Temper Sands in Prehistoric Oceanian Pottery: Geotectonics, Sedimentology, Petrography, Provenance

William R. Dickinson
Department of Geosciences
University of Arizona
Tucson, Arizona 85721
USA

THE
GEOLOGICAL
SOCIETY
OF AMERICA

Special Paper 406

3300 Penrose Place, P.O. Box 9140 ▪ Boulder, Colorado 80301-9140, USA

2006

Copyright © 2006, The Geological Society of America, Inc. (GSA). All rights reserved. GSA grants permission to individual scientists to make unlimited photocopies of one or more items from this volume for noncommercial purposes advancing science or education, including classroom use. For permission to make photocopies of any item in this volume for other noncommercial, nonprofit purposes, contact the Geological Society of America. Written permission is required from GSA for all other forms of capture or reproduction of any item in the volume including, but not limited to, all types of electronic or digital scanning or other digital or manual transformation of articles or any portion thereof, such as abstracts, into computer-readable and/or transmittable form for personal or corporate use, either noncommercial or commercial, for-profit or otherwise. Send permission requests to GSA Copyright Permissions, 3300 Penrose Place, P.O. Box 9140, Boulder, Colorado 80301-9140, USA.

Copyright is not claimed on any material prepared wholly by government employees within the scope of their employment.

Published by The Geological Society of America, Inc.
3300 Penrose Place, P.O. Box 9140, Boulder, Colorado 80301-9140, USA
www.geosociety.org

Printed in U.S.A.

GSA Books Science Editor: Abhijit Basu

Library of Congress Cataloging-in-Publication Data

Temper sands in prehistoric Oceanian pottery : geotectonics, sedimentology, petrography, provenance / by William R. Dickinson.
    p. cm.— (Special paper ; 406)
 Includes bibliographical references and index.
 ISBN-13 9780813724065 (pbk.)
 ISBN-10 0813724066 (pbk.)
  1. Clay--Additives--Oceania. 2. Sand--Oceania. 3. Sediments (Geology)--Oceania. 4. Geology, Structural--Oceania. 5. Petrofabric analysis--Oceania. 6. Oceania--Antiquities.
  I. Dickinson, William R. Special papers (Geological Society of America) ; 406.

QE471.3.D53 2006
552.5′22--dc22

                                                                                                  2006043464

Cover: View west across the head of Yalobi Bay on the south coast of Waya Island near the southern end of the Yasawa chain in western Fiji. A buried cultural horizon within the Qaranicagi rockshelter ("cave of the winds") in the cliffs above the ravine near the center of the view has yielded abundant sherds of Lapita pottery associated with anthropogenic charcoal dating to approximately 1000 BC from calibrated (calendric) radiocarbon ages (Cochrane, 2002). Photo by W.R. Dickinson.

10 9 8 7 6 5 4 3 2 1

Contents

Preface vii

Abstract 1

Introduction and Purpose 3
    Temper Terminology 3
    Temper Varieties 3
    Temper and Paste 4
    Sampling Strategy 5
    Temper Suites 5
    Topical Treatment 5
    Scientific Benefits 7

Oceanian Cultural Context 7
    Ceramic Assemblages 7
    Lapita Sherds 10
    Environmental Changes 10
    Sea-Level Highstand 11

Geotectonic Temper Classes 11
    Geotectonic Patterns 14
    Western Subregion 14
    Central Subregion 14
    Eastern Subregion 20

Temper Textural Properties 20
    Beach and Stream Sand Tempers 20
    Natural and Manually Added Tempers 21
    Epiclastic and Pyroclastic Tempers 21
    Grog and Crushed-Rock Tempers 21

Terrigenous Grain Types 21
    Grain Groups 22
    Mineral Identification 22

    Light Minerals 23
    Heavy Minerals 25
    Lithic Fragments 26

Temper Compositional Modes 27
    Grain Counting 27
    Ternary Plots 28
    Supplemental Parameters 29

Oceanic Basalt Tempers 29
    Overall Composition 30
    Caroline Tempers 30
    Rotuma-Uvea Tempers 32
    Samoan Tempers 32
    Marquesan Tempers 37

Andesitic Arc Tempers 38
    Overall Composition 39
    Bonin Temper 40
    Mariana Tempers 40
    Yap Volcanic Tempers 44
    Palau Tempers 46
    Halmahera Tempers 47
    Lease Temper 47
    Banda Temper 50
    Bismarck Arc Tempers 50
    Solomon Arc Tempers 56
    Vanuatu Tempers 61
    Fiji Platform Arc Tempers 67
    Lau Remnant Arc Temper 70
    Tongan Tempers 71

Postarc and Backarc Tempers 75
    Backarc Admiralty Tempers 75
    Postarc TLTF Tempers 76
    Backarc Vanuatu Tempers 79
    Horne Islands Tempers 79
    Postarc Fiji-Lau Tempers 80

Dissected Orogen Tempers 87
    Torres Strait Temper 88

 Aitape Coast Tempers 88
 Bismarck Archipelago 88
 Muyua (Woodlark) Temper 91
 Solomon Islands 91
 Southern Viti Levu 92

Tectonic Highland Tempers 95
 Yap Metavolcanic Temper 96
 Outer Banda Arc Tempers 97
 Aru Islands Temper 98
 D'Entrecasteaux Temper 98
 New Caledonia Tempers 99

Comparative Temper Compositions 101
 Heavy Minerals 101
 Light Minerals 104
 Lithic Fragments 104

Patterns of Ceramic Transfer 105
 Anomalous Sherds 107
 Micronesian Region 107
 Molucca-Arafura Region 109
 Bismarck-Solomon Region 112
 New Caledonia–Vanuatu Region 115
 Fiji-Lau Region 116
 Outward from Fiji 117
 Tonga Relations 118
 Transfer Distances 119

Summary and Conclusions 120

Acknowledgments 121

References Cited 122

Appendix 1: *Catalogue of Prehistoric Oceanian Potsherds Examined Petrographically*
 in Thin Section 138
Appendix 2: *Sherd Photomicrographs* 151
Appendix 3: *Index to Islands, Island Groups, and Archaeological Sites* 161

Preface

In 1965, I spent most of my first sabbatical year from Stanford University mapping volcaniclastic bedrock geology on western Viti Levu and along the Mamanuca-Yasawa island chain in liaison with the Fiji Geological Survey. During our stay in Fiji, my former wife, Peggy Dickinson, with a graduate degree in anthropology from Stanford, undertook an ethnographic study of traditional earthenware pottery-making in Yavulo village at the outskirts of Sigatoka on the southwest coast of Viti Levu (Dickinson and Sykes, 1965).

Through her, I met the late Bruce Palmer, then director of the Fiji Museum, who introduced me to Lawrence and Helen Birks, New Zealanders engaged at the time in the first excavation of the famed Sigatoka Dune archaeological site on the beachfront of the Sigatoka River delta. The Birks put to me for the first time the question of temper origins in prehistoric Oceanian potsherds. They were in a quandary as to whether the distinctive Lapita pottery recovered from their excavations in the dunes was made locally or had been imported from elsewhere within the extensive Lapita cultural region of the southwest Pacific.

Knowing from personal observation that modern potters of Yavulo add local sand as temper to clay bodies, I agreed with the Birks to investigate the question for them. My successful plan was to undertake a mineralogical and textural study of the dune sands (Dickinson, 1968a), and for comparison to examine in thin section some of their Lapita sherds to establish whether the temper sands in them are similar or different. The modern dune sands and the sherd tempers are indistinguishable in both texture and composition, demonstrating that the Lapita pottery was manufactured locally (Dickinson, 1971a, 1973b).

On my way home through Hawaii, I stopped briefly at the Bishop Museum in Honolulu, where I met Richard Shutler Jr., an archaeologist with wide Pacific experience that began in 1952 with his participation in the first excavation of the archaeological site on New Caledonia from which Lapita pottery takes its name. We immediately launched a collaboration to examine sherds from multiple Oceanian archaeological sites in thin section. Our intent was to seek direct petrographic evidence, independent of ceramic typology, for the prehistoric transfer of ceramic wares between islands or island groups.

This monograph encapsulates the fruits of nearly forty years of my subsequent investigation of Oceanian tempers from island groups spread across the expanse of the tropical Pacific Ocean from the western fringe of Micronesia on the northwest to the western fringe of Polynesia on the southeast. Over the years, 2223 prehistoric potsherds have been examined in thin section and have yielded varied insights not limited to our initial focus on evidence for ceramic transfer.

Geological Society of America
Special Paper 406
2006

Temper Sands in Prehistoric Oceanian Pottery: Geotectonics, Sedimentology, Petrography, Provenance

William R. Dickinson
Department of Geosciences, University of Arizona, Tucson, Arizona 85721, USA

ABSTRACT

Petrographic examination of temper sands in prehistoric Oceanian pottery collected by archaeologists from island groups spread across the tropical Pacific Ocean shows that the sands vary compositionally in geographic patterns that are governed by geotectonic setting and vagaries of local bedrock exposure on individual islands. The small islands serve as virtual point sources of sediment derived exclusively from the restricted array of rocks that form each island. Both natural and manually added tempers can be traced to bedrock sources by the same petrographic methodology, but independently sourcing clay bodies requires geochemical comparison of clay pastes with potential clay sources. Oceanian tempers include calcareous as well as terrigenous sands, but only the latter can be associated unequivocally with specific islands or island groups because the nature of reef tracts is similar throughout the tropical Pacific. Exotic tempers can be distinguished from indigenous tempers because their compositions are incompatible with the geology of the islands where the exotic sherds are found.

Human migration into islands of the Pacific Ocean was the last main stage in human dispersal over the planet, with no human occupation of the small islands lying beyond island Southeast Asia and Australasia until 1500 B.C. The earliest inhabitants possessed a ceramic culture, and ceramic traditions evolved over subsequent centuries to produce a varied succession of ceramic phases. Lapita pottery, which is the oldest ware in southwest Pacific island groups, is especially notable because its production was limited to a time frame short enough to allow Lapita sherds to serve a role akin to index fossils. Temper sands in Lapita and post-Lapita sherds from the same locales are indistinguishable and show that salient temper contrasts are controlled by island geology rather than habits of ancient potters. Prehistoric collecting sites for temper sand were not necessarily identical to places where modern sand accumulates because of severe environmental changes on many islands.

The compositions of terrigenous temper sands in Pacific Oceania reflect the complex pattern of circum-Pacific plate boundaries and intra-Pacific hotspot chains, and define oceanic basalt, andesitic arc, postarc-backarc, dissected orogen, and tectonic highland temper classes composed of different associations of grain types. The geographic distribution of different temper classes reflects not only the current geotectonic setting of each island group but also their paleotectonic settings when exposed rock assemblages were formed. Temper aggregates include beach, stream, and rarely dune sands, as well as grog (broken-sherd) and crushed-rock particles in some island groups.

Terrigenous grain types in Oceanian temper sands are subdivided by petrographic analysis into three main groups: light mineral grains including quartz and feldspars, heavy

Dickinson W.R., 2006, Temper Sands in Prehistoric Oceanian Pottery: Geotectonics, Sedimentology, Petrography, Provenance: GSA Special Paper 406, 164 p., doi: 10.1130/2006.2406. For permission to copy, contact editing@geosociety.org. ©2006 Geological Society of America. All rights reserved.

ferromagnesian mineral grains including opaque iron oxides and ferromagnesian silicates, and a variety of polycrystalline lithic fragments that are dominantly of volcanic derivation in most temper suites. Useful triangular compositional diagrams plot relative proportions of grain types for populations of total terrigenous grains, mineral grains exclusive of lithic fragments, ferromagnesian silicate mineral grains, all nonferromagnesian grains, only transparent mineral grains, and exclusively quartz and feldspar mineral grains. Supplemental grain parameters or indices express ratios of grain types among quartz and feldspar mineral grains, ferromagnesian grains, and lithic fragments.

Oceanic basalt tempers are mineralogically simple volcanic sands derived from basaltic to basanitic volcanic assemblages of intraoceanic hotspot chains erupted in the interior of the Pacific plate in the eastern Caroline Islands, along the northern Melanesian borderland, in Samoa and American Samoa, and in the Marquesas Islands. Andesitic arc tempers are volcanic sands displaying more compositional variability and are the most abundant tempers within the region of Oceanian ceramic cultures, occurring along island arcs flanking the Philippine Sea plate, bounding the Banda Sea in eastern Indonesia, within the Bismarck Archipelago east of New Guinea, along the reversed-polarity Solomon and Vanuatu arcs, on the Fiji platform and the Lau remnant arc, and in Tonga. Postarc and backarc volcanic sand tempers, variously displaying affinities with both oceanic basalt and andesitic arc tempers, are known from the Bismarck Archipelago, the Vanuatu backarc region, the Horne Islands of the northern Melanesian borderland, and both the Fiji platform and the Lau remnant arc. All volcanic sand tempers of Pacific Oceania are composed of phenocrystic mineral grains and volcanic lithic fragments. Most are quartz-free or quartz-poor, but quartzose variants are present locally along island arcs where silicic eruptions accompanied more typical andesitic to basaltic activity, and within backarc settings where bimodal igneous assemblages are exposed.

Most quartzose Oceanian temper sands are either dissected orogen tempers containing dominantly igneous but not exclusively volcanic detritus, or tectonic highland tempers containing recycled sedimentary detritus. Dissected orogen tempers with quartzose plutonic detritus occur in selected sherd suites from the Torres Strait Islands, the Bismarck Archipelago, and the Solomon Islands, but are especially characteristic from the south coast of Viti Levu in Fiji. Quartzose tectonic highland tempers occur in sherds from the outer Banda arc, the Aru Islands in the Arafura Sea, the D'Entrecasteaux Islands of the Solomon Sea, and New Caledonia. Nonquartzose tectonic highland tempers derived from ophiolitic rocks of uplifted oceanic crust are present in sherds from Yap and New Caledonia.

Comparisons of temper compositions among temper classes indicate that oceanic basalt and basaltic backarc tempers contain significantly higher proportions of olivine mineral grains than arc and postarc tempers, which include a varied array of temper types containing different proportions of pyroxenes and hornblendes. Dissected orogen and quartzose andesitic arc tempers display varying proportions of quartz, plagioclase, and K-feldspar within the compositional field typical for circum-Pacific orogenic sands. Tectonic highland tempers contain distinctly higher proportions of nonigneous lithic fragments than other temper classes.

The presence of exotic sherds containing temper sands incompatible with the geology of the islands from which they were recovered documents 106 instances of ceramic transfer between different islands, mostly lying within the same island groups, but also between island groups lying far apart. Two-thirds of the instances of ceramic transfer involved interisland distances of less than 200 km, and most of the remainder involved distances in the range of 200–600 km, but a few cases of ceramic transfer for 1000 km or more are known from temper analysis.

Keywords: geotectonics, hotspots, island arcs, Melanesia, Micronesia, Pacific Ocean, Polynesia, sedimentary petrography, provenance, sedimentology.

INTRODUCTION AND PURPOSE

Ceramic sherds are commonly the most abundant and durable artifacts recovered at archaeological sites in island groups of the southwestern and western Pacific Ocean. The study of temper sands by petrographic methods elucidates the tempering practices of ancient potters, and in some cases provides physical evidence for the transport of ceramic objects or raw materials between different islands. Petrographic analysis can show that materials are foreign when stylistic analysis cannot because styles can be copied (Shepard, 1965). Evidence for movement of pottery is provided by the compositions of contrasting temper sands derived from bedrock exposed on islands of differing geology. Differences in bedrock geology stem not only from vagaries of local exposure, but also from the contrasting geotectonic settings of different Pacific island groups (Dickinson and Shutler, 1968, 1971, 1979; Dickinson, 1998a).

Whether movement of pottery from one island to another resulted from commercial trade, ceremonial exchange, or human migration is indeterminate from temper studies. The neutral term *ceramic transfer* is used to denote interisland movement of pottery detected by temper analysis. Distinction between the movement of finished wares and the movement of ceramic raw material, in the form of temper sand (Best, 2002, p. 38–39), is also impossible from temper analysis alone. The connotation of pottery movement implied by the term ceramic transfer always carries the caveat that it might have been the sand, and not the pottery, that was transported between islands. Distinction between movement of finished wares and movement of raw materials is an archaeological issue that cannot be resolved by petrography alone. Nevertheless, the large volumes of ceramic transfer in some instances, and the long distances involved in others, suggests that most ceramic transfer detected by temper analysis involved movement of finished wares rather than movement of loose raw material difficult to transport in bulk.

Detection of ceramic transfer within Pacific Oceania is possible because there are no oceanic river systems to mingle sand from multiple islands, as continental river systems mix sand derived from multiple source provenances. Each island is effectively a point source of detritus, and many are small enough to expose only a restricted range of bedrock types. No sand derived from elsewhere occurs anywhere on a given island, although sand transported by turbidity currents generated on the flanks of island edifices may mingle on the deep seafloor inaccessible to island potters. Temper or clay analysis is the only sure way to detect ceramic transfer, because there are no oceanic trails along which people can leave a record of their passage in transit from one island to another. Once embarked, people leave no footprints until they reach their next landfall.

Temper Terminology

The functions of *temper* sand, as aplastic (also termed nonplastic) grit, are to enhance the workability of plastic clays during vessel fabrication, to help bind the clay during vessel drying before firing, and to strengthen finished wares by counteracting shrinkage during and after firing. Temper particles are of sand size or larger (>0.0625 mm in diameter). Almost all naturally occurring clays contain at least some fine silt particles that are not coarse enough to serve the functions of temper. Silt embedded in clay is treated as part of the clay *paste* in which the coarser temper grains are also embedded.

In archaeological parlance, the aplastic constituents of a clay body prepared for firing are often described in general as "inclusions" (within the clay), with the term "temper" reserved specifically for aplastic materials that were added deliberately to a clay body by ancient potters (Shepard, 1965). This distinction arose from recognition that some of the aplastic constituents of clay bodies occur naturally within the clays as collected, and need not be added during manipulation of ceramic raw materials.

The same distinction can be achieved, however, by referring to *natural* and *manually added* tempers. This terminology is preferable because the term "inclusion" is inherently ambiguous unless reserved for its normal mineralogical meaning as a crystal of one mineral enclosed within the crystal of another mineral species, or within noncrystalline volcanic glass. Because the identity and origin of individual temper grains making up a temper sand may rest in part upon the nature of the inclusions within them, it would be confusing to speak of inclusions within inclusions. The terminology of natural and manually added tempers has the additional advantage that the general designation as temper can be used without error in the many cases where distinction between naturally occurring and manually added tempers is difficult (Rice, 1987), requiring judgment that may be faulty.

Some fired clay bodies contain aplastic constituents that include both natural and manually added tempers. Pedogenic, colluvial, alluvial, or estuarine clays contain sandy impurities in widely variable amounts, and potters may need to add different proportions of sand manually to achieve the desired consistency for the final sand-clay mix that is worked and fired. In these cases, all the aplastic constituents can be described as temper, leaving the question of which grains are natural temper and which are manually added temper to be addressed separately. Although textural distinctions between natural and manually added temper may be equivocal, tracing either kind of temper to its bedrock provenance involves the same observations and logic.

Temper Varieties

Oceanian temper sands include *calcareous* sands of bioclastic reef debris or detritus eroded from cemented reef tracts, *terrigenous* sands derived from exposed volcanic or other silicate bedrock, and *hybrid* sands (Zuffa, 1979) of mixed terrigenous and calcareous grains. The calcareous grains are not diagnostic of geographic origin because the nature of reef tracts throughout the tropical Pacific is broadly similar. Coralgal, molluscan, and foraminiferal debris present in various proportions in different calcareous sands may be characteristic of specific depositional sites associated with reef tracts, but their regional distribution is unpredictable. Calcareous sands composed of the different biogenic components can be derived from closely spaced parts of individual reef tracts, and their variability is unsystematic even for tempers within an individual island group. Only terrigenous temper sands provide unambiguous evidence of sand origins from particular islands.

Modern sedimentary environments where usable temper sand can be collected include nearly ubiquitous island beaches, stream channels and bars on larger islands, and coastal dune tracts in rare cases. Natural temper may be present in varying proportions in residual regolith overlying weathered bedrock, in derivative colluvium transported downslope by mass wasting, or in alluvium redeposited by streams along ravines and valley floors. It is also likely that naturally tempered clay bodies can be assembled at a collecting site by kneading together intimately associated laminae of sand and clay that jointly form fluvial or estuarine deposits. Uncommon nondetrital tempers used in some island groups include *grog* tempers, in which the temper particles are pieces of broken pottery or other ceramic objects, *composite* tempers composed of mixed grog and detrital grains, and *crushed-rock* tempers derived from deliberate breakage of volcanic or other bedrock quarried on outcrop.

From a provenance standpoint, *indigenous* tempers derive from the island or cluster of closely spaced small islands where the sherds containing them were collected, whereas *exotic* tempers derive from other islands or island groups. Among indigenous tempers on larger islands or in compact island clusters, it is useful to distinguish between *local* tempers derived from near the site of sherd recovery, and *nonlocal* tempers from other parts of the same island or island cluster. A conceptual guide for applying the terms local and nonlocal in this context is to consider how far people might range afield, by walking or in paddle canoes, to bring temper back to a given locale where pottery was made. In prehistoric Oceania, there was no form of vehicular transport on land.

Temper and Paste

The firing temperatures of 600–900 °C (Intoh, 1990; Clough, 1992) reached in ceramic bonfires of Oceania were insufficient to affect the mineralogy of terrigenous temper sands, which can be studied in thin section by standard methods of sedimentary petrology (Hodges, 1963; Peacock, 1970; Bishop et al., 1982; Schubert, 1986; Gerrard, 1991). Inferences of bedrock provenance for temper sand grains can then follow the same well-developed methodology applied to sand grains in sedimentary rocks (Dickinson, 1985). Calcareous grains also persist unchanged by firing in much of the earthenware pottery recovered from Oceanian archaeological sites.

Most conclusions of this study are based upon petrographic examination of 2223 thin sections of prehistoric Oceanian potsherds prepared in the standard way, as for rocks, after vacuum-impregnation of sherds or sherd slices with epoxy resin. In cases where distinctions between quartz, K-feldspar, and plagioclase would otherwise have been uncertain, thin sections of sherds were etched with HF fumes and then stained for K and Ca (Laniz et al., 1964; Marsaglia and Tazaki, 1992). Supplemental microprobe analyses of selected mineral species were undertaken to answer specific questions raised by petrographic examination, but are unsuitable for routine surveys pursued without benefit of preliminary petrography (Freestone, 1982). Microprobe identification of individual temper grains is too laborious to allow satisfactory statistics on the multiple grain types present in temper sands to be compiled from microprobe studies alone. Microprobe analysis is ideal, however, for the conclusive identification of unusual mineral grains (Dickinson, 1971b), and for comparing the chemical compositions of a given mineral species in different sherd collections to confirm or deny a suspected temper match (Dickinson et al., 1990).

Tracing clay bodies to sources is a problem separate from temper analysis, except where all temper is natural temper, and generally requires direct chemical comparison of clay pastes with specific examples of presumed or potential clay sources (Hunt, 1988; Hunt and Graves, 1990; Kirch et al., 1991). Petrographic study of clay paste is largely unproductive because of fine grain size below the effective limit of optical resolution. As clay chemistry is influenced by varied weathering processes in local microenvironments, and even by accretion of eolian dust, the composition of residual clay cannot be predicted from knowledge of bedrock character alone.

In consequence, sourcing clays is more challenging than sourcing tempers. All the varied sands present as potential temper on a given island faithfully reflect the nature of bedrock sources on the island, although sedimentological processes may integrate or fractionate the inherent signal of bedrock provenance. By contrast, the compositions of different clay deposits on a given island may vary in ways that can only be established by empirical investigations of island clays. In selected Oceanian sherds, however, mica flakes or pyroclastic shards of volcanic glass embedded within clay pastes allow aspects of clay provenance to be inferred from petrography. Otherwise, this study is restricted to temper analysis, although chemical investigations of clay origins by others are cited wherever relevant for consideration of ceramic origins.

Chemical investigations of clay bodies must proceed with the understanding that bulk instrumental analyses of sherds (Bentley, 2000; Descantes et al., 2001) yield results that represent a composite of clay and temper constituents (Neff et al., 1988, 1989; Arnold et al., 1991; Ambrose, 1993). Combined petrographic examination and X-ray fluorescence analysis of prehistoric sherds from New Caledonia suggest strongly that variations in temper chemistry control most variations in bulk sherd chemistry (Chiu, 2003b). Only more selective microprobe or SEM techniques that focus beams on small areas within sherd slices can isolate the chemistry of clay paste apart from associated temper grains (Hunt and Erkelens, 1993; Summerhayes, 1997). Matching clay pastes compositionally with potential clay bodies requires such isolation of paste composition from the effects of temper composition on bulk sherd chemistry.

Application of laser-ablation inductively coupled plasma–mass spectrometry (LA-ICP-MS) to ceramic studies is still in its infancy (Kennett et al., 2002) but may afford advantages over more traditional geochemical techniques. A comparative LA-ICP-MS study of Tongan, Fijian, and Bismarck Archipelago sherd suites from Pacific Oceania has documented clear-cut compositional differences (Kennett et al., 2004), and an LA-ICP-MS study of compositional diversity in selected Fijian sherd suites has provided chemical data to distinguish provisionally between indigenous and exotic sherds at selected sites (Cochrane and Neff, 2006).

Sampling Strategy

It is impractical to select statistically meaningful subsets of sherds for petrographic study from the hundreds of thousands that have been collected from Oceanian archaeological sites. The petrographic work upon which this report is based involved only a small fraction of one percent of the available sherd collections. Moreover, the sherd collections themselves are but a tiny fraction of the unexcavated sherds still left in the ground at known and as yet unknown prehistoric habitation sites spread through hundreds of islands.

Sampling sherd suites for petrographic study consequently follows a nested procedure. Collaborating archaeologists first sort sherd collections megascopically by grouping sherds according to style, areal distribution by locale on individual islands or by island within island groups, stratigraphic position in excavations, and gross appearance of sand temper and clay paste. Thin sections are then prepared from representative sherds within each grouping without regard to the relative abundance of different kinds of sherds. This approach limits redundancy of results and ensures that unusual but informative temper types, reflective of cultural contacts over appreciable distances, are not overlooked. Visibly odd sherds are always examined petrographically, even if they are rare in the sherd collection from a given locale.

Unless the megascopic sorting of sherds has been misleading or ineffective for some reason not apparent before petrographic study, examination of half a dozen to a dozen sherds from each megascopic group normally proves sufficient to define the mean composition or the compositional spectrum of a given temper type within useful limits. When petrographic study raises issues of origin not appreciated during initial megascopic sorting of sherds, additional sherds are studied to address unanticipated questions.

Temper Suites

Oceanian temper sands include exclusively calcareous sands on atolls and many island beaches, volcanic sands from islands along hotspot chains built across the Pacific plate and along intraoceanic island arcs associated with subduction at parallel trenches, and mineralogically more complex sands derived from extinct and deeply dissected island arcs or from uplifted subduction complexes. The twin purposes of this monograph are to provide a primer for petrographic study of the diverse Oceanian temper suites, and to present a summary of results obtained to date from petrographic analysis. Detailed descriptions of many local temper sands have been published previously and are not repeated here, but the context and interrelations of all temper suites studied to date are discussed in comparative fashion. More descriptive detail is provided for temper types not previously treated in publications than for temper types described in previous articles that are referenced. The general implications of temper analysis for the course of Oceanian prehistory have been discussed elsewhere (Dickinson and Shutler, 2000).

Part of my intent is to provide archaeologists with the geological insights needed to address temper analysis in Pacific Oceania, and to provide geologists having limited knowledge of Pacific prehistory with the archaeological background needed to contribute effectively to Oceanian temper studies. Most temper investigators are archaeologists who have acquired minimal petrographic skills, geoscience students with limited petrographic experience when recruited by archaeologists to assist them in their work, or more mature geologists consulted by archaeologists on a temporary basis. Each of these groups of scientists must overcome inherent gaps in personal understanding to produce reliable results and reach valid conclusions.

Topical Treatment

The major topics covered in my review are (1) the different geotectonic settings of Pacific island groups and their correlation with broad temper classes, (2) the most useful petrographic criteria for the identification of salient types of temper grains, and (3) the nature of distinctive temper types in the archaeological context of Oceanian ceramic assemblages. Within each temper class, temper types from different island groups are discussed in geographic order from northwest to southeast within the region of Oceanian ceramic cultures (Fig. 1).

The prime focus is on sherds from the individually small islands of Micronesia, island Melanesia, and Polynesia that lie well out to sea from the larger islands of Australasia and Southeast Asia. Collections from comparatively large islands such as New Caledonia and Viti Levu, which occur within the region dominated by small islands, are treated as integral to the overall picture of Oceanian temper types. Also discussed are tempers in prehistoric sherds from small islands of easternmost Indonesia and the Sahul Shelf between Australia and New Guinea where the methodology of Oceanian temper analysis is applicable. Temper relationships on the large islands of Indonesia and the Philippines are not addressed, however, because sand dispersal from widespread bedrock sources to sites of deposition by extensive fluvial systems is akin to patterns on continental landmasses, rather than to relations on the smaller islands of Pacific Oceania.

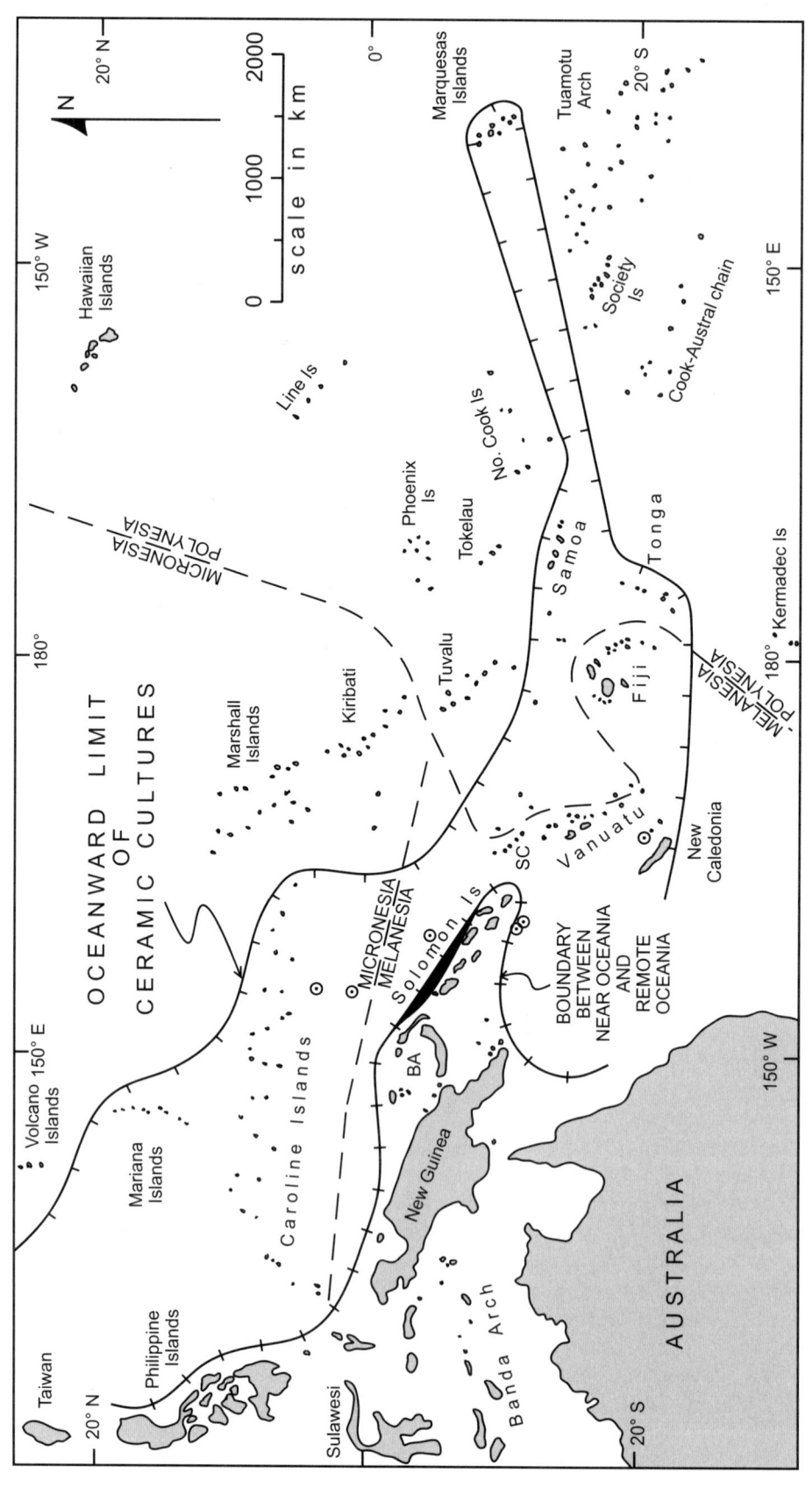

Figure 1. Extent of prehistoric ceramic cultures in equatorial Pacific Oceania (BA—Bismarck Archipelago; SC—Santa Cruz Group). Circled islands near New Caledonia and Solomon Islands are Polynesian outliers in Micronesia and Melanesia. Arch—Archipelago; Is—Islands.

Appendix 1 is a catalogue, island by island and site by site, of all Oceanian sherds examined for this report, and can be consulted to augment discussions of sherd suites in the text. Island names in the appendix and text conform to the usage of Motteler (1986). Appendix 2 is a gallery of sherd photomicrographs (Figs. A1–A30) illustrating salient temper grain types and kinds of temper aggregates in thin section. Appendix 3, an index to islands and archaeological sites of ceramic interest, provides a guide to information on Oceanian sherds and tempers from different locales in the text, figures, tables, and appendices.

Scientific Benefits

The collection of temper sands by prehistoric Oceanian potters performed a scientific service of double value. From an archaeological perspective, temper sands provide direct physical evidence of ceramic origins that otherwise could not be established with nearly as much confidence. For reconstructing cultural relations during Oceanian prehistory, the potential significance of firm evidence for ceramic origins is difficult to overestimate.

From a geological perspective, Oceanian ceramic tempers provide a broad sampling of island sands encased in matrices of fired clay that have preserved the sand samples intact and largely unweathered over many centuries spanning several millennia. Given their intimate knowledge of local sedimentary environments on far-flung islands, prehistoric potters assembled sand collections that would take modern geoscientists years or even decades to duplicate. Moreover, the ancient potters sampled Oceanian sands over a time frame of changing paleoenvironments that no longer exist in the same guise. Temper analysis thus affords a means to appraise the variability of island sands over space and time in a way that no other research avenue could offer. The study of island tempers thereby opens doors for interpretations of sand provenance that are closed to other methodological approaches.

OCEANIAN CULTURAL CONTEXT

Pacific Oceania was the last great frontier of human migration and was unoccupied until late in Holocene time. Australia was occupied well before the end of Pleistocene time, but before ca. 1500 B.C. no one had seen any islands lying seaward from the large landmasses of the Philippine and Indonesian archipelagoes. In that regard, a useful distinction has been drawn by Green (1997) between *Near Oceania* and *Remote Oceania* (Fig. 1). Within Near Oceania, where islands are intervisible or mutually visible from vessels in transit between islands, water crossings require a threshold level of seafaring ability but no navigation beyond line of sight. In Remote Oceania, however, islands or island clusters lie out of view over the horizon, and navigational skills are required to make successive landfalls.

The peopling of Remote Oceania was once thought to have been the result of happenstance from accidental voyaging in vessels lost at sea and drifting aimlessly under the influence of winds and currents. Patterns of human dispersal in time and space show unmistakably, however, that human expansion into Pacific Oceania resulted from systematic exploration and deliberate colonization (Irwin, 1992, 1993; Finney, 1996). The basic strategy of the ancient voyagers was to reconnoiter upwind, by tacking into the teeth of the easterly trade winds or by taking advantage of temporary westerly winds, allowing for safe downwind return to home base on the prevailing trade winds if no landfalls were made.

By long-standing convention, the island groups of Remote Oceania are assigned on various linguistic and cultural grounds to Micronesia ("small islands"), island Melanesia ("black islands"), and Polynesia ("many islands"). This tripartite subdivision (Fig. 1), codified almost two centuries ago (Dumont d'Urville, 1832), has subtle racial overtones and limited utility (Thomas, 1989). There are diffuse racial and cultural gradations between Melanesia and Polynesia, yet strong ethnographic contrasts between western and eastern Micronesia. There are also geographic interdigitations whereby people speaking Polynesian languages occupy so-called Polynesian outliers (Fig. 1) lying within the bounds of Micronesia and Melanesia (Kirch, 1984).

For much of prehistory, the familiar tripartite subdivision of Oceania is virtually meaningless (Green, 1991a). The earliest settlers of island Melanesia and the western fringe of Polynesia were closely related culturally, and probably spoke similar Austronesian languages (Blust, 1996; Kirch, 1997, 2000; Spriggs, 1997b; Kirch and Green, 2001). On the other hand, the earliest inhabitants of eastern Micronesia probably came from islands to the south in modern-day Melanesia, rather than from western Micronesia (Athens, 1990a; Intoh, 1997; Rainbird, 1999).

Remote Oceania was populated by surges of discovery and settlement accompanying the maritime dispersal of Austronesian peoples outward into the Pacific realm from island Southeast Asia, which Austronesian seafarers had penetrated by ca. 2500 B.C. (Bellwood, 1995; Kirch, 1995; Spriggs, 1996; Rolett et al., 2000). The motivation for the further expansion of settlement into unoccupied island groups lying beyond the previous ken remains debatable, but was perhaps the opportunity to exploit pristine environments where productive reef tracts and seabird colonies were untouched by human hand. Migration into Remote Oceania was delayed, however, until sailing technology and navigational skills evolved to meet the challenge of traversing wide expanses of ocean out of sight of land (Irwin, 1992).

Ceramic Assemblages

The oldest archaeological sites of Remote Oceania in islands beyond the Solomon Islands off New Guinea at the eastern end of the Indonesian archipelago, or beyond the Philippine Islands farther north, date from ca. 1500 B.C., and all bear evidence of ceramic cultures. This initial date applies to Early Pre-Latte wares of the Mariana Islands (Craib, 1999; Dickinson et al., 2001) on the western fringe of Micronesia in Remote Oceania, and to Lapita wares of the Mussau Group (Kirch, 2001b) in the Bismarck

Archipelago, which lies off New Guinea near the eastern limit of Near Oceania (Fig. 1). The earliest Oceanian pottery probably derived from ceramic traditions of island Southeast Asia (Shutler, 1999), although generic connections remain obscure in detail.

Lapita pottery of the southwest Pacific is typologically distinctive, decorated by characteristic dentate stamping to form elaborate anthropomorphic and more formalistic designs (Anson, 1986, 2000b; Spriggs, 1990; Sand, 2001b; Best, 2002; Chiu, 2005), and has given its name to the Lapita cultural complex (Green, 1979, 1991b, 1992; Spriggs, 1995; Smith, 1995). The comblike tools used to apply the characteristic decorative motifs were probably made from sea-turtle scutes (Ambrose, 1997, 1999). Within the Bismarck Archipelago, regarded as the Lapita homeland (Allen and White, 1989; Allen and Gosden, 1991; Gosden and Specht, 1991; Kirch, 1996), Lapita pottery was widespread by ca. 1250 B.C. (Specht and Gosden, 1997).

Makers of Lapita pottery were the first residents of the island groups in Remote Oceania lying to the southeast of the Bismarck Archipelago and Solomon Islands at the edge of Near Oceania (Fig. 1). Lapita origins are lost in the mists of time, but the Lapita cultural complex with its unique pottery probably arose from combined cultural influences (the "Triple I Model" of Green, 2000b): an *intrusion* of habits and attitudes brought by Austronesian migrants from far to the west, the *integration* of new migrant traits with practices of the previous residents of Near Oceania (Allen, 1996), and fresh *innovation* in both technology and lifestyle born from the fusion of preexisting cultures in an environment of small islands dotting the wide seas of Remote Oceania.

Within a few hundred years (1200–800 B.C.), people dispersing the Lapita cultural complex spread nearly 4000 km, at a net rate of ~10 km/yr, through multiple island groups lying as far to the east as Tonga and Samoa along the western fringe of Polynesia (Fig. 2). The path of migration was first along or past the Solomon arc to the Santa Cruz Group at the northern end of the New Hebrides island arc, thence southward to Vanuatu and New Caledonia and eastward through Fiji and Lau (Clark and Anderson, 2001a) to westernmost Polynesia (Burley and Dickinson, 2001).

In the Lapita homeland of the Bismarck Archipelago, the Lapita design tradition persisted for nearly a millennium, but was maintained for progressively shorter intervals of time in more distant island groups (Fig. 3), and perhaps for no more

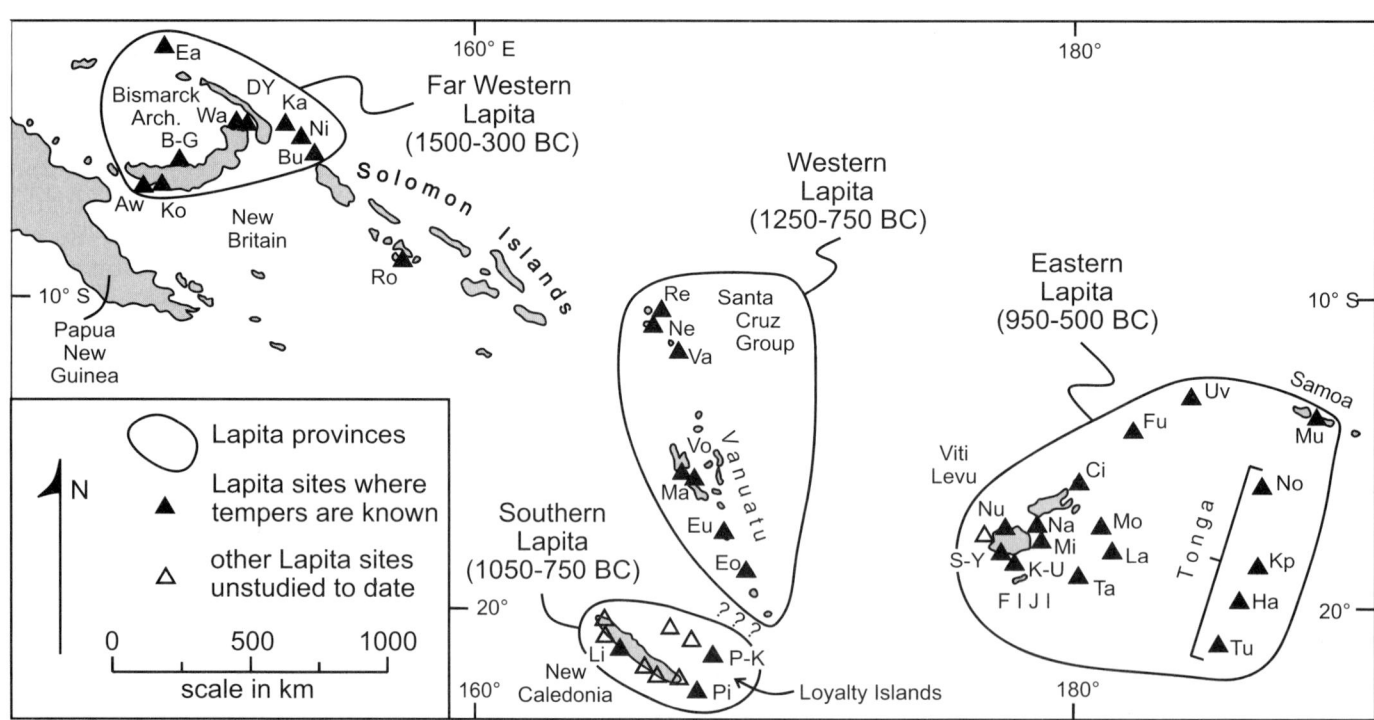

Figure 2. Distribution of Lapita wares in space and time within the southwest Pacific region, adapted after Kirch and Hunt (1988), Gosden et al. (1989), Anderson et al. (2001), Sand (2001a), and Summerhayes (2001c). Studied Lapita sites (*ms*—multiple sites on large islands; *mi*—multiple islands in island groups; island or island group of selected sites in parentheses): Aw—Arawe; B-G—Boduna and Garua; Bu—Buka; Ci—Cikobia; DY—Duke of York Islands; Ea—Eloaua (Mussau); Eo—Erromango; Eu—Erueti (Efate); Fu—Futuna; Ha—Ha'apai (*mi*); Ka—Kamgot (Babase) and nearby Balbalankin-Malekolon (Anir); Ko—Kreslo; Kp—Kapa (Vava'u); K-U—Kulu Bay (Beqa) and Ugaga; La—Lakeba; Li—Lapita (New Caledonia); Ma—Malo; Mi—Moturiki; Mo—Mago; Mu—Mulifanua (Upolu); Na—Naigani; Ne—Nendö (*ms*); Ni—Nissan; No—Niuatoputapu; Nu—Natunuku; Pi—Vatcha (Île des Pins); P-K—Patho-Kurin (Maré); Re—Reef Islands (*mi*); Ro—Roviana (New Georgia Group); S-Y—Sigatoka and Yanuca; Ta—Totoya; Tu—Tongatapu (*ms*); Uv—Uvea; Va—Vanikolo; Vo—Vao; Wa—Watom.

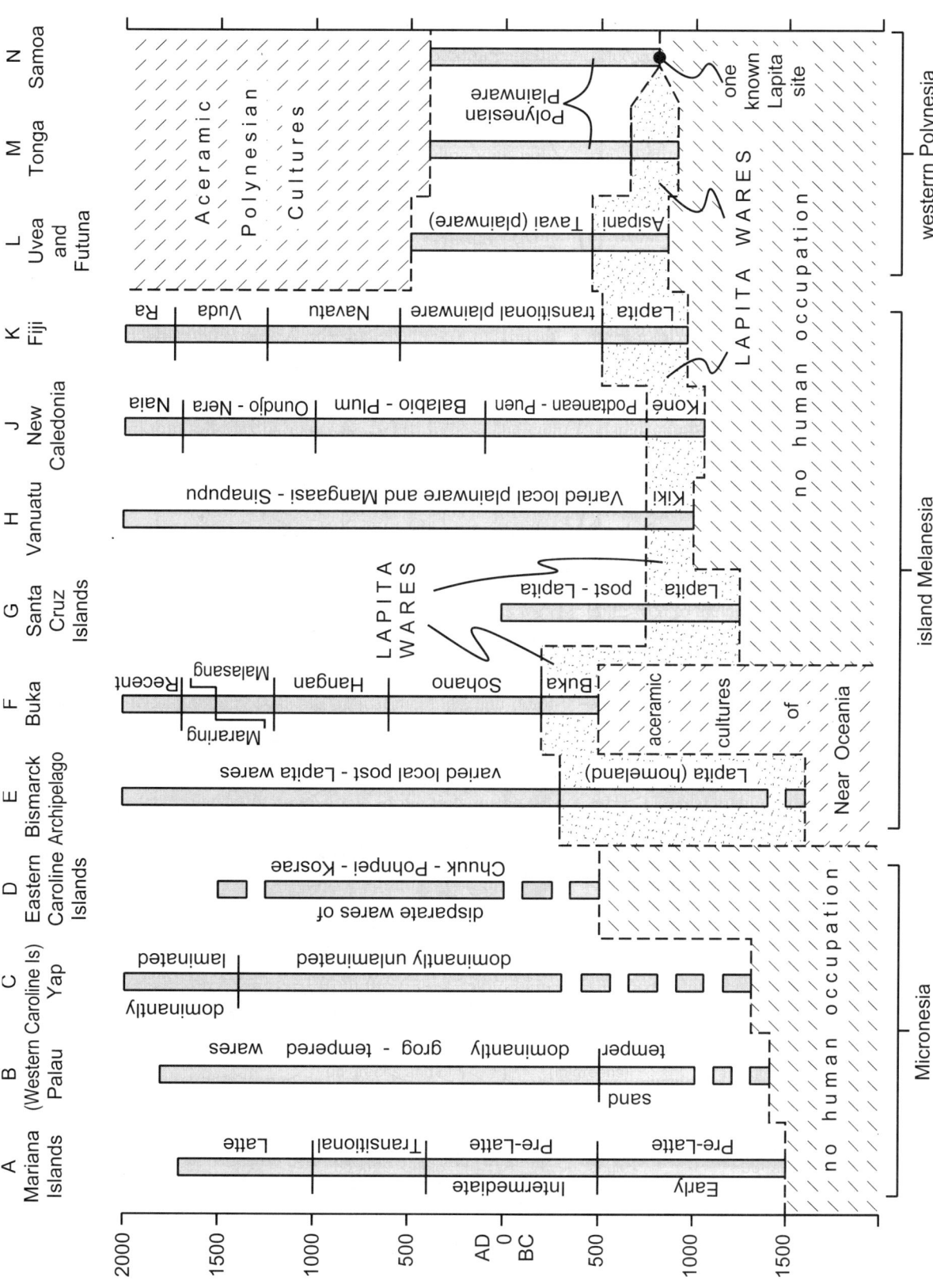

Figure 3. Duration of Lapita (stippled) and other ceramic traditions in different island groups of Pacific Oceania (Figs. 1 and 2). Continuous bars denote securely dated ceramic cultures, and broken bars denote less certain or inferred extensions in time (names of multiple locally designated ceramic phases annotated beside age bars). References: A—Kurashina and Clayshulte (1983), Moore and Hunter-Anderson (1999), Dickinson et al. (2001); B—Intoh and Leach (1985), Intoh (1990), Dodson and Intoh (1999); C—Osborne (1979), Athens and Ward (2001), Wickler (2001b, 2002), Fitzpatrick (2003), Fitzpatrick et al. (2003), Clark (2004); D—Shutler et al. (1984), Athens (1990a, 1990b), Galipaud (2001); E—Specht and Gosden (1997), Kirch (1988a, 2001b), Summerhayes (2001a); F—Wickler (1990, 2001a); G—McCoy and Cleghorn (1988), Green (1991c); H—Kirch (1984), Bedford et al. (1998), Bedford (2000, 2003), Bedford and Spriggs (2000); J—Stevenson and Dodson (1995), Sand (1996, 1997, 1999a, 2000a); K—Dickinson et al. (1998a), Clark and Anderson (1999, 2001b), Sand and Valentin (1998), Clark et al. (2001), Cochrane (2002), Kumar et al. (2003), Burley and Dickinson (2004); L—Sand (1990, 1993, 1998b, 2000b); M—Burley (1998, 1999), Burley et al. (1995, 1999, 2001), Burley and Dickinson (2001); N—Green (1974), Jennings and Holmer (1980), Clark (1993, 1996), Kirch (1993b), Petchey (2001).

than a century along the Lapita frontier in Tonga and Samoa (Burley et al., 1999; Anderson, 2001; Green, 2003). Subtle geographic and temporal variations in Lapita design motifs developed as the Lapita cultural complex evolved over time (Summerhayes, 2000a, 2001b, 2001c; Sand, 2001a) and in space (Fig. 2), but fundamental stylistic attributes were maintained despite a tendency for progressive simplification of design motifs (Sharp, 1988).

Derivative Lapitoid pottery (Golson, 1971), termed Polynesian Plainware (Green, 1974) from its lack of decoration, continued to be made in both Samoa and Tonga (Burley, 1998) for a number of centuries after the Lapita era in West Polynesia. Westward through island Melanesia (Fiji, Vanuatu, New Caledonia), diverse wares decorated by incisions and applied relief conforming to varied local traditions (Spriggs, 2000, 2003; Bedford and Clark, 2001; Clark, 2003) succeeded the Lapita ceramic phase over many centuries of subsequent ceramic evolution (Fig. 3). In parts of the Lapita homeland in the Bismarck Archipelago, transitional ceramic phases are difficult to assign to either Lapita or post-Lapita traditions, and the similarities or differences among incised and applied-relief post-Lapita wares within island Melanesia as a whole are not yet well defined (Garling, 2003). Although local and subregional post-Lapita ceramic traditions can be perceived within island Melanesia, their relationship to one another and to preceding Lapita pottery is commonly uncertain (White and Murray-Wallace, 1996). In Fiji, midway between Vanuatu and Tonga, a post-Lapita ceramic phase of Polynesian Plainware was succeeded abruptly by incised and applied-relief wares that probably reflect continued or renewed cultural ties to island groups lying farther west (Hunt, 1986, 1987; Burley, 2003; Burley and Clark, 2003).

Some red-slipped and lime-impressed Early Pre-Latte wares (1500–500 B.C.) of the Mariana Islands are decorated with punctate designs (Butler, 1994) that are somewhat reminiscent of contemporaneous Lapita ware much farther to the southeast. Mariana pottery evolved through several phases antecedent to the Latte wares associated with construction of megalithic architecture from A.D. 1000 to the time of European contact (Fig. 3). The time depth and evolutionary sequence of local ceramic traditions in the western Caroline Islands (Palau, Yap) is not as well known but may span a similar interval of prehistory. In Palau, human occupation sites and associated massive landscape terracing date to the interval 1200–800 B.C. (Athens and Ward, 2001; Welch, 2001; Wickler, 2001b, 2002; Phear et al., 2003). Paleoenvironmental indications of deforestation in Yap suggest a human presence by 1500 B.C. (Dodson and Intoh, 1999), but the oldest ceramic horizons and human burials securely dated in Palau by radiocarbon ages from charcoal do not predate 1000 B.C. (Fitzpatrick, 2003; Clark and Wright, 2003; Clark, 2004). The central and eastern Caroline Islands (Fig. 1) were not occupied until ca. 0 B.C./A.D. by people with a Lapitoid pottery tradition derived from island groups to the south (Athens, 1990b; Intoh, 1992a, 1999; Galipaud, 2001).

Lapita Sherds

Not all Lapita ceramic vessels were decorated, and there was a continuity of vessel form over many centuries, both during and after the phase of dentate stamping (Green, 2003). There are growing archaeological indications that vessels decorated with dentate-stamped motifs served a ceremonial rather than a utilitarian function (Ambrose, 1997; Terrell and Welsh, 1997; Summerhayes, 2000b; Best, 2002), and the shorter time span of the dentate-stamped tradition in island settings distant from the Lapita homeland may reflect weakening cultural ties over time as dispersal distances of human migrations increased. Moreover, only the shoulders and rims of Lapita ceramic vessels were typically decorated, so that both decorated and undecorated sherds may derive from breakage of the same Lapita vessels. For petrographic analysis, however, decorated Lapita sherds are unique in providing the only unambiguous horizon marker of initial ceramic production in widespread island groups, and in that sense serve a role akin to index fossils in stratigraphic successions. Establishing the ages of undecorated sherds recovered from various island locales requires understanding the internal stratigraphy of excavated archaeological sites, whereas decorated Lapita sherds can be dated to short time intervals even where found reworked on the surface.

Temper analysis reveals no systematic contrasts between Lapita and post-Lapita, or between Pre-Latte and Latte, tempers from the same places, except for a decline over time in the use of calcareous sand as temper, and a concomitant growing preference for terrigenous sand tempers instead. The gradual tendency for disuse of calcareous temper probably reflects the proclivity of calcareous grains to calcine in the upper part (750–900 °C) of the earthenware firing range, with deleterious effects on ceramic wares (Intoh and Leach, 1985, p. 84–91; Intoh, 1990, p. 49). Successful use of calcareous temper in many island groups during early phases of Oceanian prehistory suggests that initial Oceanian firing technology was unable to achieve temperatures above the lower part (600–750 °C) of the earthenware range.

Environmental Changes

Exact collecting sites for prehistoric temper sands are difficult to specify from examination of current island landscapes because of significant changes in island environments over the past few millennia. The most widespread effect has been a drawdown in regional sea level throughout the tropical Pacific since the peak of a mid-Holocene highstand (Dickinson, 2001b, 2003). More local influences on sedimentary environments include tectonic uplift or subsidence of selected islands (Dickinson and Green, 1998; Dickinson, 2001b), the spread of lava flows or tephra blankets over islands near eruptive centers (Monzier et al., 1994; Robin et al., 1994; Torrence et al., 2000; Green and Anson, 2000a, 2000b; Anson, 2000a; Torrence, 2001; Green, 2002), alluviation of lowland valleys in response to enhanced erosion triggered by deforestation that accompanied expanding human impact on inland vegetation (Spriggs, 1986, 1997a; Clark and Michlovic, 1996; Dickinson et al., 1998a; Anderson et al., 2006), accretion of successive beach ridges or spits to form enlarged coastal strands (Kirch, 1988c, 1993a; Kirch et al., 1990; Dickinson et al., 1994; Gosden and Webb, 1994; Amesbury et al., 1996; Hunt and Kirch, 1997), and progressive erosion of many coastlines exposed to continued wave attack.

All these processes have altered the distribution of island sand deposits suitable for temper. Except for instances of pottery manufacture in recent centuries, detailed local paleoenvironmental investigations are required for an understanding of potential temper sources during prehistory. Regardless of detailed sedimentological relationships on a given island through time, however, the mineralogy and petrologic character of all local terrigenous sands are controlled by the nature of bedrock sources on the same island. Even where specific prehistoric collecting sites cannot be identified, indigenous and exotic temper sands can be differentiated.

For evaluation of temper sources through time, the effects of volcanic or alluvial cover and of coastal aggradation or erosion vary from island to island, and from site to site on each island, but effects of a mid-Holocene highstand in regional sea level were similar throughout Pacific Oceania. Exceptions are restricted to locales where island subsidence has counteracted the emergence of island paleoshorelines (Dickinson and Green, 1998), or where tectonic uplift has further enhanced island emergence (Dickinson, 2000b).

Sea-Level Highstand

The mid-Holocene highstand in sea level, varying regionally from 1.1 m to 2.6 m above modern sea level (Dickinson, 2001b), was produced by an early Holocene eustatic rise in global sea level followed by a late Holocene hydro-isostatic decline in regional sea level. The pre-mid-Holocene eustatic rise stemmed from augmentation of seawater volume by meltwater released from wasting Pleistocene glaciers, whereas the post-mid-Holocene decline in tropical sea level reflects *equatorial ocean siphoning* (Mitrovica and Peltier, 1991). The latter hydro-isostatic process is a delayed response to deglaciation as compensatory mantle flowage gradually adjusts the geoid to the transfer of mass from circumpolar ice sheets to the global ocean. Water withdrawn from tropical seas by equatorial ocean siphoning fills space created by postglacial collapse of submerged annular arches surrounding formerly glaciated regions, and by postglacial tilt of drowned continental shelves downward toward adjacent ocean basins loaded by the excess mass of meltwater (Mitrovica and Milne, 2002).

The mid-Holocene hydro-isostatic highstand peaked no later than ca. 2000 B.C., well before any human presence in Remote Oceania, but highstand conditions persisted within the island groups that nurtured ceramic cultures until ca. 1200 B.C. or later, overlapping the earliest phases of human occupation (Dickinson, 2003). Consequently, the oldest coastal habitation sites yielding Early Pre-Latte pottery in the Mariana Islands at the northwest limit of Remote Oceania, and Lapita pottery in Tonga at the southeast limit of Oceanian ceramic cultures, lie on mid-Holocene paleoshorelines perched well above and inland from modern shorelines (Butler, 1994; Dickinson et al., 1999b; Burley et al., 2001). Intervening island groups experienced a similar prehistory that is not yet as well documented archaeologically.

Modern coastal geomorphology is accordingly not a reliable guide to the distribution of potential temper sands available to prehistoric potters during initial cultural phases in Pacific Oceania. As the gradual post-mid-Holocene hydro-isostatic decline in regional sea level was monotonic and approximately linear (Dickinson, 2001b, 2003), a slow approach to modern conditions developed over time, but for many centuries after sea level began to fall coastal environments were in a state of flux.

In places where detailed paleolandscape analysis has been undertaken, fossil sources of temper sand can be identified. On the island of 'Uiha in the Ha'apai Group of central Tonga, for example, a mid-Holocene paleobeach ridge at least a kilometer long is composed of black placer sand that was reworked from Pleistocene tephra (Dickinson et al., 1999b). The tephra deposits form a blanket over the interior of the island and doubtless extended seaward, during synglacial eustatic drawdowns in sea level, over ground now forming submerged shoals along the flanks of the island. The crest of the paleobeach ridge, at present covered by soil and dense vegetation, stands 2 m higher in elevation than the berm crest topping the modern beach face of white calcareous sand along the present shoreline. When the paleobeach ridge was still an active geomorphic feature, prehistoric potters would have had access to an attractive supply of temper sand that is unavailable under present landscape conditions. Analogous fossil placers, or other sand deposits not related to active sedimentological processes, are probably widespread in Pacific Oceania. Fossil sand deposits, rather than modern sand accumulations, were the sources of temper in prehistoric times.

GEOTECTONIC TEMPER CLASSES

The petrologic affinities of terrigenous temper sands in Pacific Oceania are governed by the distribution of geologic provinces controlled by the geotectonic framework of the various island groups. Five fundamental temper classes (Fig. 4; Table 1) can be identified (modified after Dickinson and Shutler, 1968, 1971, 1979, and Dickinson, 1998a):

1. *Oceanic basalt* tempers occur along hotspot chains of island shield volcanoes that grew from seamounts built as volcanic edifices atop the Pacific plate of lithosphere as it drifted across magma hearths within the underlying mantle. Detritus was derived mainly from mafic (basaltic to basanitic) lavas and breccias, with minor debris from subordinate differentiates of the mafic magmas (such as hawaiite and trachyte). Minor detritus was derived also from local exposures of eroded subvolcanic intrusions forming feeder dikes, sills, necks, and plug domes

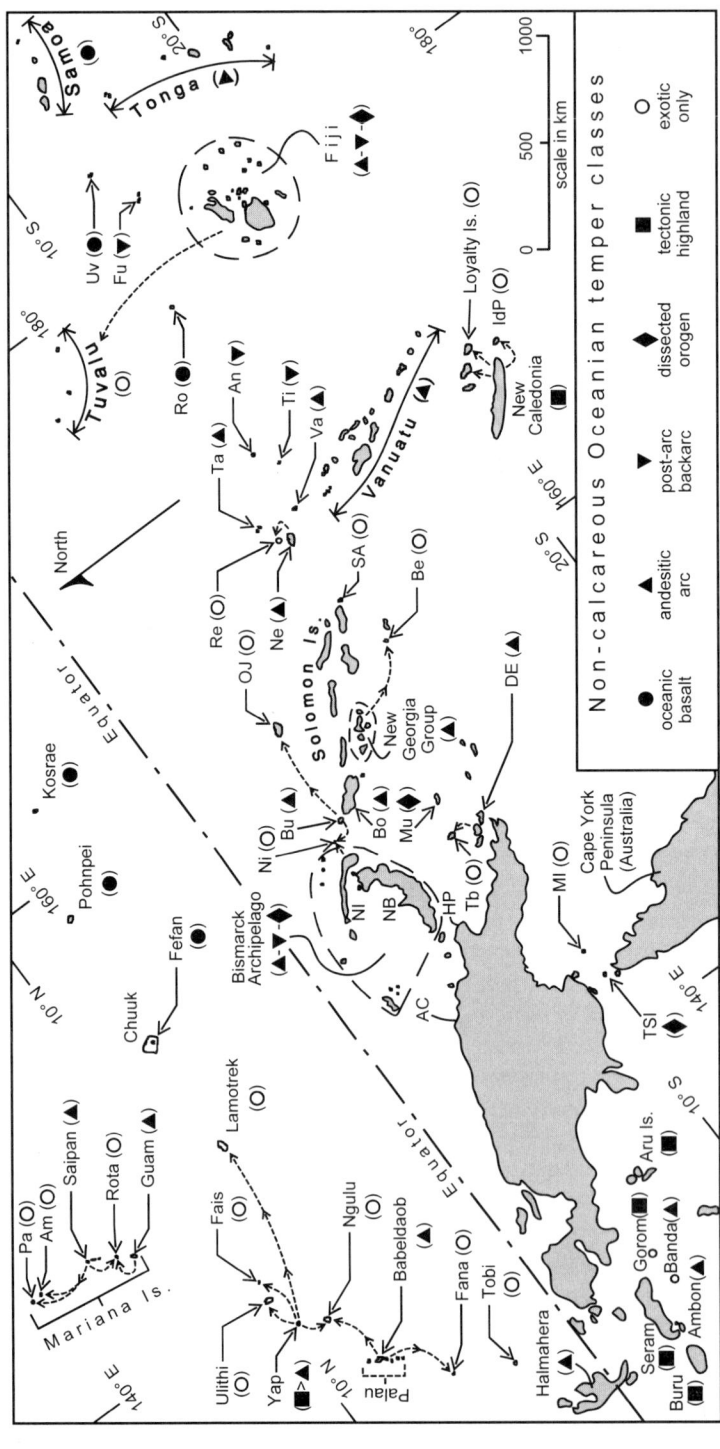

Figure 4. Islands and island groups of Pacific Oceania showing the known distribution of noncalcareous (terrigenous) temper classes (smaller islands lacking known ceramic sites not shown). Unshaded islands are atolls or clusters of calcareous islets. Dashed arrows denote inferred ceramic transfer to sites yielding sherds with exotic tempers only. Selected islands and island clusters: Am—Alamagan; An—Anuta; Be—Bellona; Bo—Bougainville; Bu—Buka; DE—D'Entrecasteaux Islands; Fu—Futuna (and nearby Alofi); IdP—Île des Pins; Ml—Murray Islands; Mu—Muyua (Woodlark); Ne—Nendö; Ni—Nissan; NB—New Britain; NI—New Ireland; OJ—Ontong Java; SA—Santa Ana; Pa—Pagan; Re—Reef Islands; Ro—Rotuma; Ta—Taumako; Tb—Trobriand Islands; Ti—Tikopia; TSI—Torres Strait Islands; Uv—Uvea (Wallis); Va—Vanikolo. Locales on New Guinea: AC—Aitape coast; HP—Huon Peninsula.

TABLE 1. SAND GRAIN TYPES IN DIFFERENT GEOTECTONIC TEMPER CLASSES

Temper class	Monominerallic grains	Polyminerallic grains
Oceanic basalt	CLINOPYROXENE, hornblende, PLAGIOCLASE, OLIVINE	BASALT or BASANITE, dolerite, trachyte
Andesitic arc	biotite, PYROXENE, HORNBLENDE, PLAGIOCLASE, olivine, oxyhornblende, quartz	ANDESITE, BASALT, BASALTIC ANDESITE, dacite, dolerite, rhyodacite
Postarc/backarc	CLINOPYROXENE, hornblende, PLAGIOCLASE, olivine	BASALT, dolerite, rhyodacite (or rhyolite), SHOSHONITE
Dissected orogen	biotite, pyroxene, epidote, hornblende, K-feldspar, PLAGIOCLASE, QUARTZ	argillite-hornfels, chert-metachert, MICROGRANITIC, slate-phyllite, VOLCANIC-METAVOLCANIC
Tectonic highland (quartzose)	feldspars, glaucophane, hornblende, QUARTZ	argillite, CHERT-METACHERT, quartzite, SLATE-PHYLLITE
Tectonic highland (ophiolitic)	chrome-spinel, PYROXENE, epidote, olivine, PLAGIOCLASE	BASALT-METABASALT, microgabbro, serpentinite

Notes: all-capitals, abundant grain types; lower-case, subordinate grain types; some sand grains derived from volcanic rocks (right column) are vitric grains composed in whole or in part of volcanic glass.

composed of igneous rock more coarsely crystalline than eruptive products but similar in mineralogical composition. Magma sources were generated by pressure release on upwelling mantle without crustal influences.

2. *Andesitic arc* tempers occur along intraoceanic island arcs where subduction of oceanic lithosphere at trenches triggers mantle melting to build magmatic arcs along the edges of overriding plates of oceanic lithosphere. Magmas incorporate volatile elements and other materials from the descending plates of lithosphere, and feed heterogeneous eruptive suites that include lavas and volcaniclastic strata ranging from basaltic through andesitic to dacitic or rhyodacitic in composition, with proportions of rock types varying from arc to arc and from place to place along individual arc axes. Detritus was derived mainly from volcanic edifices, but also from eroded subvolcanic intrusions emplaced as hypabyssal stocks with associated swarms of dikes and sills.

3. *Postarc-backarc* tempers occur along extinct island arcs where arc magmatism has been extinguished, along remnant arcs stranded behind active frontal arcs by backarc seafloor spreading that has formed intervening marginal seas, and within marginal seas where backarc eruptions have locally formed isolated islands. Detritus was derived principally from volcanic assemblages either dominated by mafic lavas and breccias, or bimodal with mafic and felsic end members but only rare rocks of intermediate composition. Postarc volcanic assemblages overlie older arc assemblages, either on outcrop or at depth, whereas backarc assemblages are erupted through oceanic crust. Postarc igneous suites are varied but include shoshonitic and other alkalic lavas and breccias unfamiliar along active island arcs. Magma origins are complex and poorly understood but involve the rise of mantle melts through rifted island arcs, with the opportunity for interaction of mantle-derived magmas with crustal materials.

4. *Dissected orogen* tempers are mineralogically complex sands that occur where uplift and deep erosion of island arcs have exposed the plutonic roots of island-arc structures. Detritus was derived from granitic and more mafic intrusions, metamorphic envelopes surrounding the intrusive bodies, and both volcanic and sedimentary strata capping ancient island arcs. Dissected orogen tempers are potentially gradational to andesitic arc tempers where the level of erosion along an arc trend is intermediate in depth, but the presence of coarsely crystalline plutonic detritus in significant proportions serves as a criterion for recognition of dissected orogen tempers. Neither plutonic nor metamorphic detritus is present in the volcanic sands of oceanic basalt, andesitic arc, and postarc-backarc tempers.

5. *Tectonic highland* tempers are mineralogically varied sands derived from uplifted and eroded subduction complexes assembled by tectonic processes of thrusting and folding at oceanic trenches or along crustal suture belts. Principal variants include (1) quartzose detritus from metasedimentary rocks that represent deformed seafloor strata or underthrust continental strata and (2) ophiolitic detritus from mafic and ultramafic igneous and metamorphic rocks of the ophiolite succession that forms the crust and uppermost mantle of oceanic lithosphere. In different instances, the metasedimentary and ophiolitic assemblages have been overthrust above either undeformed seafloor or continental crustal profiles. Sand grains of radiolarite (chert), serpentinite, chrome-spinel, or blueschist in temper sands are especially diagnostic of tectonic highland tempers.

Geotectonic Patterns

The western and southwestern fringes of the Pacific Ocean are adorned by festoons of island arcs controlled by intricate plate interactions (Fig. 5). Trenches, spreading centers, and transform faults are linked in complex geometric patterns to define the boundaries of multiple microplates aligned regionally along junctures between much larger Pacific, Indo-Australian, and Eurasian plates of lithosphere. The configurations and evolution of the plate boundaries has controlled the distribution of different geotectonic temper classes (Fig. 4) in the varied island groups lying within the region of Oceanian ceramic cultures (Fig. 1).

Although Figure 5 provides a regional overview of the geotectonic framework of all the islands of interest, the scale is inadequate to depict geotectonic features significant for temper analysis shown on subregional geotectonic maps for western (Fig. 6A), central (Fig. 7), and eastern (Fig. 8) segments of the region of Oceanian ceramic cultures. Because the present distributions of provenance types that control geotectonic temper classes are contingent in part upon the past course of geologic history (Dickinson, 1973a), subregional tectonic reconstructions of past geotectonic patterns (Figs. 6B and 9) provide insights for understanding provenance relations of Oceanian temper sands. Many islands of interest expose ancestral geologic assemblages that pertain to previous geotectonic patterns not readily apparent from present plate configurations.

Western Subregion

Ceramic sites extending southward from the Mariana Islands to the Banda arc of eastern Indonesia (Fig. 6A) lie along the eastern edge of the Philippine Sea plate and southward into an area of complex interaction between the Pacific, Eurasian, and Indo-Australian plates near the western end of New Guinea. The Pacific seafloor is subducted downward to the west beneath the Philippine Sea plate along the Izu-Bonin-Mariana arc-trench system, and the seafloor of the Philippine Sea plate is in turn subducted downward to the west beneath the Eurasian plate along the Philippine and Ryukyu Trenches (Fig. 5). Farther south, the continental block of Australia and New Guinea on the Indo-Australian plate is being drawn into a subduction zone linked to the Banda arc on the extremity of the Eurasian plate west of the Sahul Shelf (Fig. 6A). North of the Banda arc, lateral motion between the Pacific and Indo-Australian plates is accommodated by strike slip along the Sorong fault system of New Guinea (Figs. 5 and 6A), with easterly segments locally termed the Yapen and Bewani faults (not labeled separately on Figs. 6A and 7). The Sorong fault system is currently the principal active plate boundary between the Pacific and Indo-Australian plates north of the Australia–New Guinea continental block, with only limited seismicity along the New Guinea and Manus-Mussau Trenches.

The Mariana island arc along the eastern edge of the Philippine Sea has counterparts to the south in Yap and Palau (Fig. 6A), but subduction at the Yap and Palau Trenches involves the Caroline plate rather than the Pacific plate (Fig. 5). As plate motions along northern and eastern edges of the Caroline plate are apparently slower than plate motions along western and southern edges, the Caroline plate can be regarded as a dislocated appendage of the Pacific plate to the northeast. Farther south, multiple plate boundaries in the Maluku (=Molucca) area of eastern Indonesia reflect much more intricate disruption of the Indo-Australian and Eurasian plates near their juncture with the Philippine Sea plate north of the Banda Sea (Figs. 5 and 6A).

A tectonic reconstruction for 30 Ma (mid-Oligocene) clarifies the key ancestral relationships of geotectonic elements fringing the modern Philippine and Banda Seas (Fig. 6B). Closure of post–30 Ma interarc basins of the Mariana arc-trench system restores the Paleogene basement of islands lying within the forearc belt of the active Mariana arc-trench system to the flank of the Palau-Kyushu Ridge, a remnant arc that extends southward to include the Paleogene volcanic assemblage of Palau. Halmahera and an accreted arc now forming the northern fringe of New Guinea (West Irian) were linked along strike as a south-facing arc-trench system (with subduction downward to the north). The distance at 30 Ma between the southern end of the north-south Mariana-Yap-Palau arc complex and the east-west Halmahera–north New Guinea arc complex is uncertain because of Neogene subduction (downward to the south) of oceanic lithosphere at the New Guinea Trench and Manokwari Trough (Fig. 6A). Those Neogene subduction zones formed by reversal of arc polarity after the north New Guinea arc accreted in middle Miocene time (15–12 Ma) to continental crust underlying southern New Guinea. Similarly, both the distance and the azimuthal position of the passive continental margin along the northern edge of the Australia (–New Guinea) continental block at 30 Ma are uncertain, both because of subduction of seafloor (downward to the north) beneath the north New Guinea arc while it was still in an intraoceanic position (Fig. 6B) and because of strike slip of uncertain net amount along the Sorong fault (Fig. 6A) after arc accretion. The Banda arc-trench system (Figs. 5 and 6A) of eastern Indonesia does not appear on the reconstruction for 30 Ma because the Banda arc is a Neogene geotectonic feature linked to development of the backarc basin forming the Banda Sea (Honthaas et al., 1998). Backarc Banda seafloor is Miocene-Pliocene (6.5–3.5 Ma) in age (Hinschberger et al., 2005).

Central Subregion

East of New Guinea (Fig. 7), the modern geotectonic pattern is dominated by active magmatic arcs of so-called reversed polarity along the southern edge of the Bismarck Sea and along the Solomon island chain where seafloor of marginal seas adjacent to the Australia–New Guinea continental block is subducted downward to the north beneath the Pacific plate. (The term reversed polarity stems from the perspective that normal plate

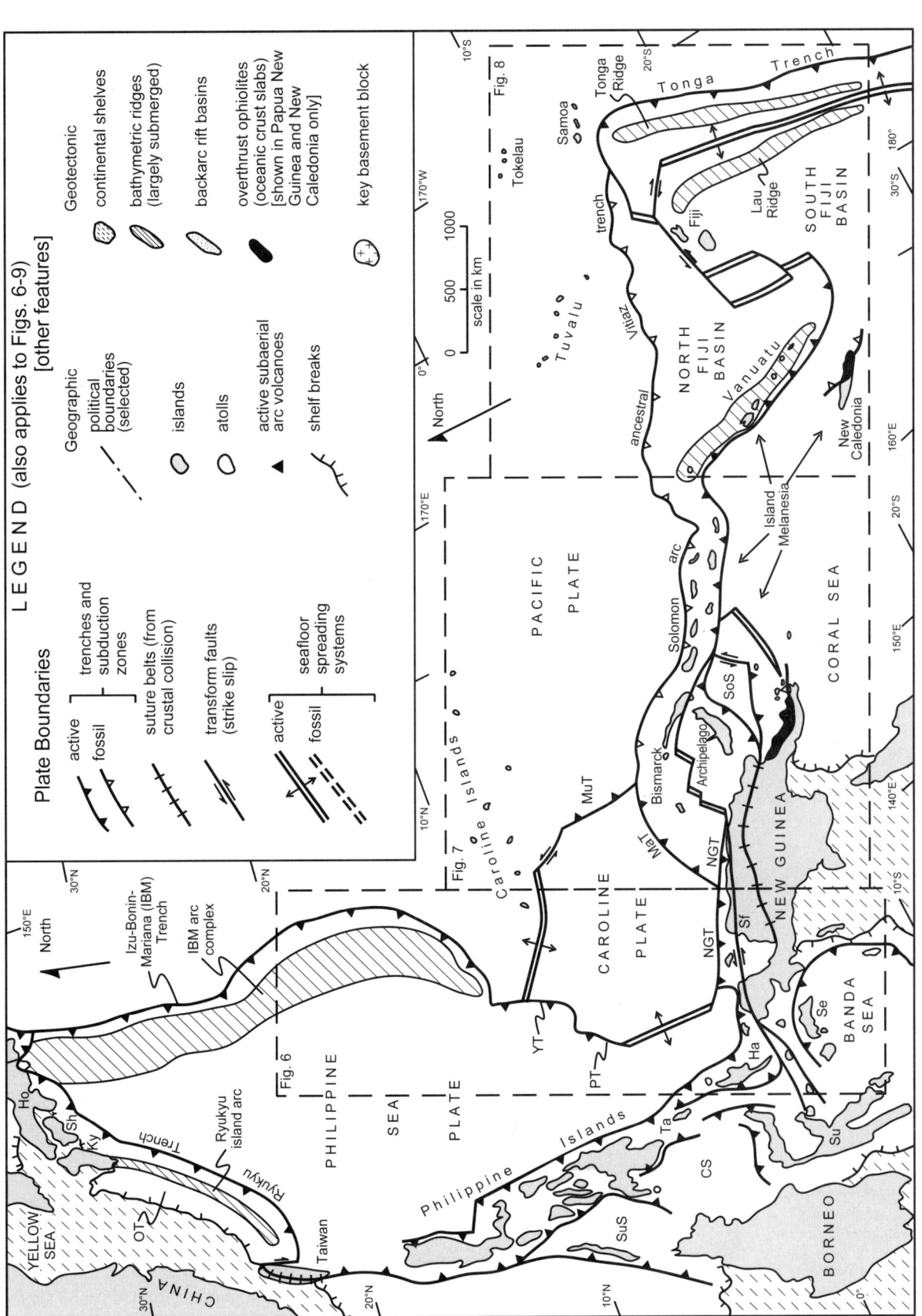

Figure 5. Geotectonic framework of the western and southwestern margins of the Pacific Ocean including the small islands of Remote Oceania within the ceramic culture region (Fig. 1) and adjoining continental blocks and larger islands of Near Oceania (legend applies also to Figs. 6–9). Geotectonic features: CS—Celebes Sea; MaT—Manokwari Trench; MuT—Mussau Trench; NGT—New Guinea Trench; OT—Okinawa Trough (extensional backarc basin); PT—Palau Trench; Sf—Sorong fault; SoS—Solomon Sea; SuS—Sulu Sea; YT—Yap Trench. Islands: Ha—Halmahera; Ho—Honshu; Ky—Kyushu; Se—Seram; Sh—Shikoku; Su—Sulawesi; Ta—Talaud. Adapted after Silver and Moore (1978), Hamilton (1979), Lee and Lawver (1995), Bautista et al. (2001), Hall (2002), Hu et al. (2002), and Sibuet and Hsu (2004) (approximate outlines shown).

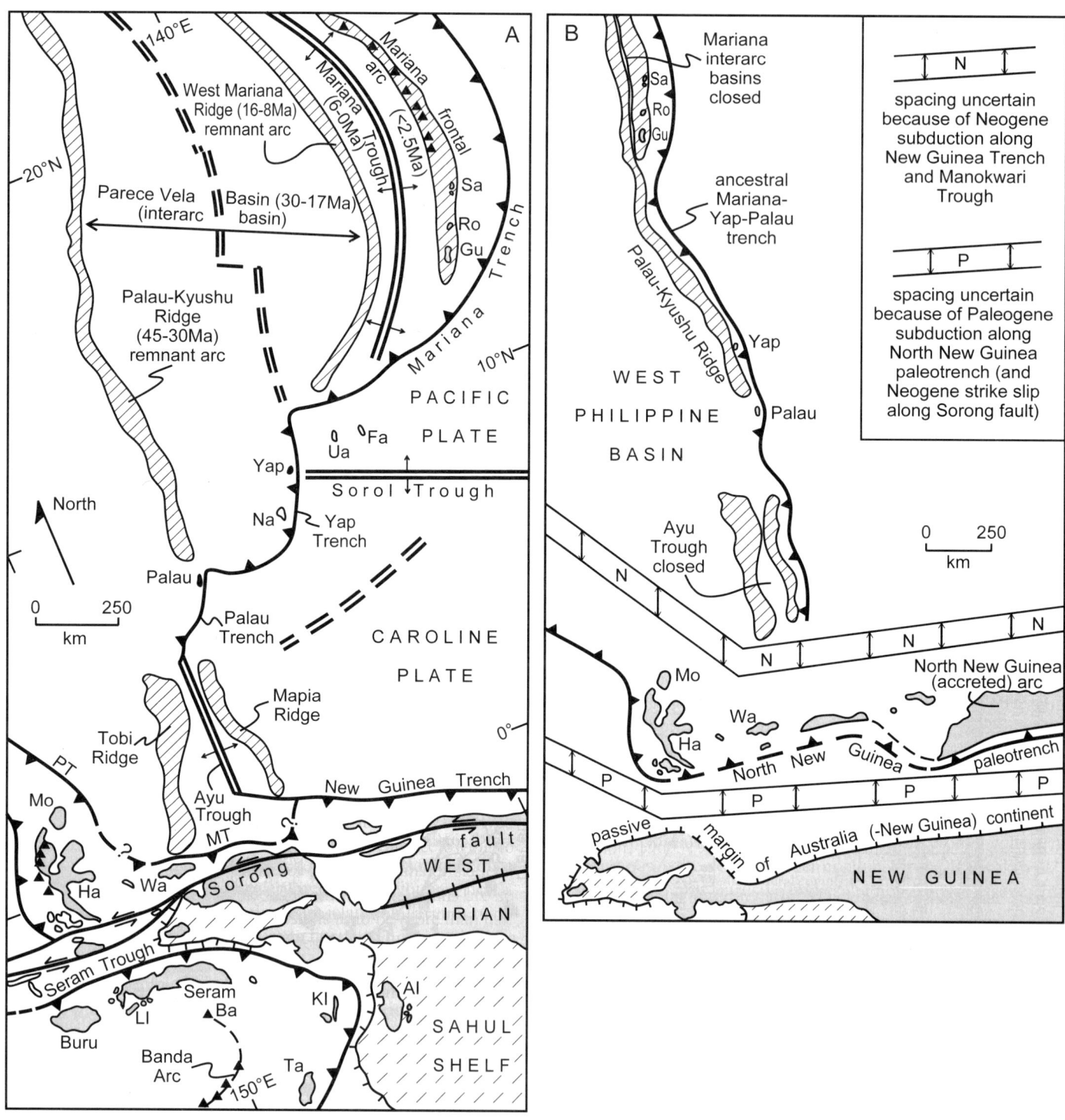

Figure 6. Geotectonic pattern of the Mariana-Yap-Palau arc-trench system along the eastern edge of the Philippine Sea plate and the Maluku (Molucca) region of eastern Indonesia to the south (see Fig. 5 for location and legend). A: modern configuration. B: mid-Oligocene reconstruction (ca. 30 Ma). Geotectonic features: PT—Philippine Trench; MT—Manokwari Trough. Islands: Ba—Banda; Gu—Guam; Ha—Halmahera; Mo—Morotai; Ro—Rota; Sa—Saipan; Ta—Tanimbar, Wa—Waigeo. Atolls: Fa—Fais; Na—Ngulu; Ua—Ulithi. Island clusters: AI—Aru Islands; KI—Kai Islands; LI—Lease Islands. Adapted after Bracey (1975), Hamilton (1979), Meijer et al. (1983), Hegarty et al. (1983), Hegarty and Weissel (1988), Milsom et al. (1992), Lee and Lawver (1995), Hall (1996, 2002), Dickinson (2000b), Charlton (2000), Hawkins (2003), and Hinschberger et al. (2005).

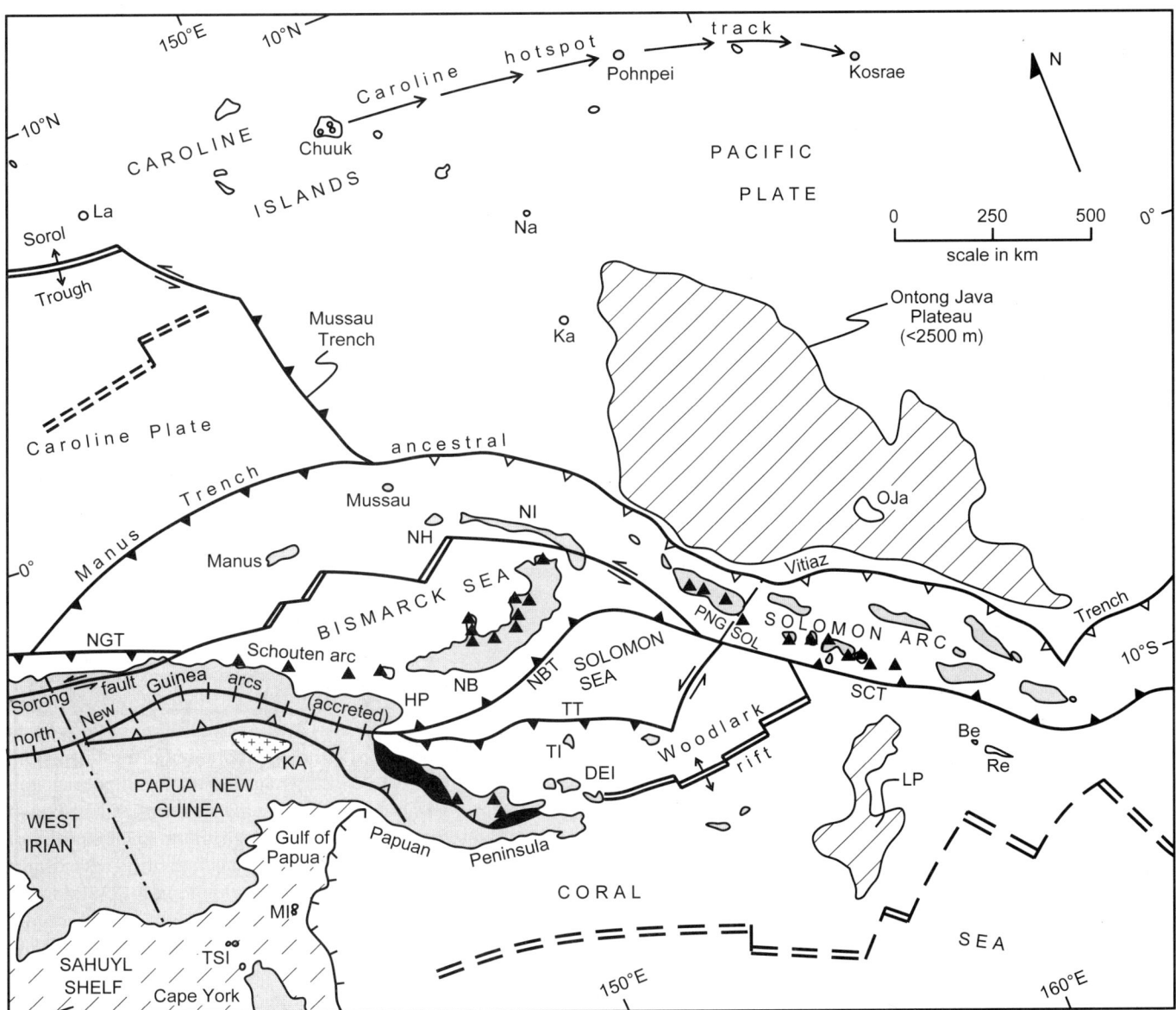

Figure 7. Geotectonic pattern of the Bismarck Archipelago, Solomon island arc, and adjoining regions (see Fig. 5 for location and legend). Political entities (at interisland boundary): PNG—Papua New Guinea; SOL—Solomon Islands. Geotectonic features: KA—Kubor Anticline; LP—Louisiade Plateau (microcontinental fragment); NBT—New Britain Trench; NGT—New Guinea Trench; SCT—San Cristobal Trench; TT—Trobriand Trough. Islands: Be—Bellona; HP—Huon Peninsula (of Papua New Guinea); NB—New Britain; NH—New Hanover; NI—New Ireland; Re—Rennell. Atolls: Ka—Kapingamarangi; La—Lamotrek; Na—Nukuoro; OJa—Ontong Java. Island clusters: DEI—D'Entrecasteaux Islands; MI—Murray Islands; TI—Trobriand Islands; TSI—Torres Strait Islands. Adapted after Jaques and Robinson (1977), Hamilton (1979), Weissel et al. (1982), Hegarty et al. (1983), Davies and Jaques (1984), Hegarty and Weissel (1988), Abbott et al. (1994a), Dickinson (2000a, 2002a), and Quarles van Ufford and Cloos (2005).

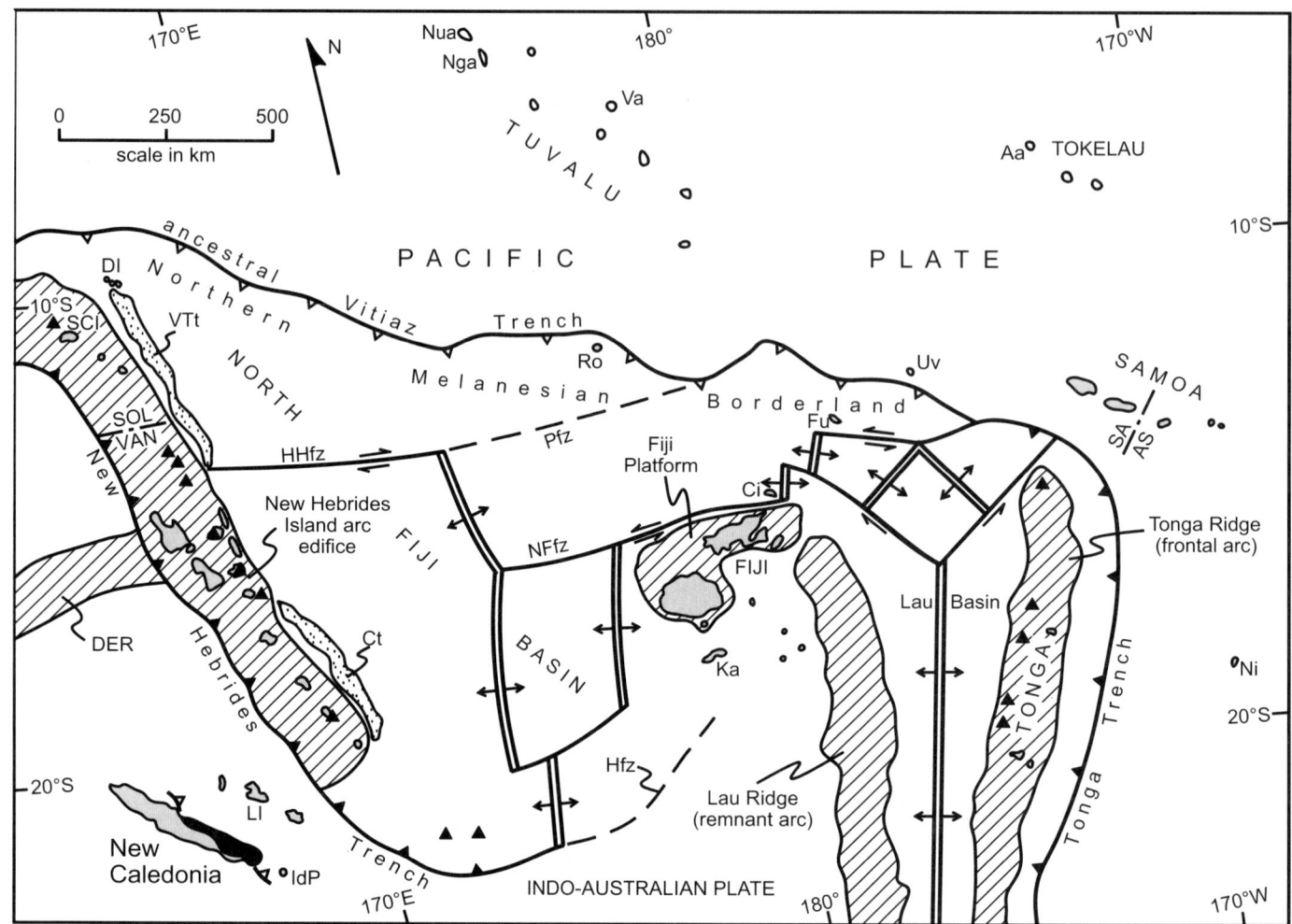

Figure 8. Geotectonic pattern of southwest Pacific island groups east and southeast of Solomon island arc (see Fig. 5 for location and legend). Complex patterns of backarc seafloor spreading schematic within Lau Basin and North Fiji Basin. Political entities (at interisland boundaries): AS—American Samoa; SA—Samoa (formerly Western Samoa); SOL—Solomon Islands; VAN—Vanuatu. Backarc rift troughs: Ct—Coriolis; VTt—Vot Tande. Other geotectonic features: DER—D'Entrecasteaux Ridge (aseismic); Hfz—Hunter fracture zone (dormant); HHfz—Hazel Holme fracture zone; NFfz—North Fiji fracture zone; Pfz—Pandora fracture zone (dormant). Islands: Ci—Cikobia; Fu—Futuna-Alofi; IdP—Île des Pins; Ka—Kadavu; Ni—Niue; Ro—Rotuma; Uv—Uvea (Wallis). Atolls: Aa—Atafu; Nga—Nanumanga; Nua—Nanumea; Va—Vaitupu. Island clusters: DI—Duff Islands; LI—Loyalty Islands; SCI—Santa Cruz Islands. Adapted after Weissel (1977), Louat and Pelletier (1989), Auzende et al. (1994), Maillet et al. (1995), Pelletier et al. (1998, 2000, 2001), and Dickinson (2001a, 2002b).

behavior involves subduction of the Pacific plate beneath continental margins and island arcs surrounding the Pacific Ocean.) Active seafloor spreading within the Bismarck Sea and along the Woodlark rift system, subordinate subduction of normal polarity along the Trobriand Trough of the Solomon Sea, and transform faults linking local spreading centers and trenches are in the process of modifying the morphology of both marginal seas and island arcs east of New Guinea. A westward extension of the reversed-polarity New Britain arc has accreted to the margin of the Papua New Guinea continental block, by consumption of intervening oceanic crust, as the Finisterre Range of the Huon Peninsula (Fig. 7). The crustal suture belt is exposed along an intramontane valley to the south of the Finisterre Range (Abbott et al., 1994a). The offshore Schouten arc (Fig. 7) reflects continuing subduction of oceanic lithosphere into the mantle beneath the accreted arc structure.

Until Miocene time (Fig. 9), Paleogene volcanogenic assemblages of the Bismarck Archipelago including the accreted Finisterre arc segment of Papua New Guinea, and of the Solomon arc, formed a linked island-arc complex of normal polarity associated with subduction of the Pacific plate downward to the southwest along the ancestral Vitiaz paleotrench flanking the Solomon arc on the northeast. Reversal of arc polarity to form the modern system of reversed-polarity arc-trench systems was induced by arrival of the Ontong Java Plateau, composed of overthickened oceanic crust (Gladczenko et al., 1997; Phinney et al., 1999),

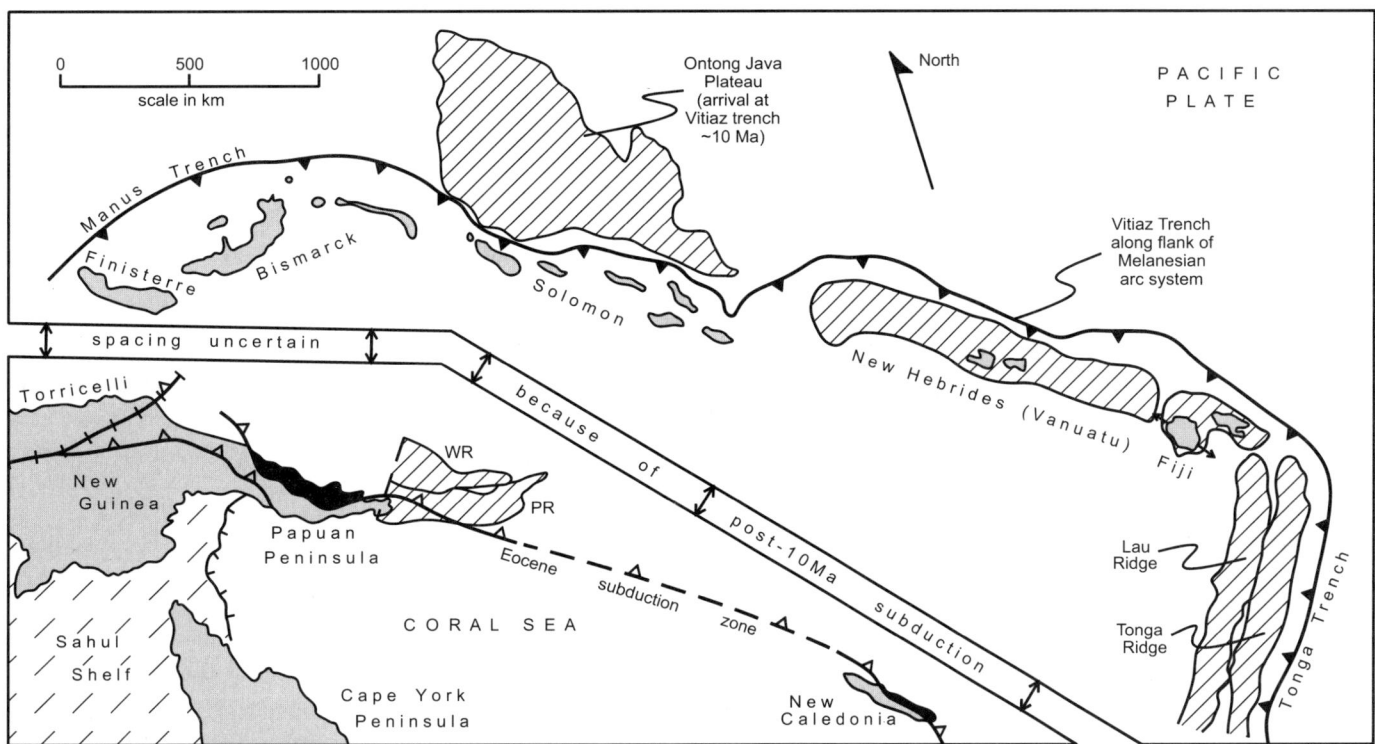

Figure 9. Tectonic reconstruction (ca. 10 Ma) of island Melanesia and associated island groups (Figs. 7 and 8) before polarity reversal of the Vitiaz or ancestral Melanesian arc including Finisterre, Bismarck, Solomon, Vanuatu (New Hebrides), and Lau segments (see Fig. 5 for legend). Configuration and orientation of Vitiaz (-Manus) paleotrench arbitrarily held fixed (Figs. 7 and 8), except for restoration of sinistral offsets of eastern segments along North Fiji and Pandora fracture zones (Fig. 8) of northern Melanesian borderland west of north end of Tonga Trench (limited indentation of the Vitiaz arc by attempted subduction of the Ontong Java Plateau opposite the Solomon segment of the Melanesian arc system is suggested by local deflection of the Vitiaz trend). Bismarck Sea, Lau Basin, North Fiji Basin, and Woodlark Basin closed by restoration of post–7.5 Ma seafloor spreading. Fiji platform back-rotated (45° clockwise) to bring the axis of the Wainimala orogen (double-headed arrow) of Viti Levu into alignment with the ancestral Vitiaz arc trending northwest from the Lau Ridge to the prereversal New Hebrides arc (back-rotated 45° counterclockwise to close the North Fiji Basin). Flanks of closed Woodlark Basin: PR—Pocklington Rise; WR—Woodlark Rise. Exotic Torricelli arc accreted to northern New Guinea ca. 12 Ma. Adapted after Falvey (1978), Kroenke (1984), Davies and Jaques (1984), Cullen and Pigott (1989), Wells (1989a, 1989b), Falvey et al., (1991), Hill and Raza (1999), Dickinson (2000a, 2001a, 2002b), Cluzel et al. (2001), and Quarles van Ufford and Cloos (2005). See Yan and Kroenke (1993), Lee and Lawver (1995), and Hall (2002) for regional context.

at the Vitiaz paleotrench (Dunkley, 1983; Kroenke, 1984). The immense crustal bulk of the world's largest oceanic plateau (Petterson, 2004) choked subduction at the Vitiaz paleotrench to force reversal of arc polarity as a means of continuing plate convergence between the Pacific and Indo-Australian plates.

Intra-Pacific plate reconstructions imply that the Ontong Java Plateau arrived at the Vitiaz paleotrench in Late Miocene time, not more than ~2.5 million years before or after 10 Ma (Wells, 1989a, 1989b; Musgrave, 1990). The southwestern edge of the oceanic Ontong Java Plateau was subsequently incorporated tectonically into the northeastern flank of the Solomon arc (Tejada et al., 1996, 2002; Petterson et al., 1997) to form the basement of Malaita and Santa Isabel. Structural interaction between the plateau and the arc may have indented the ancestral Vitiaz arc trend by ~175 km (Fig. 9). The oldest postreversal volcanic assemblages of the reversed-polarity Solomon arc date to ca. 7.5 Ma (Petterson et al., 1997). Older volcanogenic assemblages exposed in the Bismarck Archipelago and along the Solomon arc were erupted along the ancestral Vitiaz island arc, rather than along the modern reversed-polarity island arcs.

After Vitiaz arc reversal, the originally intraoceanic Finisterre arc of the Huon Peninsula was accreted to Papua New Guinea beginning in Pliocene time (Abbott et al., 1994b). Before Vitiaz polarity reversal at ca. 10 Ma, however, originally intraoceanic island arcs were accreted to the northern flank of New Guinea in the Torricelli Range (Fig. 9), and westward along the northern coast (Fig. 6), by ca. 12 Ma (Hill and Raza, 1999; Quarles van Ufford and Cloos, 2005). The geotectonic relation of these westerly accreted arcs of New Guinea to the Vitiaz arc remains uncertain (Fig. 9). Also uncertain are the regional geotectonic affinities of an ophiolitic slab of oceanic crust thrust westward over the eastern flank of the Papuan Peninsula (Figs. 7 and 9) along a Paleocene subduction zone (Lus et al., 2004) prior to mid-Eocene time (Davies and Jaques, 1984). Emplacement of the Papuan ophiolite may have been coeval with wedging of the Papuan Peninsula away from the Australian continental margin by late Paleocene to early Eocene seafloor spreading (Dickinson, 2002b) in the Coral Sea (Fig. 7).

Eastern Subregion

East of the Bismarck-Solomon area, modern geotectonic elements include the reversed-polarity New Hebrides (Vanuatu) island arc and the normal-polarity Tongan island arc (Fig. 8). The two arc-trench systems, each displaying backarc spreading to form marginal seas in the North Fiji and Lau Basins, are linked geotectonically by a geometrically complex transform system crossing intervening Fiji. Along its principal strand at the North Fiji fracture zone (Fig. 8), the transform accommodates lateral motion between the edge of the Pacific plate lying along the northern Melanesian borderland to the north and the edge of the Fiji platform lying at the northeastern extremity of the Indo-Australian plate to the south.

Before polarity reversal of the Vitiaz island arc in Late Miocene time (ca. 10 Ma), the crustal underpinnings of the New Hebrides island arc, the Fiji platform, and Tonga were aligned along strike as a unified segment of the Vitiaz arc subducting Pacific seafloor downward to the southwest (Fig. 9). Including its Finisterre-Bismarck-Solomon segments farther west, the ancestral Vitiaz arc formed an integrated Melanesian island arc of normal polarity extending for ~5000 km eastward from New Guinea (Fig. 9). The Tongan segment of the Vitiaz arc included both the modern Tonga Ridge frontal arc and the modern Lau Ridge remnant arc now separated by the interarc Lau Basin.

Restoration of the New Hebrides island-arc structure to an initial position along the northern Melanesian borderland parallel to the trend of the ancestral Vitiaz arc requires counterclockwise back-rotation by ~45° (Fig. 9), compatible with the 52° ± 12° implied by paleomagnetic data (Musgrave and Firth, 1999). Restoration of the Fiji platform requires clockwise back-rotation of ~45° to bring structural trends of the Paleogene Wainimala orogen of Viti Levu into alignment with the ancestral Vitiaz arc. Suggested greater rotation of the Fiji platform (Musgrave and Firth, 1999) is discounted here because it would produce an unlikely tectonic reconstruction with Wainimala tectonic trends lying athwart the Vitiaz arc trend (Dickinson, 2001a).

On New Caledonia, a disrupted fragment of the Gondwanan continent (Dickinson, 2002b), Paleogene subduction thrust an ophiolitic slab of oceanic crust across the island from the northeast late in Eocene time (Cluzel et al., 1997, 2001). Emplacement of the ophiolite occurred when New Caledonia was in an intraoceanic position as an elongate continental sliver or ribbon continent after Late Cretaceous to Early Eocene separation from the Gondwanan margin. Eocene overthrusting of ophiolite on the Papuan Peninsula and on New Caledonia were related events (Fig. 9) separated by only a few million years in time, but the regional relations of the Paleogene subduction zone that was responsible remain uncertain (Dickinson, 1973a; Parrot and Dugas, 1980; Aitchison et al., 1995). The volcanic underpinnings of the coralline Loyalty Islands northeast of New Caledonia (Fig. 8) represent a remnant, however, of an island arc of reversed polarity that was paired geotectonically with the subduction zone that overthrust forearc ophiolite over the ribbon continent of New Caledonia (Dickinson, 1973a; Milsom, 2003).

TEMPER TEXTURAL PROPERTIES

Oceanian temper sands include terrigenous aggregates composed of rock-forming silicates and oxides, aggregates of exclusively calcareous grains composed of calcite or aragonite, and mixtures of the two in hybrid temper sands. Terrigenous tempers are mostly beach sands, but include stream sands and rarely coastal dune sands. Calcareous and hybrid tempers containing modern reef debris are beach sands, but some hybrid tempers are stream sands in which the calcareous grains are detrital limeclasts reworked from limestone bedrock. Rare nondetrital sand tempers include both broken-sherd (grog) and crushed-rock aggregates. Terrigenous natural temper sands residual within impure clay bodies accompany grog particles in some composite tempers. The textures of temper aggregates in relation to clay paste and the internal textures of nonterrigenous or nondetrital temper grains allow distinctions to be drawn between tempers of different origins.

Beach and Stream Sand Tempers

Island beach sands are well sorted but variably rounded, most commonly subrounded to subangular rather than well rounded (Figs. A4, A6, and A15). Associated coastal dune sands, where present, are texturally similar and cannot be distinguished from beach sands by petrographic study. The limited abrasion of island beach sands, as compared to many continental beach sands, is probably a function both of limited residence time on island beaches and of subdued wave energy on shorelines protected from fair-weather surf by offshore fringing or barrier reefs.

Island stream sands are generally only moderately sorted (Figs. A8, A16, and A22), with subangular to subrounded grains, but discrimination between tempers composed of beach or stream sand is inferential and not always conclusive. Strong placer concentrations of so-called heavy minerals of high specific gravity are most commonly indicative of beach origin (Dickinson, 1994), but stream placering has been detected for some temper sands, and placering is also characteristic of the rare dune sands used as temper. Placer sand tempers are both common and widespread in Oceanian sherd suites (Figs. A23–A27).

Admixtures of calcareous grains to form hybrid terrigenous-calcareous sands are diagnostic of beach origin except where detrital limeclasts reworked from emergent limestone terraces occur locally in stream sands. To date, prominent detrital limeclasts have been observed only in sherds from the widely separated islands of Saipan in the Mariana Islands and Erromango in Vanuatu. Debris derived from modern reefs can be distinguished from detrital limeclasts by the internal microscopic textures and fabrics of calcareous grains. Modern reef debris consists of a mixture of skeletal grains (bioclasts) displaying oriented internal crystalline structures produced by coralline or molluscan organisms and of cryptocrystalline micrite either of algal origin or composed of cemented lime mud. By contrast, reworked detrital limeclasts are composed dominantly of microspar with more coarsely microgranular textures of diagenetic origin.

In many Oceanian sherd suites, postburial leaching of calcareous grains by groundwater percolating through the subsurface of archaeological sites has produced vacuoles (secondary pores) of sand size and shape that can impart the illusion of untempered wares if the significance of the vacuoles is not appreciated (Dickinson, 1995, 2000a). Where calcareous grains are only partially leached, or leached only from parts of a sherd, their former presence is not in doubt, but wholesale removal of calcareous grains from a sherd suite presents a more severe interpretive challenge. Nevertheless, the rounded shape of vacuoles produced by dissolution of calcareous grains is distinctive, both megascopically and microscopically. When sherds are well impregnated with epoxy resin before slicing, plucking of grains during the grinding of thin sections can be excluded as the origin of sand-sized vacuoles visible in thin section. Any ambiguity on that score can be removed entirely by impregnating with blue-dyed epoxy, which will fill vacuoles produced by dissolution of calcareous grains but would not fill any vacuoles produced by plucking of grains during sawing or grinding to make a thin section.

Natural and Manually Added Tempers

Beach sand tempers are all manually added tempers, but stream sand tempers may be either natural or manually added. Distinct contrasts in grain size between the finest sand grains of temper and the coarsest silt particles embedded in surrounding clay paste imply manually added temper, whereas gradation in grain size between the finest temper grains and the coarsest silt within clay paste suggests natural temper. Where a limited amount of natural temper embedded within clay bodies as collected was augmented by manually added temper during ceramic processing, the distinction between natural and manually added temper is inherently ambiguous. The natural tempers easiest to recognize as such are composed of poorly sorted, angular, and strongly weathered grains enclosed within impure clay collected from residual regolith or colluvial soil.

Epiclastic and Pyroclastic Tempers

Almost all temper sands derived from volcanic sources in Oceanian sherd suites are epiclastic volcanic sands produced by weathering, erosion, and subsequent sedimentary transport of detritus from exposed lava or pyroclastic rocks and sediments. Grain shapes are those of sedimentary particles rather than pyroclastic tephra. In scattered instances, however, tephra aggregates collected from unconsolidated or only slightly reworked pyroclastic blankets of volcanic ash were apparently used as temper. Textural criteria for the distinction between epiclastic and pyroclastic volcanic sands may be ambiguous, however, where sedimentary reworking of pyroclastic debris was limited. Unabraded tephra particles can be identified most readily by the presence of smoothly curved concave reentrants and sharply pointed convex projections on the margins of grains composed in whole or in part of volcanic glass. The concave reentrants represent the edges of burst gas bubbles formed by the vesiculation that disrupted magma during pyroclastic eruptions, and the pointed projections reflect the shredding of magma by vesiculation as gas bubbles expanded. Tephra particles formed by individual mineral crystals do not display comparably diagnostic shapes, although mineral grains in tephra tend to be markedly angular and are commonly broken crystals reflecting the explosivity of pyroclastic eruptions.

Grog and Crushed-Rock Tempers

The identification of grog (broken-sherd) particles (Fig. A2) is dependent upon recognition of discrete sand-sized (or larger) domains of fired clay displaying different microscopic fabrics or textures, or coloration, than the surrounding clay paste forming the bulk of a sherd (Whitbread, 1986). The actual origin of grog particles in Oceanian tempers is uncertain, but most may have been derived from disaggregation of dried clay blocks or from the breakage of prefired ceramic objects that were specially prepared to produce grog, rather than from the breakage of finished ceramic vessels. Usage of grog as temper may well reflect cultural tradition, but probably also indicates a paucity of local noncalcareous sands suitable for temper. To date, grog and composite grog-terrigenous tempers are known only from the widely separated localities of Palau in the western Caroline Islands, Pohnpei in the eastern Caroline Islands, and sites on Vitiaz Strait between New Guinea and New Britain.

Manually added temper aggregates derived from artificially crushed rock are common globally in continental settings (Hodges, 1963; Rice, 1987), but rare among Oceanian tempers. Crushed-rock tempers composed of wholly unabraded angular fragments of volcanic rock of uniform composition and texture are known only from selected sites in Samoa and American Samoa. Extreme angularity, with ragged grain edges, is characteristic of the homogeneous lithic fragments in Samoan crushed-rock tempers (Fig. A3). Along the margins of some fragments of crushed volcanic rock, tiny plagioclase microlites protrude from the edges of the fragments in a manner never seen on the edges of abraded lithic fragments in natural sand aggregates.

TERRIGENOUS GRAIN TYPES

The classification of terrigenous grain types in temper sands is based on precepts developed for the treatment of sandstone detrital modes as used for provenance analysis in sedimentary geology (e.g., Dickinson, 1970; Wolf, 1971; Dickinson and Rich,

1972; Graham et al., 1976; Dickinson et al., 1979; Critelli and Ingersoll, 1995; Garzanti et al., 2000, 2002). In continental settings, quartz is commonly the most abundant mineral in ceramic temper sands (Peacock, 1970; Bishop et al., 1982; Williams, 1983), but quartz is typically absent in Pacific island tempers and is subordinate to other grain types in nearly all the tempers in which it does occur (Dickinson, 1998a). Criteria for the identification of different kinds of mineral grains and lithic fragments presented here are intended as a guide for study rather than as a definitive treatment of petrographic methodology. Inexperienced petrographers will need to consult standard textbooks on optical mineralogy to learn how to perform diagnostic operations with the petrographic microscope, and petrography texts for full accounts of rock textures and fabrics.

Grain Groups

Various generic groups of grain types are important for the classification and analysis of grain aggregates. Most fundamental is the distinction between monocrystalline silicate or oxide mineral grains and polycrystalline lithic fragments. The term lithic fragment is preferable to rock fragment, for both mineral grains and lithic fragments are derived from island bedrock. Monocrystalline sand grains are derived from mineral constituents of rocks having a crystal size coarser than silt, within the sand range or larger (>0.0625 mm in diameter). These include phenocrysts in volcanic rock, the constituent crystals of coarser-grained igneous and metamorphic rocks, and preexisting sand grains in sandstone. Lithic fragments, on the other hand, are derived from source rocks so fine-grained that the derivative sand grains are composed of multiple internal crystals. Monomineralic but multicrystalline sand grains are termed aggregate grains, which mainly include polycrystalline quartz from quartz veins and epidote grains of either hydrothermal or metamorphic origin.

The rock components that yield lithic fragments are typically aphanitic in the sense of having an internal grain size too fine for individual crystals or sediment particles to be resolvable with a standard ten-power hand lens. Lithic fragments composed of multiple internal crystals can also be derived from finer-grained varieties of phaneritic rock, in which individual crystals are discernible through a hand lens. The latter types of lithic fragments can conveniently be termed microphanerite and are most commonly derived from hypabyssal igneous rocks forming small stocks, dikes, or sills that cool at rates intermediate between those that pertain to plutonic and volcanic rocks.

Among the igneous rock assemblages of prime importance as sources for most Pacific temper sands, the cooling rate of magma is the prime control on crystallinity. In plutonic rocks such as granite and gabbro, which cool slowly at depth within the crust, there is time for various chemical constituents that enter the lattices of given minerals to diffuse through magma over distances great enough to allow crystals to grow to large size, and thereby to form uniformly phaneritic crystalline aggregates. By contrast, eruptive volcanic rocks cool so rapidly that crystals are forced to grow too fast to gather chemical constituents from appreciable volumes of magma, and the resulting crystalline aggregates are aphanitic. Where the cooling rate of a volcanic rock is fast enough, some or all of the parent magma is chilled as amorphous volcanic glass that congeals without the growth of any crystalline materials. By convention, pyroclastic or sedimentary particles of volcanic glass are regarded as lithic fragments.

Many volcanic rocks are porphyritic, containing crystals of two distinct size populations, and can yield both mineral grains and lithic fragments to derivative sediment. The larger crystals termed phenocrysts, or microphenocrysts when only apparent microscopically, grow as intratelluric crystals within slowly cooling magma chambers at depth beneath volcanoes before the rise of magma to the surface. Phenocrysts are set in a groundmass of aphanitic or glassy materials that formed during chilling of magma after eruption.

Also important for grain classification is the distinction between light and heavy minerals with specific gravities below or above ~2.8. The dominant light minerals in Pacific temper sands are feldspars, and secondarily quartz, whereas the heavy minerals are mainly ferromagnesian silicates and oxides. Heavy minerals are locally concentrated relative to light minerals during sedimentological reworking by placering on beaches, or in the bed load of streams. The specific gravity of lithic fragments depends upon internal mineralogy, but is typically intermediate between light and heavy minerals.

Opaque grains (Figs. A26 and A27) cannot be studied closely in transmitted light using a standard petrographic microscope, and can be identified mineralogically with confidence only by special techniques using reflected light. Gross distinctions can be achieved, however, by shining light off the surfaces of standard thin sections. Most opaque grains in Oceanian tempers are dominantly equant grains of magnetite (iron oxide spinel) or typically more elongate ilmenite (iron-titanium oxide), both of which are gray in reflected light and derive from igneous and metamorphic rocks. Some tempers contain pedogenic particles of iron oxide reworked from ferruginous nodules of weathering horizons. These are dominantly hydrous limonite, dull or white in reflected light, but some are anhydrous hematite, which is bright red in reflected light. For all the sand tempers studied, opaque grains are reported as opaque iron oxide without attempts at mineralogical differentiation, but most are probably magnetite.

Mineral Identification

Identifications of mineral grains and individual mineral constituents of lithic fragments in temper sands are based on observations of optical properties in transmitted light with the aid of standard accessories to the petrographic microscope. Optical properties are discussed as they appear in thin sections of standard thickness (0.03 mm), but charts in standard textbooks indicate the differences observed for some optical properties in overthick thin sections.

The standard observational modes are as follows (denoted by letters keyed to Table 2): (A) in plane-polarized light passing through only one Nicol prism (note that the plane light of petrography is not the same as unpolarized "plain" light); (B) with Nicols crossed to produce complete extinction of light passed through isotropic materials; (C) with both the Bertrand-Amici lens and the substage condenser engaged to produce interference figures, either uniaxial or biaxial, in convergent polarized light with the Nicols crossed; (D) using either the gypsum plate or the quartz plate to determine the optic sign of interference figures, whether positive or negative.

Standard optical properties most useful for the identification of minerals in Pacific temper sands are listed in Table 2, which is provided in part to indicate for readers who are unfamiliar with petrographic techniques the variety and instrumental sophistication of petrographic observations relevant for temper studies. Petrographic methodology involves much more than simply viewing sand grains at high magnification. Bear in mind, moreover, that full pleochroic schemes, maximum birefringence, true extinction angles, crystallographic twins, and centered interference figures are observable only for grains transected in favorable crystallographic orientations. This constraint can be surmounted by searching thin sections for grains of minerals that are in the correct orientation for each optical property sought. Pleochroism and extinction angles are best displayed in grains displaying maximum birefringence, whereas centered optic-axis interference figures are obtained only for grains displaying minimum birefringence. Supplemental but laborious optical techniques that can be worthwhile to resolve especially thorny questions include (1) the quantitative determination of refractive indices by immersion of separated mineral grains in oils of known refractive index and (2) the study of thin sections on the universal stage. This device allows a mineral to be placed in any arbitrary crystallographic orientation but requires manipulative skills that are arduous to acquire.

Light Minerals

Light minerals, all with low refringence or relief, are clear or white to pale gray megascopically and colorless (wholly transparent) in thin section. Most are quartz or feldspar (Figs. A12 and A19) of low birefringence displaying white to palest yellow interference tints under crossed Nicols. By contrast, carbonate grains (calcite or aragonite) display high relief and extreme birefringence producing pale high-order interference tints (Fig. A1). Both calcite and aragonite also reverse relief (from positive to negative) as the microscope stage is rotated (a unique diagnostic trait).

Feldspar is distinguished from quartz by the presence of cleavage and by interference figures that are biaxial rather than uniaxial. Moreover, little-abraded feldspar grains are commonly subhedral, whereas quartz grains are most commonly anhedral. Quartz from intrusive igneous rocks commonly displays undulatory extinction and curvilinear trains of minuscule inclusions or vacuoles, but phenocrystic quartz derived from volcanic rocks (Fig. A16) typically displays sharp and uniform extinction. Volcanic quartz is also commonly euhedral, of a bipyramidal habit that appears as equant to distorted diamond shapes in thin section. Some volcanic quartz, however, has embayed grain margins marked by rounded concave reentrants formed by partial resorption of quartz back into surrounding magma. Volcanic feldspars and quartz may both contain inclusions of volcanic glass.

Plagioclase commonly displays closely spaced polysynthetic twinning that makes for easy discrimination from quartz and K-feldspar. Internal lamellae that pass into extinction for different rotations of the microscope stage produce a distinctively banded visual display. Some plagioclase is untwinned, however, and polysynthetic twinning is invisible where the twin planes lie subparallel to the plane of a thin section. In these cases, reliance on polysynthetic twinning to identify plagioclase leads to error. Where polysynthetic twinning is not present or not visible, obtaining biaxial interference figures for plagioclase or uniaxial interference figures for quartz is the most direct means to distinguish plagioclase grains from quartz grains.

Establishing the composition (Na/Ca ratio) of the plagioclase grains in a temper sand is difficult from petrographic observations alone, and is commonly not worth the effort involved in using ancillary mineralogical techniques. Most detrital plagioclase in Pacific temper sands is andesine or labradorite, although oligoclase from plutonic rocks and bytownite from phenocrystic plagioclase in mafic volcanic rocks are present in some temper sands. Methodology that relies upon maximum extinction angle for discrimination among these plagioclase varieties is useful for igneous rocks in which all the plagioclase has equilibrated with the same magma during crystallization, but is not appropriate for sedimentary aggregates derived from multiple source rocks. In general, each plagioclase grain in a given sand aggregate may have a somewhat different composition. Moreover, curves relating maximum extinction angle to Na/Ca ratio differ appreciably for plutonic and volcanic plagioclases that are in different states of internal crystalline order.

Volcanic plagioclase commonly displays monotonic or oscillatory zoning that is reflective of internal variations in composition. The compositional zoning is revealed by concentric domains within a plagioclase crystal that pass into extinction at different orientations as the microscope stage is rotated. Diagenetic albite is commonly distinctive from the presence of multiple inclusions of hydrous Ca-bearing alteration minerals of high relief and moderate birefringence, but may display little or no twinning. In rare cases, unmixed anorthoclase derived from phenocrysts in volcanic rock may occur in temper sands as a solid solution of albite and K-feldspar. Exceedingly narrow intersecting twin bands and incipient exsolution lamellae are typically visible microscopically, but identification of anorthoclase is inherently equivocal from petrographic observations alone.

K-feldspar with slight negative relief, as opposed to the low positive relief of quartz and plagioclase, is rare in Oceanian temper sands because K-feldspar phenocrysts are absent from nearly

TABLE 2. OPTICAL PROPERTIES USEFUL FOR IDENTIFICATION OF MINERAL SPECIES

Property	Description
Color (A)	Tones of coloration (commonly faint) in transmitted light
Pleochroism (A)	Contrasting colors displayed by some anisotropic minerals as viewed in different crystallographic orientations (by rotating microscope stage)
Refringence (A)	Apparent relief against mounting medium governed by refractive index (the *Becke line*, a bright highlight parallel to a grain margin, moves into the material of higher refractive index when the thin section is taken out of sharp focus by slightly lengthening the distance between the eyepiece and the microscope stage)
Habit (A or B)	Crystal shape in the sense of equant, prismatic, or acicular (needle-like), and crystal form in the sense of euhedral (bounded entirely by crystal faces), subhedral (some crystal faces), or anhedral (no crystal faces)
Cleavage (A or B)	Characteristic planes of inherent breakage along selected crystallographic directions (visible as planar seams or cracks internal to mineral grains); minerals may have none, one, or more cleavage directions
Inclusions (A or B)	Internal domains of one or more mineral crystals (or volcanic glass) enclosed within host crystals of another mineral species
Birefringence (B)	Varied colorations defined by interference tints (akin to segments of the rainbow spectrum) observed with nicols crossed for anisotropic minerals placed in different crystallographic orientations by rotation of the microscope stage because of different refractive indices in different crystallographic directions (spectrum colors range from bright and intense for first-order tints to progressively pale pastel tints for higher orders)
Extinction angle (B)	Angular rotation of microscope stage, as measured relative to dominant cleavage direction or elongation in habit, required to bring an anisotropic mineral into extinction (black); extinction angles vary from zero or parallel (*straight extinction*) to diversely inclined (extinction direction coincides with change in pleochroic coloration if present)
Zoning (B)	Gradients in chemical composition within crystals, as reflected by internal domains with gradational boundaries that pass into extinction at different rotations of the microscope stage (common in volcanic plagioclase)
Twinning (B)	Partitioning of crystals by twin planes into discrete segments (with different crystallographic orientations) that pass into extinction at different rotations of the microscope stage (multiple internal twinning is termed *polysynthetic*, and is common in plagioclase and microcline)
Interference figures (B+C)	Uniaxial with isogyres that remain in crossed position throughout stage rotation, or biaxial with isogyres that separate during stage rotation (for off-axis interference figures, uniaxial isogyres remain straight during stage rotation as they sweep across the field of view, but biaxial isogyres curve during passage across the field of view); best viewed in mineral orientations displaying the lowest birefringence
Optic axial angle or 2V (B+C)	Determined for biaxial minerals by the amount of curvature of isogyres in optic-axis (centered) interference figures (line of view along an optic axis), or by the amount that isogyres separate during stage rotation for acute bisectrix interference figures (line of view along a direction exactly intermediate between the orientations of paired optic axes)
Optic sign (B+C+D)	Positive (+) or negative (−), whether uniaxial or biaxial, as determined from interference figures

Notes on observational modes designated by letters in parentheses (see text for explication): A, Nicols uncrossed; B, Nicols crossed; C, Bertrand-Amici lens and substage condenser both engaged; D, gypsum plate or quartz plate employed.

all eruptive products of either hotspot chains or intraoceanic island arcs. K-feldspar is common, however, in detritus derived from the exposed granitic plutons of dissected orogens. Plutonic K-feldspar is commonly perthite containing exsolution lamellae or bleblike inclusions of albite, the sodic end member of the plagioclase solid solution, which pass into extinction for different rotations of the microscope stage than the host K-feldspar. Plutonic K-feldspar may in part be triclinic microcline with polysynthetic grid twinning having a crisscross appearance unlike the lamellar polysynthetic twinning of plagioclase. Unmixed and untwinned K-feldspar grains of plutonic derivation are dominantly orthoclase, but volcanic K-feldspar of phenocrystic origin is commonly sanidine, with straight and uniform extinction and with a low optic axial angle that makes interference figures difficult to distinguish from those of uniaxial quartz.

Reliable distinctions among quartz, plagioclase, and K-feldspar grains at the level of confidence required to establish relative proportions quantitatively by counting grains may require etching sherd thin sections by exposure to HF fumes. The etch residues are then stained for K to produce yellowish etch pits on K-feldspar, and for Ca to produce pinkish etch pits on plagioclase. Quartz does not etch in HF, so it remains unstained. Albite etches but does not stain for either K or Ca. Uniformity of either etching or staining is difficult to achieve, and practical experience is required to interpret etch stains correctly.

Heavy Minerals

Common heavy minerals include opaque iron oxides, not specifically identifiable by standard optical methods, and dominantly ferromagnesian silicates with high relief or refringence and moderate to strong birefringence producing high first-order and low second-order interference tints. The interference tints under crossed Nicols are generally comparable to the tints in the first and second light spectra of a double rainbow. Typical transparent heavy minerals in Oceanian temper sands are variably prismatic pyriboles (pyroxenes and amphiboles), blocky olivine, flakes of mica, and aggregate epidote. Isotropic garnet and chrome-spinel (Fig. A28), the latter translucent rather than transparent, occur in selected temper sands, as do rarer heavy minerals discussed as appropriate in descriptions of individual temper types.

Megascopic inspection of tempers with a hand lens can sort placer tempers with abundant dark grains from nonplacer tempers with abundant pale grains. Megascopic distinctions among jet-black to dark green ferromagnesian heavy minerals are generally not feasible, however, even though the shiny metallic luster of opaque magnetite and ilmenite is somewhat different from the vitreous luster of ferromagnesian silicate minerals. The similar megascopic appearance of optically opaque and transparent ferromagnesian minerals as sand grains precludes reliable discrimination with a hand lens, but microscopic examination readily distinguishes the two kinds of heavy mineral grains.

Distinctions among olivine, orthopyroxene (typically hypersthene), clinopyroxene (typically augite), and hornblende (or related amphiboles) are critical for Oceanian temper analysis because differing relative proportions of these ferromagnesian silicate minerals as phenocrysts are diagnostic of contrasting volcanic assemblages. In placer temper sands, they are commonly the dominant mineral grains. Although all display comparably high refringence or relief, moderate to strong birefringence, and variably developed cleavage, they can be distinguished by their respective optical properties (Table 3). Because different varieties of hornblende, pyroxene, and olivine display slightly different habits and optical properties, identification of the specific variants present in each temper sand is generally required before recognition by inspection is possible.

The combination of deep coloration, strong pleochroism, moderate extinction angle, nonorthogonal cleavages (60°–120°), and elongate prismatic habit are diagnostic of hornblende (Fig. A21). Comparatively pale coloration, weak pleochroism, and orthogonal cleavages distinguish pyroxenes from hornblende (Figs. A17, A19, and A22). Clinopyroxene (Figs. A23–A25)

TABLE 3. OPTICAL PROPERTIES OF COMMON FERROMAGNESIAN SILICATE MINERALS

Optical property	Hornblende	Clinopyroxene (augite)	Orthopyroxene (hypersthene)	Olivine
Color	green to brown	pale green	colorless to pale green	colorless
Pleochroism	strong (in greens and/or browns)	none or weak (in greens)	weak (faint green to pale mauve)	none
Habit	elongate prismatic	stubby prismatic	variably prismatic	blocky anhedral
Cleavages	60°/120° angle	~90° angle	90° angle	weak
Extinction angle	15°–25° (low)	~45° (high)	0° (straight)	N/A
Birefringence	moderate (but variable)	moderate (1st order tints)	weak	strong (2nd order tints)
2V (optic axial angle)	variable	~60° (+)	>75° (−)	near 90° (−)

in addition displays stubby prismatic to equant habits and high extinction angles, whereas orthopyroxene has straight extinction and weaker birefringence than either hornblende or clinopyroxene. Recognition of orthopyroxene, as opposed to clinopyroxene, can also be based in part on a characteristic faint pleochroism (green to mauve) in orthopyroxene (hypersthene), and on contrasting optic axial angles (2V) that are ~60° optically positive for augite (clinopyroxene) but ≥75° optically negative for hypersthene (orthopyroxene).

Olivine is colorless or very pale green in thin section, with weak cleavage and strong birefringence that produces interference tints of higher order than pyroxenes display. Olivine grains are also typically equant and anhedral, lacking the prismatic habit that is common for hornblendes and pyroxenes. Weak coloration distinguishes olivine from hornblende, and distinctly higher birefringence as well as lack of any coloration or pleochroism distinguishes olivine from orthopyroxene. Distinction between olivine and clinopyroxene requires close attention to the paler color of olivine (Fig. A20), its somewhat higher birefringence, and its high and optically negative optic axial angle (2V) approaching 90°. Helpful for identifying olivine in many temper sands is a tendency for alteration to coatings of hydrous red iddingsite along grain margins and internal fracture surfaces during the deuteric (hydrothermal) latest phase of igneous crystallization (Fig. A20). Epidote (including clinozoisite) can also be confused with clinopyroxene, but generally occurs as multicrystalline aggregate grains with a yellowish green cast, and commonly displays anomalous (nonspectral) interference tints produced by strong dispersion of light of different wavelengths.

Micas occur as subhedral platy grains with flat faces and ragged edges displaying straight extinction parallel to a prominent cleavage. Both colorless muscovite (hydrous aluminosilicate) and pleochroic green or brown biotite (hydrous ferromagnesian silicate) occur in Oceanian temper sands (Figs. A14 and A18), but the latter is by far the more abundant. Excellent closely spaced cleavages separating flexible leaves impart a characteristic blotchy pattern of extinction to all micas. In a few temper sands of volcanic derivation, oxyhornblende or lamprobolite is present as acicular prismatic grains displaying extreme pleochroism, from yellowish brown to intense red, and an extinction angle low enough (5°–10°) to invite confusion with biotite.

Lithic Fragments

Generic classification of lithic fragments is based jointly on their internal mineral composition, texture, and fabric. Texture pertains to the grain size of constituent crystals or clasts, whereas fabric pertains to their morphology and spatial arrangements. Lithic fragments are composed of varying mixtures of the same minerals that occur as separate temper grains, as well as variable proportions of amorphous (optically isotropic) volcanic glass in the case of most volcanic lithic fragments. Whereas different kinds of mineral grains represent discrete and mutually exclusive mineral species, different categories of lithic fragments are inherently gradational in nature. The petrographic challenge for classifying lithic fragments in a reproducible manner lies in developing operational criteria for distinguishing between generic categories that have geological significance (Table 4). Discussions of certain rare kinds of lithic fragments, notably serpentinite and blueschist, are deferred to discussions of particular temper types in which they occur.

The internal fabrics of lithic fragments derived from igneous, metamorphic, and sedimentary rocks differ in ways that reflect disparate rock-forming processes. Igneous lithic fragments tend to display subhedral crystals set either in volcanic glass or in microcrystalline aggregates formed by devitrification of volcanic glass, or mutually interlocking in the case of lithic fragments from intrusive rocks. Metamorphic lithic fragments typically display more interdigitated boundaries between largely anhedral crystals, and commonly display the oriented planar fabrics (foliation) characteristic of slate and phyllite. Most sedimentary lithic fragments have a fragmental internal fabric of detrital clasts, although chert is composed of microcrystalline mosaics of chalcedonic quartz formed during diagenesis.

Igneous lithic fragments span a gradational spectrum in texture from wholly aphanitic volcanic lithic fragments through hypabyssal lithic fragments derived from subvolcanic intrusions to plutonic lithic fragments in which multiple crystals of sand size remain attached as segments of individual lithic fragments. The hypabyssal and plutonic lithic fragments are so gradational in texture that the term microphanerite (Fig. A11) is used to designate them all. Note, however, that most detritus from deep-seated plutons is composed of individual mineral grains of quartz and feldspar, rather than polycrystalline lithic fragments. Volcanic lithic fragments deserve special attention because of their prevalence in many Oceanian temper sands, and include four principal compositional-textural types based on the nature of the groundmass in which variable proportions of microphenocrysts are set: vitric (glassy), felsitic, microlitic, and lathwork.

Vitric lithic fragments range widely from (1) silicic glass (obsidian), which is pale tan or colorless in thin section and has negative relief (Fig. A10), to (2) mafic glass (sideromelane), which is brownish or reddish in thin section and has positive relief (Fig. A8), and to (3) tachylite, which is black and opaque or semiopaque in plane light from submicroscopic inclusions of

TABLE 4. CATEGORIES OF LITHIC FRAGMENTS IN OCEANIAN TEMPER SANDS

Igneous and meta-igneous
  microphaneritic (microgranular internal texture and fabric)
  volcanic (vitric, felsitic, microlitic, lathwork)
  metavolcanic (greenschist and amphibolite)
Sedimentary and metasedimentary
  chert-metachert (microcrystalline chalcedony)
  microgranular (fragmental internal texture) siltstone
  argillite (including shale with mass extinction)
  tectonite (foliated internal fabric)
  limeclasts (reworked carbonate rock)
Polycrystalline quartz (quartzite and vein quartz)

iron oxides. Some otherwise isotropic volcanic glass displays faint birefringence from the presence of submicroscopic cryptocrystalline alteration products.

Felsitic lithic fragments (Fig. A7), derived mainly from dacite or felsic andesite, are composed of anhedral mosaics of intergrown feldspar and quartz formed by pervasive devitrification of volcanic glass. Microlitic lithic fragments, derived mainly from andesite or basaltic andesite, contain tiny subhedral but untwinned plagioclase microlites, which may be either tightly packed (pilotaxitic texture of Fig. A5) or set in a glassy matrix (hyalopilitic texture of Fig. A6), and in some cases are oriented into fluidal fabrics produced by flowage of lava before congealing. Lathwork lithic fragments (Fig. A4), derived mainly from basalt or basanite, are composed of subhedral and twinned plagioclase laths separated by interstitial glass (intersertal texture) or by grains of ferromagnesian minerals (intergranular texture).

Gradations between vitric and microlitic, felsitic and microlitic, and microlitic and lathwork categories of volcanic lithic fragments introduce operator subjectivity into the classification of volcanic lithic fragments, but in few instances are specific types of volcanic lithic fragments diagnostic of a particular temper type. Many volcanic lithic fragments are microporphyritic and can be described as phyric or aphyric according to the presence or absence of microphenocrysts, respectively. The nature of phenocrysts or microphenocrysts can be specified by modifiers (feldspar-phyric, pyroxene-phyric, etc.). The term vitrophyric is useful to describe lithic fragments in which microphenocrysts are set in volcanic glass or tachylite. The metavolcanic lithic fragments present in some Oceanian temper types range from actinolite-chlorite-albite greenschist to hornblende-plagioclase amphibolite, both with typically foliate internal fabrics.

Sedimentary lithic fragments include a range of fine-grained clastic rocks from claystone through mudstone to siltstone. Metasedimentary lithic fragments range from aphanitic (hornfels and slate) to microphaneritic (metaquartzite and phyllite) varieties. Because degree of metamorphism is difficult to gauge from the internal character of individual sand grains, both sedimentary and metasedimentary lithic fragments (Fig. A13) are grouped jointly as (1) chert-metachert (microcrystalline quartz mosaics), (2) microgranular (quartz-rich siltstone or hornfels), (3) argillite (clay-rich mudrock or shale, the latter displaying mass extinction but not visibly foliated crystals), (4) tectonite (foliated quartz-mica slate or phyllite), and (5) limeclasts (reworked limestone).

A degree of operator subjectivity also enters into the recognition of categories of sedimentary-metasedimentary lithic fragments because of gradations from argillite to more silicic chert or to coarser-grained microgranular lithic fragments, and from microgranular to tectonite fabrics. Moreover, distinction between foliated and unfoliated internal fabrics of temper grains depends in part upon the orientation of the foliation plane with respect to the plane of a thin section. Polycrystalline quartz grains range from lithic fragments of quartzite, either granular or foliated internally, to aggregate grains of vein quartz. Chert-metachert lithic fragments are partly indeterminate in origin because microcrystalline chalcedony may form through the diagenesis of siliceous sediment (e.g., radiolarite), as replacement nodules in limestone, as vesicle fillings (amygdules) in volcanic rock, or by intense recrystallization of silicic volcanic rocks affected by hydrothermal activity.

TEMPER COMPOSITIONAL MODES

The qualitative recognition of key grain types forms the basis for geotectonic classification of temper types, and for appreciation of relationships between placer and nonplacer tempers from the same localities, but quantitative representations of temper compositions are necessary for full temper comparisons. The primacy of qualitative data on sand tempers as a general principle for ceramic studies (Freestone, 1991) does not negate the supplementary utility of quantitative data.

Grain Counting

The standard quantitative methodology in sedimentary petrology is to count the grain types falling beneath the intersection points of a rectilinear geometric grid superimposed on a thin section using a mechanical stage to manipulate the thin section. Point counting is inefficient, however, for temper analysis because a majority of grid points fall on clay paste (Stoltman, 1989). Establishing the relative proportions of clay paste and sand temper has minimal value because experience has shown that ancient Pacific potters tended to attain approximately the same optimal mix of temper and paste. Except for selected sparsely tempered wares, temper grains amount volumetrically to between a quarter and a third of the clay bodies in nearly all Oceanian earthenware.

For systematic compositional analysis of sherd tempers, frequency counts of numbers of grains in thin section are much less time-consuming than point counts and yield equally reproducible, though not identical, results. Frequency counts and point counts differ for poorly sorted sands in which different grain types differ systematically in median grain size. Frequency counts treat all grains of varied sizes as equivalent, whereas the larger grains are encountered preferentially during point counting. Where comparative data are available, however, percentages of grain types derived alternately from point counts and frequency counts are not different enough to alter interpretations of temper origins (Dye and Dickinson, 1996; Dickinson, 2000a; Dickinson et al., 2001).

Frequency counts may be complete areal counts for which all temper grains within a sherd slice are counted, ribbon counts for which the grains lying within bands spaced at systematic intervals across a sherd slice are counted, traverse counts for which the only grains counted are those encountered beneath the microscope crosshairs along traverses spaced systematically across a sherd slice, or partial areal counts for which only the grains lying within selected evenly spaced areas of the sherd slice are counted. The different methods of frequency counting yield statistically indistinguishable results (Middleton et al., 1985). Unless sherd slices are too small or temper is too sparse, frequency-counting methodology (whether areal, ribbon, or traverse counts) was adjusted to yield counts of ~400 temper grains per sherd. This number best balances time demands with the statistics of counting error. The standard deviation of counting error is then less than 2.5 percentage points for all percentage values, and less than 1.5 percentage points for percentages less than ten (Van der Plas and Tobi, 1965).

One significant difference between the results of point counts and frequency counts arises because the Gazzi-Dickinson convention (Ingersoll et al., 1984) for conversion of raw point counts to summary compositional modes cannot be applied. By the Gazzi-Dickinson convention, sand-sized crystals or clasts within polycrystalline lithic fragments are recalculated as mineral grains to restrict percentages of points reported as lithic fragments to points falling exclusively on aphanitic constituents of sand grains. For summations of frequency counts, however, lithic fragments are perforce summed in their entireties as lithic fragments, without reapportionment of any of their internal constituents to various categories of accompanying mineral grains. This practice is inherent for frequency counting because whole individual grains are counted, rather than discrete grid points beneath the microscope crosshairs.

Ternary Plots

The standard compositional plots used for analysis of sandstone provenance are QtFL and QmFLt diagrams (Dickinson, 1985), each with three poles where Qm is monocrystalline quartz grains, F is monocrystalline feldspar grains, Qt is total quartzose grains (both monocrystalline and polycrystalline), Lt is total lithic fragments (including quartzose lithic fragments), and L is labile lithic fragments (excluding quartzose lithic fragments). Common supplementary triangular plots for partial grain populations (Dickinson, 1985) include the QmPK diagram for monocrystalline mineral grains, where P is plagioclase and K is K-feldspar, and the QpLvLs (or QpLvmLsm) diagram for lithic fragments, where Lv (= Lvm) is volcanic-metavolcanic grains and Ls (= Lsm) is sedimentary-metasedimentary grains. None of the standard triangular diagrams are attractive, however, for plotting compositions of Pacific island tempers because none include heavy ferromagnesian mineral grains. Heavy minerals are rare in most sandstones, particularly on continents, but are major constituents of many Pacific island tempers.

For Oceanian tempers, the following ternary (triangular) compositional diagrams (each with three poles as specified in Table 5A) are used instead for summations of frequency counts (exclusive of calcareous grains or grog particles after Dickinson, 1998a):

1. LF-QF-FM: full terrigenous grain population where LF is total polycrystalline lithic fragments, QF is the sum of monocrystalline quartz and feldspar mineral grains (Q + F = QF), and FM is total ferromagnesian mineral grains (both opaque and nonopaque).
2. QF-FS-OP (same as QF-FS-OO of Dickinson, 1998a): partial grain population exclusive of lithic fragments where FS is ferromagnesian silicate mineral grains (nonopaque), and OP is opaque iron oxide grains (FS + OP = FM).
3. Q-F-LF: partial grain population exclusive of ferromagnesian grains (FM); the poles for Q-F-LF diagrams of temper sands are ostensibly the same as for the QmFLt diagrams of sedimentary petrology, but data points do not plot in identical positions because the Gazzi-Dickinson convention for point counts cannot be applied to grain frequency counts.
4. Q-F-FS: partial grain population exclusive of lithic fragments (LF) and opaque grains (OP), restricting the grain population plotted to transparent mineral grains.

Placering effects within a suite of related temper sands are shown by trends in the arrays of plotted points for individual temper sands toward the FM pole in LF-QF-FM space and toward

TABLE 5. POLES OF TERNARY COMPOSITIONAL DIAGRAMS FOR TEMPER SANDS

Diagram	Grain population	Top center pole	Lower left pole	Lower right pole
	A. Standard Plots			
LF-QF-FM	all	LF	QF	FM
QF-FS-OP	non-lithic	QF	FS	OP
Q-F-LF	non-ferromagnesian	Q	F	LF
Q-F-FS	non-lithic and non-opaque	Q	F	FS
	B. Supplemental Plots			
hbl-pyx-olv	ferromagnesian silicate	hbl	pyx	olv
Q-P-K	quartz and feldspar	Q	P	K
Mph-Lvm-Lsm	lithic	Mph	Lvm	Lsm

Notes: F, monocrystalline feldspar mineral grains; FM, all ferromagnesian mineral grains (silicates and oxides); FS, ferromagnesian silicate mineral grains; hbl, hornblende and oxyhornblende mineral grains (summed); K, monocrystalline K-feldspar mineral grains; LF, all lithic fragments (summed); Lsm, sedimentary and metasedimentary lithic fragments (summed); Lvm, volcanic and metavolcanic lithic fragments (summed); Mph, all microphaneritic igneous (plutonic and hypabyssal) lithic fragments (summed); olv, olivine mineral grains; OP, opaque ferromagnesian mineral grains; P, monocrystalline plagioclase feldspar mineral grains; pyx, pyroxene mineral grains (both clinopyroxene and orthopyroxene as summed); Q, monocrystalline quartz mineral grains; QF, monocrystalline quartz and feldspar mineral grains (summed) the FS pole in Q-F-FS space, or sequentially toward the FS pole and then toward the OP pole in QF-FS-OP space. The observed compositional shifts from placering hold true because the specific gravity of FM grains is uniformly greater than QF and LF grains, and OP grains have markedly higher specific gravity than FS grains. Quartzose tempers are differentiated from nonquartzose tempers by the Q-F-LF and Q-F-FS diagrams, whereas tempers containing a high ratio of lithic fragments to mineral grains emerge primarily from the LF-QF-FM plot, and secondarily from the Q-F-LF plot, which avoids the influence of placering on grain populations. Ratios of plagioclase to K-feldspar do not figure in any of the four standard compositional plots, but are uniformly high in Pacific temper sands.

Megascopic examination, with a hand lens or binocular microscope, can detect some of the salient trends shown on triangular compositional diagrams based on microscopic study of temper grains. Quartz and feldspar grains (QF) are pale (white to light gray), ferromagnesian grains (FM) are dark (black or nearly so), and lithic fragments (LF) display varied intermediate tones of gray (or red). In favorable instances, quartz can be distinguished from feldspar by lustrous curving surfaces of conchoidal fracture, whereas feldspar grains tend to be bounded partly by degraded but flat crystal faces or cleavages. Ferromagnesian silicate (FS) mineral grains display vitreous luster, whereas most opaque iron oxide grains (OP) display metallic luster, although as noted previously the two kinds of luster are difficult to distinguish on rounded sand grains. Megascopic sorting of Oceanian sherd collections by temper type can be improved markedly, however, by correlation of the visual display of different temper grains with the results of microscopic study of representative sherds containing different temper types.

Supplemental Parameters

As none of the four triangular compositional diagrams can display all the compositional differences among temper types, and none highlight the qualitative differences important for interpretations of temper origins, eight supplemental grain parameters (Dickinson, 2001a) are useful for codifying generic differences among the various Pacific temper types (Table 6).

Quartz (QZi) and feldspar (PFi) indices reflect differences among light mineral grains by indicating the quartzose nature of selected tempers and relative proportions of different feldspars. Volclithic (VLi) and sedlithic (SLi) indices show relative proportions of volcanic and nonigneous lithic fragments. The oxide index (OXi) is sensitive to degrees of placering. Olivine (OLi), pyribole (PYi), and pyroxene (PXi) indices reflect relative proportions of key ferromagnesian silicate mineral grains of diagnostic value for gauging temper origins.

For temper comparisons, three supplemental triangular compositional diagrams are also used (Table 5B). Proportions of key ferromagnesian mineral grains are plotted on the hbl-pyx-olv diagram where *hbl* is hornblende plus oxyhornblende, *pyx* is total pyroxenes, and *olv* is olivine (Dickinson, 1998a). Proportions of key light mineral grains in temper sands of complex mineralogy are plotted on the QPK diagram where Q is quartz, P is plagioclase, and K is K-feldspar (Dickinson, 1982c). Proportions of generic types of lithic fragments in wholly or partly nonvolcanic temper sands are plotted on the Mph-Lvm-Lsm diagram where *Mph* is microphaneritic igneous lithic fragments derived from plutonic and hypabyssal intrusions, *Lvm* is volcanic-metavolcanic lithic fragments, and *Lsm* is sedimentary-metasedimentary lithic fragments.

OCEANIC BASALT TEMPERS

Oceanic basalt tempers are mineralogically the simplest of Oceanian temper sands because all are exclusively volcanic sands derived from islands exposing mainly basaltic eruptive assemblages. Parent magmas are exclusively mantle melts produced by the fusion of upwelling asthenosphere as pressure is released on mantle materials, and igneous activity is not directly influenced

TABLE 6. SUPPLEMENTAL GRAIN PARAMETERS FOR OCEANIC TEMPER SANDS

Parameter	Symbol	Definition	Explanation
Quartz index	QZi	$100 \times Q/(Q + F)$	quartz (Q)/feldspar (F) ratio
Feldspar index	PFi	$100 \times P/(P + K)$	plagioclase (P)/K-feldspar (K) ratio
Volclithic index	VLi	$100 \times VRF/TRF$	ratio of volcanic lithic fragments (VRF) to total lithic fragments (TRF)
Sedlithic index	SLi	$100 \times SMR/TRF$	ratio of sedimentary-metasedimentary lithic fragments (SMR) to total lithic fragments (TRF)
Oxide index	OXi	$100 \times OP/FM$	ratio of opaque iron oxide grains (OP) to total ferromagnesian mineral grains (FM)
Olivine index	OLi	$100 \times olv/FS$	ratio of olivine mineral grains (olv) to total ferromagnesian silicate mineral grains (FS)
Pyribole index	PYi	$100 \times pyx/(pyx + hbl)$	pyroxene (pyx)/hornblende (hbl) ratio
Pyroxene index	PXi	$100 \times cpx/(cpx + opx)$	clinopyroxene (cpx)/orthopyroxene (opx) ratio

by continental crust or subducted lithosphere. Felsic differentiates produced by fractional crystallization of mafic magmas are exposed locally but are volumetrically minor.

In the northwestern and southwestern segments of the Pacific Ocean basin where Oceanian ceramic traditions evolved, nearly all exposed volcanic assemblages of intra-Pacific island chains are alkalic, with transitional and marginally tholeiitic basalts rare on outcrop where present at all (Keating et al., 1984b; Dickinson, 1998b). The detritus in oceanic basalt tempers was derived dominantly from alkalic olivine basalt or basanite, which is comparably mafic lava less saturated geochemically with silica than basalt and containing feldspathoidal minerals as well as feldspars in the groundmass of lava. Because lithic fragments of mafic groundmass material derived alternately from basalt and basanite cannot be distinguished in thin section, both are treated as basaltic detritus.

More felsic differentiates of the mafic basaltic and basanitic magmas include hawaiite, mugearite, benmoreite, and trachyte (in order of increasingly felsic character). Lithic fragments of hawaiite are distinguishable from basalt only by the composition of plagioclase laths in the groundmass, and are treated as basaltic. Trachytic lithic fragments can be distinguished from basaltic lithic fragments by groundmasses with distinctly higher feldspar contents (Fig. A5). Lithic fragments derived alternately from trachyte, benmoreite, and mugearite cannot be distinguished reliably in thin section, and all are treated as trachytic.

Most islands yielding oceanic basalt tempers are generally interpreted as hotspot chains composed of islands that supposedly age monotonically downstream as plate motions carry Pacific lithosphere over magma sources in underlying asthenosphere. Many intra-Pacific island chains, however, involve multiple hotspots, eruptions out of sequence along a chain, or rejuvenated volcanism that develops atop long-dormant volcanic edifices (Dickinson, 1998b; Dickinson and Green, 1998). Variations in geologic history along hotspot chains, and in the consequent petrology of eruptive assemblages, do not alter the general correlation of oceanic basalt tempers with geotectonic setting, but do influence temper analysis in subtle ways. Only four island groups yielding oceanic basalt tempers occur within the region of Oceanian ceramic cultures: (1) the high volcanic islands of the eastern Caroline Islands in the western Pacific region (Fig. 7), (2) isolated islands (Rotuma, Uvea) along the northern Melanesian borderland to the west of Samoa (Fig. 8), (3) the Samoan chain (Fig. 8), and (4) the Marquesan cluster of islands in the central Pacific region to the east (Fig. 1).

Overall Composition

Oceanic basalt tempers are composed exclusively of volcanic detritus ($VLi = 100$; $SLi = 0$) derived either from basalt or basanite lavas and breccias or from trachyte plugs. Gradations between the two variants of oceanic basalt temper have not been encountered, suggesting that the trachytic tempers were derived from restricted bedrock sources and that trachytic detritus is generally diluted by much more abundant basaltic detritus at most depositional sites where sand accumulates on intra-Pacific islands. Most volcanic lithic fragments are lathwork (Figs. A4 and A15), or vitric with mafic glass, except for microlitic (pilotaxitic) and felsitic lithic fragments derived from trachytes in Samoa and American Samoa (Fig. A5).

Oceanic basalt tempers, apart from trachytic variants and some Marquesan placer sands, have greater olivine contents ($OLi > 25$) than any other Oceanian temper types, and this facet of their ferromagnesian mineralogy is diagnostic. Even so, the ratio of olivine to pyroxene in oceanic basalt tempers is commonly less than in basaltic bedrock sources because olivine is more susceptible to destruction by weathering on outcrop. Quartz and K-feldspar mineral grains are absent from all known oceanic basalt tempers ($QZi = 0$; $PFi = 100$), although sanidine microphenocrysts occur in lithic fragments of trachyte in trachytic tempers from Tutuila in American Samoa. Quartz derived from quartz-bearing trachyte plugs (Stearns, 1944; Macdonald, 1944, 1968; McDougall, 1985) might be present in some derivative sands on Tutuila, but has not yet been observed in temper sands. Hornblende is generally absent ($PYi = 100$) in oceanic basalt tempers, again except for trachytic variants ($PYi \approx 0$), and orthopyroxene is uniformly absent or rare ($PXi \approx 100$). In most placer sands, opaque iron oxide grains are distinctly subordinate to ferromagnesian silicate mineral grains ($OXi < 25$). Their low abundance, in comparison to many other Oceanian tempers, probably reflects a paucity of magnetite microphenocrysts in intra-Pacific lavas as opposed to the Pacific-margin lavas of island arcs and associated geotectonic elements.

Some clinopyroxene grains in oceanic basalt tempers are titanaugite, which displays weak pleochroism in pale green and brown to lavender tones that invite potential confusion with orthopyroxene (hypersthene). Either extinction angles or optic axial angles must be observed closely for conclusive identification of titanaugite as opposed to orthopyroxene. The lack of titanaugite in other volcanic sand tempers of Oceania can be attributed to the relative depletion of titanium in magmas of all island arcs along plate margins, as opposed to the intraplate magmas of intra-Pacific island chains.

Caroline Tempers

Sherds have been studied from a single key archaeological site on each of the three volcanic edifices of the Caroline hotspot track (Fig. 10) composed dominantly of alkalic basalt suites (Mattey, 1982). The tempers in sherds from the three islands differ systematically in texture and composition, reflecting differences in the local ceramic traditions (Shutler et al., 1984; Athens, 1990a, 1990b). All are interpreted as oceanic basalt tempers indigenous to the respective volcanic islands.

Chuuk Temper

Tempers in sherds excavated from Fefan Island, which rises from the lagoon of the Chuuk (=Truk) almost-atoll, are hybrid

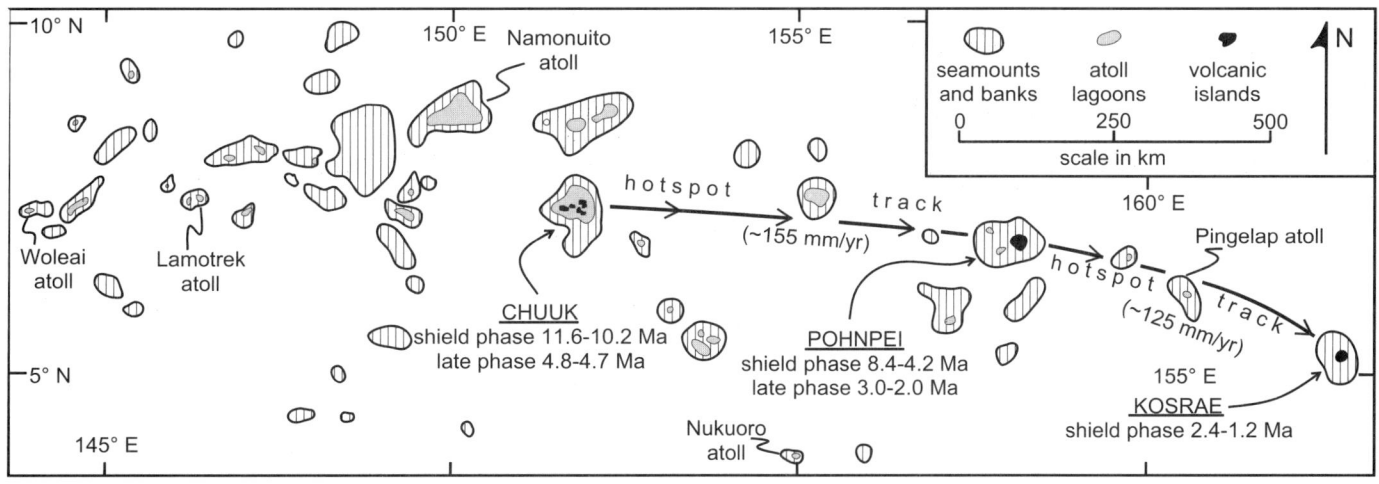

Figure 10. Central and eastern Caroline Islands showing geologic relations of Chuuk (Truk)–Pohnpei (Ponape)–Kosrae (Kusaie) hotspot track amid submerged seamounts, banks, and atoll pedestals of unknown ages. Ages of volcanic islands after Dixon et al. (1984) and Keating et al. (1984a, 1984b). Rate of hotspot migration as inferred from shield phases of volcanism (ignoring more alkalic late phases of volcanism) decreasing from ~155 mm/yr (Chuuk to Pohnpei) to ~125 mm/yr (Pohnpei to Kosrae).

beach sands composed in varying proportions of calcareous debris and basaltic detritus (Dickinson, 1984; Intoh and Dickinson, 2002). Calcareous reef debris, or vacuole pores produced by dissolution of calcareous grains, are predominant in all sherd thin sections. Weathered basaltic detritus is present only sparingly, but ubiquitously, as the following grain types: plagioclase and clinopyroxene mineral grains, opaque iron oxide grains, abraded fragments of mafic volcanic glass with brownish to reddish hues, and lathwork volcanic lithic fragments, some of which contain microphenocrysts of olivine largely altered to iddingsite. The nature of the terrigenous detritus is compatible with derivation from the alkalic basalt volcanic assemblage of Chuuk (Stark and Hay, 1963; Mattey, 1982). Fefan Island itself is composed dominantly of olivine basalt bedrock (Stark and Hay, 1963, p. 7), which is a suitable source for the terrigenous fraction of Fefan tempers.

Pohnpei Temper

Surface sherds derived from early cultural horizons of the Nan Madol megalithic site on Pohnpei (=Ponape) contain composite tempers composed of basaltic detritus mixed in a majority of thin sections with varying proportions of grog particles (Intoh and Dickinson, 2002). The most common grains of volcanic derivation are lathwork lithic fragments, most deeply weathered, many vesicular or amygdaloidal, and some containing olivine microphenocrysts largely altered to iddingsite. Subordinate mineral grains of plagioclase, clinopyroxene, and opaque iron oxide are also present in selected sherds, as are calcareous grains of reef debris. The lithic fragments are typically rounded or subrounded, suggestive of beach origin, whereas the grog particles are angular or subangular and display an internal particulate character unlike the internal volcanic textures of the lithic fragments. The character of the terrigenous detritus is compatible with derivation from the alkalic basalt volcanic assemblage of Pohnpei (Yagi, 1960; Mattey, 1982; Dixon et al., 1984). Addition of supplemental grog temper to clay bodies may reflect the paucity of sand deposits either within the deeply weathered and thickly vegetated interior or along the mangrove-clad shorelines of Pohnpei.

Kosrae Temper

Tempers in sherds excavated from early horizons of the Lelu site on a tiny islet off Kosrae (=Kusaie) are exclusively calcareous beach sands providing no temper evidence for indigenous or exotic origin (Dickinson, 1995). Postburial dissolution of calcareous grains from Lelu sherds led to initial megascopic description of many sherds as untempered. The clay pastes are all poorly sorted silty clay, probably of local stream origin, which contains sparse sand grains compatible with derivation from the basaltic bedrock of Kosrae. Lithic fragments of mafic volcanic glass, both plagioclase and clinopyroxene mineral grains, opaque iron oxide grains, and rare olivine grains are all identifiable as apparently natural temper present in the clay bodies before addition of calcareous temper grains. All the clay pastes contain tiny thin flakes of strongly pleochroic brown phlogopitic mica diagnostic of derivation from Kosrae bedrock (Dickinson, 1995). Mica is rare in most alkalic basalt assemblages of Pacific islands, but phlogopite occurs as phenocrysts in the main lava series of Kosrae (Mattey, 1982) and indicates an indigenous origin for the Lelu sherds. As part of sparse natural temper in selected sherds, phlogopite grains of sand size are also present, as are grains of pleochroic (nearly colorless to red-brown) titaniferous hornblende (kaersutite), which also occurs as phenocrysts in Kosrae lavas (Mattey, 1982).

Rotuma-Uvea Tempers

The isolated islands of Rotuma and Uvea (= Wallis), which cap seamounts that rise abruptly from the seafloor along the northern Melanesian borderland, are composed dominantly of Pleistocene (younger than 1.5 Ma) to Holocene alkalic olivine basalt lava shields and associated pyroclastic cones (Stearns, 1945; Macdonald, 1945; Duncan, 1985; Woodhall, 1987). Uvea lies just north of the Vitiaz paleotrench on Pacific lithosphere, but Rotuma lies south of the Vitiaz paleotrench where its basaltic eruptions were probably related to opening of the oceanic North Fiji Basin as the Vanuatu island arc rotated away from the Vitiaz paleotrench after polarity reversal of the ancestral Vitiaz arc (Figs. 8–9).

Rotuma Tempers

Indigenous prehistoric sherds from Rotuma contain two local variants of oceanic basalt temper (Ladefoged et al., 1998). One variant derived from the erosion of subaerial scoria deposits or clinker capping aa lava flows is composed mainly of lithic fragments (68%–95%) of tan to brown volcanic glass that is characteristically microvesicular and contains microphenocrysts of olivine. The subordinate mineral grains are exclusively olivine and clinopyroxene (OLi = 80–85). These glass-rich Rotuma tempers range from moderately sorted and subrounded aggregates of probable beach origin to poorly sorted and subangular aggregates of alluvial or colluvial origin. The other variant of Rotuma temper, present in one sherd as subangular to subrounded beach or stream sand, was derived from the erosion of crystalline lava. The dominant lithic fragments (71%) display lathwork internal textures, and mineral grains include minor plagioclase (6%) and opaque iron oxide grains (OXi = 4) in addition to olivine and clinopyroxene (OLi = 45).

Olivine grains and microphenocrysts in the lava-derived Rotuma temper have a distinctive internal appearance owing to the presence of minute festoonlike or gridlike chains of skeletal but conjoined magnetite inclusions. Some olivine grains are so highly charged with magnetite inclusions as to be more than half opaque in plane light. This unique internal structure within olivine crystals has not been observed in any other Oceanian tempers. Woodhall (1987, p. 12) notes, however, that magnetite in lavas of Rotuma typically occurs as intergrown inclusions within olivine phenocrysts, a description that fits the unusual relations between magnetite and olivine in the temper sand. The close similarity of unusual internal features of olivine crystals in Rotuma bedrock and in derivative temper sand provides evidence for indigenous origin of the temper.

Uvea Temper

The temper sands in six sherds from Uvea (Sand, 1998b) are homogeneous populations of basaltic lithic fragments containing microphenocrysts of plagioclase and olivine, the latter tinged red from alteration to iddingsite, set in a groundmass of opaque (tachylitic) volcanic glass. The temper grains are commonly microvesicular, and many have the curved cuspate margins typical of basaltic tephra. As the distinctive temper type is unique for Oceanian sherds studied to date, it is presumed to be indigenous to Uvea. The temper was probably collected from pyroclastic deposits on the flanks of local cinder cones, for unaltered basaltic ash at eruptive centers on Uvea is known to be composed dominantly of opaque black volcanic glass (Macdonald, 1945, p. 567). Lapita sherds from the Utuleve site (Sand, 2000b) on Uvea have not as yet been examined in thin section and may or may not contain the same distinctive temper type.

Samoan Tempers

Tempers in abundant sherd collections from Samoa and American Samoa are sedimentologically varied (Table 7), dominantly of basaltic composition but including subordinate trachytic variants in selected sherds from multiple sites (Fig. 11). Beach sands, some hybrid with admixtures of calcareous reef debris, are the most abundant tempers, but colluvial debris and apparently alluvial sands are also present as temper in sherds from selected sites, and aggregates composed of crushed volcanic rock occur in sherds from several sites. Ferromagnesian basaltic tempers (Dickinson, 1969, 1974a, 1976b) are the most characteristic temper sands.

My previous interpretation (Dickinson, 1993) that some Samoan sherds contain composite tempers including grog particles is not sustained by more experienced reappraisal. The temper grains once thought to be grog particles are weathered basaltic lithic fragments that were originally mafic volcanic glass but are now composed largely of clay minerals. The distinction

TABLE 7. INDIGENOUS TEMPER SANDS FROM SITES ALONG THE SAMOAN ISLAND CHAIN

Group	Sav	Mul	Jan	Par	Vai	Sas	Tat	Mal	Aoa	Aun	Vat	Tog	sum
Beach sand	-	10	4	-	6	3	-	-	3	-	1	20	47
(Hybrid)		(5)	(2)								(1)	(7)	(15)
Stream sand	2	-	-	-	-	-	-	-	-	-	1	7	10
Colluvium	-	-	-	-	-	-	2	2	3	1	-	-	8
Crushed rock	-	-	4	1	4	1	1	5	7	-	-	3	26
Trachytic	-	-	2	-	4	2	-	-	3	-	-	-	11

Notes: Sites (Fig. 11) tabulated from west to east: Aoa, 'Aoa Valley; Aun, Aunu'u Island; Jan, Jane's Camp; Mal, Malae'imi Valley; Mul, Mulifanua (ferry berth) Lapita site; Par, Paradise; Sas, Sasoa'a (Falefa Valley); Sav, Savai'i (Pulemelei); Tat, Tataga-matau; Tog, To'aga; Vai, Vailele; Vat, Va'ota.

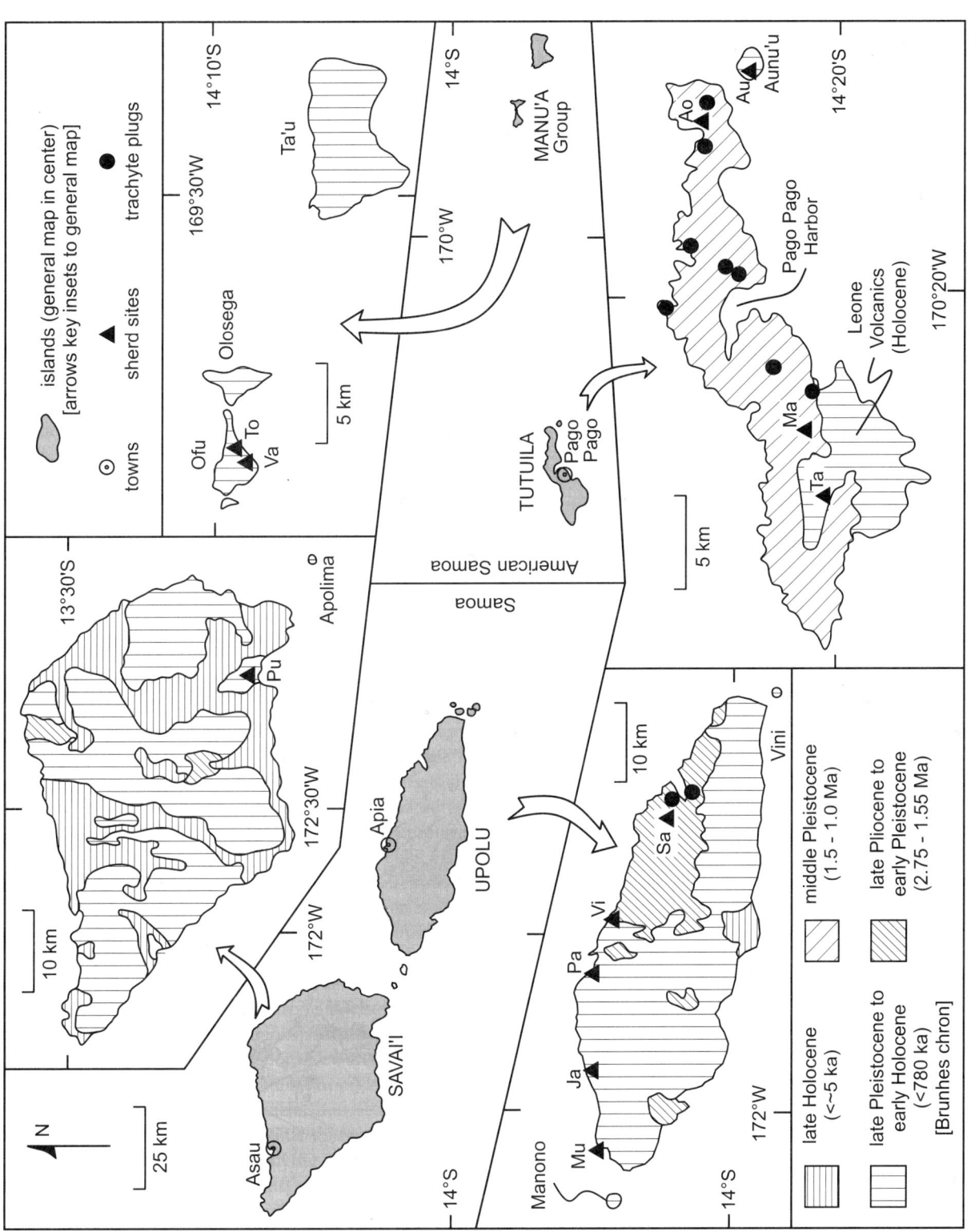

Figure 11. Distribution of bedrock geologic units and archaeological sites yielding studied sherds in the Samoan islands. Sherd sites: Ao—'Aoa Valley; Au—Aunu'u; Ja—Jane's Camp; Ma—Malae'imi Valley; Mu—Mulifanua (ferry berth Lapita); Pa—Paradise; Pu—Pulemelei; Sa—Sasoa'a; Ta—Tataga-matau; To—To'aga; Va—Va'ota; Vi—Vailele. Ages of eruptive assemblages after Stearns (1944), Kear and Wood (1959), Kear (1967), Natland (1980), McDougall (1985), and Natland and Turner (1985).

between grog particles and weathered vitric fragments is subtle, based on the presence or lack of an internal clastic texture, and is commonly not conclusive for individual temper grains.

The only decorated Lapita sherds known from Samoa derive from the submerged Mulifanua site near the western end of Upolu. Island subsidence under the growing volcano load of nearby Savai'i, where voluminous Holocene eruptions have been frequent, has carried mid-Holocene paleoshorelines below modern sea level (Dickinson and Green, 1998). The Mulifanua site was discovered during dredging for a ferry berth, and other Lapita sites may be present elsewhere beneath undredged shoals surrounding Upolu (Green, 2002). Radiocarbon geochronology indicates, however, that the oldest cultural horizons yielding Polynesian Plainware sherds from sites on Tutuila and Ofu in American Samoa are virtually indistinguishable in age from the Mulifanua Lapita site, suggesting that Lapita decoration was abandoned shortly after the initial occupation of Samoan islands (Fig. 3).

Ferromagnesian Basaltic Sands

Volcanic sands composed exclusively of basaltic lithic fragments and ferromagnesian mineral grains derived from phenocrysts in basalt are widespread as temper in Samoan sherds and are dominant at most archaeological sites that have yielded large sherd collections (Table 7). This observation is expected because olivine basalt is the dominant lithology of Samoan eruptive assemblages of all ages on all islands (Macdonald, 1944, 1968; R.N. Brothers in Kear and Wood, 1959; Stice and McCoy, 1968; Hubbard, 1971; Hawkins and Natland, 1975). The most typical grain aggregates are well-sorted and subrounded beach sands (Fig. A4).

The degree of placering, to concentrate ferromagnesian mineral grains over volcanic lithic fragments, is highly variable along a gradational spectrum of temper compositions (Fig. 12A), but is not diagnostic of sherd collections from specific sites. Ferromagnesian silicate grains are predominant in the placer aggregates, with opaque iron oxide grains distinctly subordinate (Fig. 12B), as reflected by oxide indices (OXi) that average only 5–10 in the sherd suites from various sites and are consistently less than 15–20 (Table 8). A large range in the relative proportions of mineral grains and lithic fragments in the sherd tempers from a single site was confirmed by an independent petrographic study of sand tempers in 35 additional Mulifanua sherds not used for this study (Petchey, 1995). Her study also revealed no systematic differences in temper type between decorated and undecorated Lapita sherds from Mulifanua.

The sand tempers in some sherds from To'aga on Ofu are more placered than other ferromagnesian basaltic tempers from elsewhere in Samoa (Fig. A20), but the tempers in other To'aga sherds overlap broadly in composition with tempers from several sites on Upolu (Fig. 12A). Selected sherds from both Va'ota and To'aga on Ofu contain variable minor proportions of feldspar mineral grains that do not occur in comparably ferromagnesian tempers from Upolu (Fig. 12C). In general, however, the gross compositions of Samoan ferromagnesian tempers can neither support nor deny the possibility of ceramic transfer within the Samoan islands. Moreover, the internal textures of lithic fragments in ferromagnesian basaltic tempers in sherds from multiple sites are generally comparable and provide no basis for distinction among sherds from different sites. Lathwork grains typically form 75%–95% of the lithic fragments, with mafic glass fragments of brownish to reddish hues forming most of the remainder.

The olivine index (OLi) serves, however, to subdivide the ferromagnesian basaltic tempers of Samoa and American Samoa into two areally distinct mineralogical groups. Tempers of beach origin from sites in western Upolu have consistently high olivine indices near 80, whereas tempers of comparable sedimentology from Sasoa'a in eastern Upolu, and still farther east on Ofu in American Samoa (sparse basaltic temper and mixed temper sand of Dickinson, 1993), have consistently lower olivine indices that average 35–40 and are uniformly <50 (Table 8). This difference can be attributed to petrological differences in potential source rocks for derivative sands exposed along the Samoan island chain. Basaltic assemblages erupted during the Brunhes Chron (younger than 780 ka) on western Upolu (Fig. 11) tend to contain olivine phenocrysts to the exclusion of clinopyroxene phenocrysts, whereas older basalts of the Pliocene-Pleistocene Fagaloa Volcanics in eastern Upolu (Fig. 11) tend to contain both olivine and clinopyroxene in variable proportions as phenocrysts.

The contrast in petrology and consequent mineralogy arises because the Fagaloa Volcanics were erupted during the shield phase of volcano construction on Upolu and nearby Savai'i, whereas the younger lavas represent more alkalic volcanism that resumed on partly eroded shields (Natland, 1980; Natland and Turner, 1985; Wright and White, 1986/87; Hart et al., 2004). The post-Fagaloa flows and breccias of Late Pleistocene and Holocene age on both Upolu and Savai'i have poured like lava icing over the eroded remnants of late Pliocene to early Pleistocene shield edifices (Dickinson and Green, 1998).

The young volcanic islands in the Manu'a Group of American Samoa (Fig. 11) expose eruptive suites that are typical of the shield-building phase of Samoan volcanism and commonly contain both clinopyroxene and olivine phenocrysts (Stice, 1968). The lowest average olivine index (OLi) yet recorded for Samoan temper sands pertains to tempers in selected sherds from Ofu interpreted as local stream sands (profuse basaltic temper of Dickinson, 1993). The olivine index (average 28; range 22–32) overlaps, however, with that of temper sands in Ofu sherds interpreted as beach sands (Table 8). The OLi value for the terrigenous fraction of modern hybrid beach sand on Ofu is intermediate between OLi values for beach and stream sand tempers in sherds from Ofu (Table 8), but the modern beach sand is more strongly placered than any Ofu temper sands (Fig. 12A).

The consistent contrasts in olivine content afford presumptive evidence against prehistoric ceramic transfer of wares

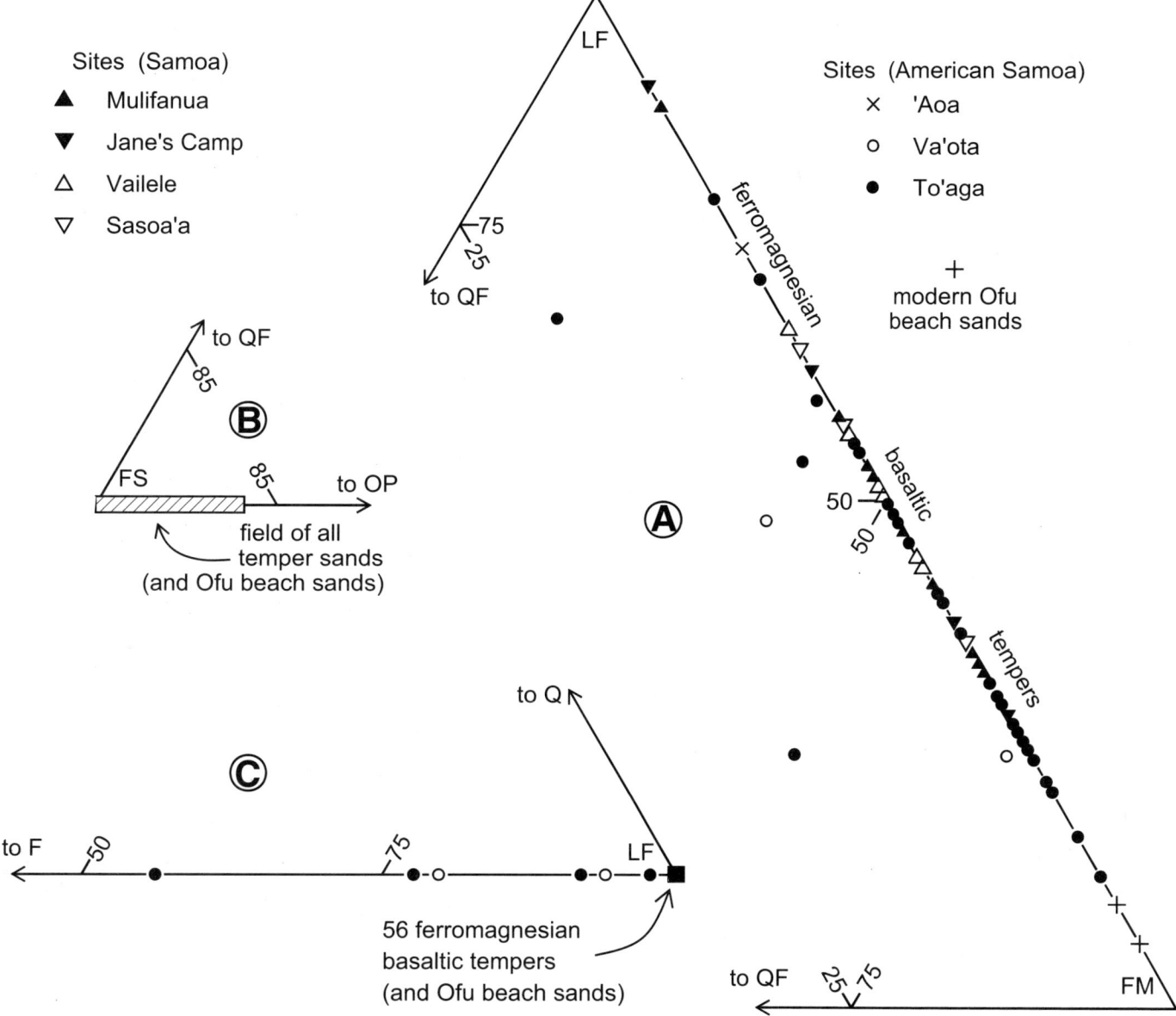

Figure 12. Compositions (grain frequency percentages) of beach and stream sand tempers in Samoan sherds. A: LF-QF-FM diagram. B: QF-FS-OP diagram. C: Q-F-LF diagram. See Table 5A for poles.

TABLE 8. SUPPLEMENTAL GRAIN PARAMETERS OF FERROMAGNESIAN BASALTIC TEMPER SANDS FROM SAMOA AND AMERICAN SAMOA

Site, island, nation or territory	Olivine index (OLi)	Oxide index (OXi)
Mulifanua, western Upolu, Samoa (*n*=10)	80 (77–85)	5 (1–11)
Jane's Camp, western Upolu, Samoa (*n*=3)	81 (76–86)	7 (5–11)
Vailele, western Upolu, Samoa (*n*=6)	80 (77–84)	7 (0–12)
Sasoa'a, eastern Upolu, Samoa (*n*=3)	37 (31–40)	9 (6–14)
To'aga-Va'ota beach sand tempers, Ofu, American Samoa (*n*=18)	38 (17–50)	9 (3–17)
To'aga-Va'ota stream sand tempers, Ofu, American Samoa (*n*=9)	28 (22–32)	3 (1–6)
Terrigenous (placer) fraction of modern hybrid beach sand (beach face between Va'ota and To'aga), Ofu, American Samoa (*n*=2)	32 (31–32)	11 (11–11)

Notes: tabulated values are means with ranges in parentheses; *n* is number of sherds used in tabulation.

tempered with ferromagnesian basaltic sands over any appreciable distances within the Samoan islands. No sherds examined from western Upolu contain tempers with the comparatively low OLi indices characteristic of tempers in sherds from eastern Upolu and Ofu, nor vice versa. The possible movement of pottery containing ferromagnesian basaltic tempers among sites in western Upolu cannot be evaluated closely because the tempers in sherds from all studied sites are similar, but the distinct contrast in OLi index for sherds from western and eastern Upolu suggests that sands available in close proximity to each Upolu site were used as temper. The geologic significance of the similarity in OLi index for sherds from Ofu in the Manu'a Group and from eastern Upolu is difficult to judge because so few sherds containing beach sand tempers have been examined to date from intervening Tutuila. A similarity in island geology is more attractive, however, as an explanation for the temper similarity than wholesale ceramic transfer over a distance of ~200 km (Fig. 11).

The limited abrasion and poor sorting of the constituent grains in the Ofu tempers interpreted as stream sands are atypical of Samoan tempers interpreted as beach sands, and the high ratio of temper to clay (~60:40) in sherds containing the stream sands as temper suggests that the fluvial sand may be natural temper embedded in alluvial deposits of sandy clay (the more typical ratio of temper to clay in most Oceanian sherd suites is <40:60). A preponderance of the sherds containing profuse basaltic temper (Dickinson, 1993) of inferred alluvial origin for which clay paste has been studied by SEM techniques cluster compositionally with Ofu colluvial clays from near the To'aga site where the sherds were excavated (Hunt and Erkelens, 1993, p. 145).

The clay pastes in many sherds from the To'aga site on Ofu cannot be matched closely with any known Ofu clays, but at least some sherds containing each of the varied tempers observed in To'aga sherds (Table 7) resemble local colluvial clays present near the To'aga site (Hunt and Erkelens, 1993, p. 145). The unfamiliar clay bodies may derive from untested clay sources located elsewhere on Ofu, on adjacent Olosega, or possibly from the nearby larger island of Ta'u within the Manu'a Group (Fig. 11) where prehistoric pottery was also made (Hunt and Kirch, 1988).

Two recently studied sherds from the Pulemelei site (Table 7) on southeastern Savai'i (Fig. 11) contain, as manually added temper, moderately sorted and subrounded to rounded stream sands composed almost exclusively of basaltic lithic fragments displaying dominantly intergranular (lathwork) internal textures. Microphenocrysts in the lithic fragments include olivine (tinged on grain margins by reddish deuteric iddingsite) and less common clinopyroxene. Rare monocrystalline grains of clinopyroxene are also present, as are vitrophyric and vitric grains probably derived from chilled lava surfaces or basaltic tephra. The grain aggregates were probably derived from Holocene alkalic olivine basalt widely exposed along the elevated crest of Savai'i in the hinterland of Pulemelei (Fig. 11). Despite generic similarities, the basaltic Pulemelei temper does not closely resemble any other known Samoan tempers in either texture or mineralogy.

Colluvial Debris

In some sherds from Tutuila, the only temper present is unabraded and deeply weathered basaltic detritus interpreted as natural temper embedded as residual grit within colluvial clay paste (Table 7). Most temper grains are altered volcanic lithic fragments now composed largely of clay minerals produced during weathering, but sparse weathered mineral grains are also present in some cases. The colluvial tempers in sherds from different sites vary slightly in appearance, but generally afford no means to infer local or nonlocal origins. In the case of a sherd examined from Aunu'u islet offshore from Tutuila (Fig. 11), however, the best-preserved temper grains are tan to brown fragments of altered palagonitic glass, and several display reentrants in grain margins typical of vesicle walls partly bounding ash particles in hyaloclastite produced by phreatomagmatic eruptions of the type that constructed the palagonitic Aunu'u tuff cone (Stearns, 1944; Macdonald, 1944). This evidence for local origin suggests that Tutuila sherds containing colluvial clays with natural temper also derive from near the sites of their respective recovery.

Crushed-Rock Tempers

The only temper grains in numerous sherds from multiple sites in Samoa are unabraded and poorly sorted aggregates composed dominantly of unweathered fresh lithic fragments (Fig. A3) of identical internal texture, fabric, and composition. Associated smaller mineral grains are species present as microphenocrysts within the lithic fragments. Because some of the lithic fragments have partly cuspate margins resembling the grain outlines of microscoria particles in basaltic ash, the crushed-rock tempers were initially misinterpreted as fresh volcanic ash derived by inference from active cinder cones in eruption during human occupation of Samoa (feldspathic basaltic temper of Dickinson, 1969, 1976b, 1993). In many instances, however, the lithic fragments are too coarsely crystalline internally to be credible as fragments of basaltic tephra, and the locally cuspate grain margins can be understood instead as arising from the breakage of vesicular lava.

Crushed-rock tempers are most common in sherd collections from Tutuila (Table 7) where prehistoric adze quarries are more extensive and prevalent than elsewhere in Samoa (Best et al., 1992; Clark et al., 1997). My previous suggestion (Dickinson and Shutler, 2000) that ancient Samoan potters may have acquired crushed-rock temper from wastage in adze quarries is countered, however, by the observation that fresh vesicular lava is generally unexposed in the adze quarries. A more likely source of crushed vesicular basalt is afforded by jagged aa on the surfaces of Holocene lava flows, which are widespread on southern Tutuila (Stearns, 1944) but are also present on other Samoan islands (Fig. 11) where crushed-rock tempers are known (Table 7).

Although the lithic fragments of crushed-rock temper are lithologically identical in each sherd, there is wide variability in the lithology represented from sherd to sherd, as well as from site to site, and there is no presumption that all came from the same island. In principal, crushed-rock tempers could be traced to specific collecting sites using either petrographic or geochemical methods, but the sourcing problem is akin to the challenge of sourcing specific clay bodies. Petrologically, all the crushed-rock tempers are fully compatible with Samoan origins, but matching each to a specific outcrop of volcanic rock would be daunting, given the large size of individual Samoan islands and the multiplicity of feasible bedrock sources on them.

The subordinate mineral grains and the microphenocrysts within lithic fragments of most crushed-rock tempers are plagioclase, rather than ferromagnesian clinopyroxene or olivine, suggesting that many may have been derived from hawaiite rather than basalt or basanite. In crushed-rock tempers from sites on Upolu and Tutuila, the internal textures of lithic fragments are most commonly fine-grained lathwork (intergranular), broadly similar petrologically to more heterogeneous lithic fragments that are dominant in the ferromagnesian basaltic sand tempers of Samoa. On Ofu, however, all known crushed-rock tempers were derived from microlitic vitrophyre with a tachylitic groundmass (Dickinson, 1993). This consistent contrast between Ofu and other Samoan crushed-rock tempers suggests that all the crushed-rock tempers were derived from somewhere on the islands where sherds containing them were recovered, and do not reflect interisland ceramic transfer.

Trachytic Tempers

At several sites on Upolu (Dickinson, 1969, 1976b) and Tutuila, subordinate numbers of sherds contain temper sands of trachytic (Fig. A5) rather than basaltic composition (Table 7). The trachytic tempers are poorly sorted aggregates that contain largely homogeneous populations of lithic fragments evidently derived from a restricted bedrock source in each case. Derivation from wastage in adze quarries can be ruled out because no Samoan adzes contain more than ~50% silica (Best et al., 1992; Clark et al., 1997), whereas mugearites and trachytes of Upolu contain 56%–61% silica (Natland, 1980), and trachytes and quartz trachytes of Tutuila contain 66%–71% silica (Macdonald, 1944). The limited variability and poor sorting of the lithic fragments in the trachytic tempers suggest origins for the grain aggregates from taluslike colluvial aprons flanking trachyte plugs, which form prominent steep-sided peaks on the Samoan landscape. The slight abrasion of temper grains and minor variations in the lithology of lithic fragments are compatible with that inference.

Trachytic tempers in sherds from sites on Upolu (Table 7) are all compositionally similar sands in which feldspar-rich lithic fragments have microlitic (pilotaxitic) internal textures and display fluidal internal fabrics with oriented plagioclase microlites bundled into parallel arrays. Subordinate mineral grains include anorthoclase as well as plagioclase, minor oxyhornblende, and rare clinopyroxene. As trachyte plugs occur on Upolu only at the crests of steep ridges rising abruptly from Fagaloa Bay east of Falefa Valley (Fig. 11), all the closely similar trachytic tempers in Upolu sherds probably derive from that general locale. If so, they provide evidence for intraisland (or alongshore) ceramic transfer (or temper transport) of at least 40 km to archaeological sites on western Upolu (Fig. 11). Lithic fragments in the trachytic tempers of sherds from multiple Upolu sites are similar enough to derive from the same plug or closely related plugs.

Trachytic tempers in sherds from 'Aoa Valley on Tutuila (Table 7) are distinctly different lithologically from the Upolu trachytic tempers. Nearly all the temper grains, which are more poorly sorted than those in Upolu trachytic tempers, are lithic fragments of felsitic internal texture containing blocky sanidine microphenocrysts set in a densely microcrystalline groundmass of mosaic quartz and feldspar. Rare sanidine mineral grains are present in the tempers as well. No close lithologic matches with trachytes described from plugs exposed in highlands rising abruptly from 'Aoa Valley can be inferred with present information, but available petrographic descriptions of the numerous trachyte plugs on Tutuila (Fig. 11) are sketchy and not exhaustive (Macdonald, 1944).

The contrast between Upolu and Tutuila trachytic tempers provides no evidence for interisland ceramic transfer of sherds containing the trachytic tempers. An indigenous, if not entirely local, origin for all the trachytic tempers is supported by the circumstance that no trachyte plugs are present in the Manu'a Group (Stice and McCoy, 1968), and no trachytic tempers have been encountered in sherd collections from Ofu (Table 7).

Marquesan Tempers

The five indigenous Marquesan sherds discovered to date are Polynesian Plainware recovered from three sites on three different Marquesan islands (Atuona Valley on Hiva Oa, Ho'oumi on Nuku Hiva, and Hane on Ua Huka). All three islands are composed dominantly of tholeiitic to alkalic olivine basalt and basanite containing phenocrysts of olivine and clinopyroxene, and minor plagioclase, with phenocrysts of brown hornblende present in areally restricted exposures of more felsic differentiates of the mafic magmas (Brousse et al., 1978a, 1978b; Brousse and Sevin, 1978; Le Dez et al., 1996; Ielsch et al., 1998). The temper sands in Marquesan sherds are composed exclusively of terrigenous detritus presumably derived from local island bedrock. The olivine index (OLi < 5) of Marquesan tempers is distinctly less than in oceanic basalt tempers from farther west (Samoa, Uvea, Rotuma), and the presence of hornblende in several of the temper sands (Table 9) is also distinctive.

Three variants of Marquesan temper (Table 9) are represented in the few sherds available for study (Dickinson and Shutler, 1974; Dickinson et al., 1998b). Sherds containing subrounded and well-sorted placer sands as temper derive from coastal sites (Ho'oumi, Hane) where beach placers were probably available nearby. A less placered but equally well sorted

TABLE 9. GENERAL CHARACTERISTICS OF INDIGENOUS MARQUESAN TEMPER SANDS

	Placer beach sand	Non-placer beach sand	Alluvial sand
Sherd number(s)	(1) 85-1061 (2) MUH1-186-21	MUH1-J86-22	#4 and #5
Site(s) of recovery	(1) Ho'oumi on Nuku Hiva (2) Hane on Ua Huka	Hane on Ua Huka	Atuona Valley on Hiva Oa
Temper sand texture	well sorted; rounded to subrounded	well sorted; subrounded to subangular	poorly sorted; subangular to subrounded
Dominant grain types	ferromagnesian mineral grains (91%-94%): $OLi=2-3$; $OXi=49-66$ $PYi=PXi=100$	plagioclase (45%) and ferromagnesian mineral grains (45%): $OLi=5$; $OXi=11$ $PYi=PXi=100$	lithic fragments (microlitic and microporphyritic)
Subordinate grain types	lithic fragments (4%-7%); plagioclase (2%)	lithic fragments (10%); brown hornblende (1%)	clinopyroxene, plagioclase, opaques, minor brown hornblende

Note: Sherd numbers after Kirch et al. (1988) and Dickinson et al. (1998b).

feldspar-rich temper in an additional Hane sherd is probably also beach sand. More poorly sorted and less abraded lithic-rich sands of probable fluvial origin occur as temper in sherds from an inland site (Atuona Valley) where only alluvial deposits would have been readily available for collection as temper. The textural and compositional variety of Marquesan tempers suggests that local pottery was made at multiple locales, rather than being dispersed from a common origin within or outside the Marquesas Islands.

The recognition of indigenous Marquesan pottery is important because it indicates that Polynesian migrants still making pottery reached the Marquesas Islands before pottery manufacture was abandoned in West Polynesia (Kirch et al., 1988). Most Marquesan prehistory was aceramic, and no indigenous pottery is known from any other island group of East Polynesia. Arrival of the first Marquesan settlers during waning phases of the Polynesian ceramic tradition is implied by the paucity of Marquesan sherds that have been discovered despite multiple Marquesan excavations (Sinoto, 1983).

Until recently, the youngest sherds in Samoa and Tonga were thought to date from the interval A.D. 200–400 (Fig. 3), and initial settlement of the Marquesas Islands was thought to have occurred during the interval A.D. 300–600 (Spriggs and Anderson, 1993; Dickinson, 2003). The sherds from Atuona Valley derive from an undated stratigraphic horizon (Kirch et al., 1988), but the sherds from Hane were recovered from a horizon that lies beneath a rock pavement originally dated to the interval A.D. 300–600 (Sinoto, 1966). Reappraisal of the radiocarbon chronology for the Hane site indicates, however, that its occupation dates instead from as late as A.D. 900–1200 (Anderson and Sinoto, 2002). Initial human settlement of the Marquesas Islands probably did not significantly predate A.D. 1000 (Conte and Anderson, 2003; Allen, 2004). This revised timing provides support for interpretations that pottery manufacture in American Samoa continued until A.D. 800–1200 (Ayres et al., 2001).

Otherwise, Polynesian migrants to the Marquesas Islands would have had no ceramic tradition to perpetuate.

ANDESITIC ARC TEMPERS

Andesitic arc tempers are volcanic sands more complex mineralogically than oceanic basalt tempers, and commonly contain a greater variety of lithic fragments. Parent magmas were erupted along festoons of island arcs that fringe the western margin of the Pacific Ocean basin (Fig. 5) where subduction of oceanic lithosphere induces melting of mantle fluxed by volatile elements and contaminated to varying degrees by crustal materials carried to mantle depths on subducted oceanic plates. Eruptive assemblages include a spectrum of lavas and pyroclastic products ranging from subalkalic basalt or basaltic andesite at the mafic end, through both mafic and felsic andesites, to dacite and more rarely rhyodacite or rhyolite at the felsic end. The proportions of igneous rocks falling along this spectrum vary widely among magmatic arcs. Andesitic rocks are most prevalent overall, but some arcs erupt mainly basalt and basaltic andesite, and lack any felsic products. Distinctions among adjacent compositional members of the gradational spectrum of rock types require geochemical analysis and cannot be made for individual temper grains. In particular, the nature of detritus from more silicic end members of arc suites (dacite, rhyodacite, rhyolite) is broadly similar from parent rocks spanning the full range of silicic petrology.

The volcanic assemblages of island arcs vary petrochemically in a manner reflected most directly by the ratio of potassium to silica in eruptive suites (Dickinson, 1975a). The most common distinction is between arc tholeiitic (low K) and calc-alkalic andesitic suites, but in actuality a spectrum of K/Si ratio is exhibited by different island arcs, giving rise to a range in overall arc petrology and associated mineralogy (Gill, 1981; Thorpe, 1982). Along any given arc, however, lavas of varying silica content have analogous K/Si ratios.

The variable petrochemistry of different arcs can be specified by the $K_{57.5}$ value, which is derived from geochemical variation diagrams of K_2O versus SiO_2 for arc lavas and is the typical percentage of K_2O for an SiO_2 percentage of 57.5 (Dickinson, 1975a; Gill, 1981). The silicic end members of eruptive suites in arcs with low $K_{57.5}$ (<1.3), including Tonga and New Britain, are K-poor dacites. Silicic lavas in arc suites with intermediate $K_{57.5}$ (1.4–1.7) locally erupt rhyodacite as well as dacite, and include arc assemblages in Vanuatu and the Solomon and Mariana arcs. The Halmahera and Banda arcs with high $K_{57.5}$ (>1.8) in the Molucca Islands (Maluku) of eastern Indonesia locally expose rhyolite with K-feldspar phenocrysts, which are lacking in the rhyodacites and dacites of other island arcs within the region of Oceanian ceramic cultures.

Island arcs are alignments of active and dormant volcanic islands, or extinct but persistent volcanic edifices, each of which may expose somewhat different components of the igneous assemblage that is characteristic of a given island arc, and in varying proportions locally. There are commonly variations in temper type from island to island along an island arc in addition to systematic variations from arc to arc. In some cases, dissection of dormant volcanic edifices has exposed subvolcanic dikes, sills, and other small intrusive bodies that contribute hypabyssal microphanerite lithic fragments to the dominantly volcanic sands locally available for use as temper. Segments of remnant arcs stranded within oceanic realms by backarc seafloor spreading, but as yet largely undissected and uncovered by postarc or backarc igneous assemblages, yield derivative detritus that is indistinguishable from the volcanic detritus present along active island arcs.

Arc volcanism includes more explosive eruptions than occur in other Oceanian geotectonic settings. In addition to lavas and breccias emplaced near volcanic centers, intermittent ash plumes and episodic ash clouds associated with caldera collapse spread pyroclastic debris over wide areas surrounding the eruptive centers. The explosivity of arc activity stems from the injection of volatile elements into subjacent mantle by plate subduction. Two results are important for ceramic analysis. First, temper sands may be derived from reworking of widespread ash blankets as well as from erosion of volcanic edifices. Second, some clay pastes in potsherds contain a petrographically visible component of volcanic ash in the form of curvilinear and branching glass shards of silt size. The shards are made of generally clear felsic glass that contrasts with the murky coloration of the clay matrix in which the ash is embedded, and impart a characteristic *vitroclastic* texture (as seen in tuff) to the clay pastes containing them.

Andesitic arc tempers are the most widespread of all Oceanian temper classes, and occur along all the following island arcs and closely associated arc segments: (1) the Volcano Islands of the Bonin arc between the Mariana Islands and Japan (Fig. 1); (2) an ancestral Paleogene segment (Fig. 6B) of the still-active Mariana arc, which is now exposed in the forearc belt of the Mariana chain where the Pacific plate is subducting beneath the Philippine Sea plate (Fig. 6A); (3) extinct Miocene and Paleogene components of the Yap and Palau arcs aligned southward from the Mariana arc along the western fringe of Micronesia at the eastern edge of the Philippine Sea plate (Fig. 6A); (4) the arcuate and partly contorted Halmahera and Banda arcs in the Molucca Islands of eastern Indonesia (Fig. 6A); (5) both active modern and disrupted ancestral arc segments in the Bismarck Archipelago east of New Guinea (Figs. 7 and 9); (6) the reversed-polarity Solomon (Fig. 7) and Vanuatu (Fig. 8) arcs subducting the seafloor of marginal seas downward to the northeast beneath the Pacific plate; (7) remnant arcs on the Fiji platform (Fig. 8) related both to Miocene subduction before and to Pliocene subduction after polarity reversal of the ancestral Vitiaz arc (Fig. 9); (8) the Miocene Lau remnant arc between the Fiji platform and Tonga (Figs. 8 and 9); and (9) the active Tonga (or Tofua) arc where Pacific seafloor is subducting downward to the west beneath the easternmost extremity of the Indo-Australian plate (Fig. 8).

Overall Composition

Andesitic arc tempers are composed almost exclusively of volcanic detritus (VLi ≈ 100; SLi = 0), with only rare lithic fragments of hypabyssal microphanerite present as nonvolcanic detritus. Volcanic lithic fragments are most typically microlitic (Figs. A6 and A15), derived from basaltic andesite and andesite, but lathwork grains derived from basalt are sparingly present, and felsitic grains derived from felsic andesite and silicic dacite-rhyolite are abundant in selected tempers (Fig. A7). Microscopic distinctions between microlitic or felsitic lithic fragments derived alternatively from felsic andesite or more silicic parent sources are generally not feasible unless quartz microphenocrysts are present as an indication of rhyodacitic composition. Vitric or vitrophyric lithic fragments of varied coloration are dominant in some temper types, but provide no clear indication of petrological affinity. Mafic volcanic glass most commonly displays shades of brown or dull red, whereas felsic volcanic glass is clear or pale tan in thin section. Mafic (Fig. A8) and felsic (Figs. A10 and A16) volcanic glass particles are present in temper sands of basaltic (or mafic andesitic) and felsic (andesitic or dacitic to rhyolitic) parentage, respectively, and opaque tachylitic glass is common in lithic fragments of microporphyritic vitrophyre. In rare instances, different populations of volcanic lithic fragments allow empirical distinctions between temper sands from different islands, but diverse and undiagnostic populations of lithic fragments are more common.

The olivine content of andesitic arc tempers is consistently lower (OLi < 25) than in oceanic basalt tempers, because basaltic end members of igneous assemblages in island arcs are typically subalkalic, with a petrochemistry giving rise to lower olivine contents. Either pyroxene or hornblende, which commonly occur in association, may be dominant along a given arc trend or segment (variable PYi), and orthopyroxene may be present as well as clinopyroxene (variable PXi). Pleochroic schemes for hornblendes range from green through green-brown to brown, and contrasting varieties of hornblende have diagnostic value in some cases for discrimination between different temper types. Both biotite and oxyhornblende are prominent and even dominant, either separately or jointly, in selected andesitic arc tempers. Systematic variations in populations of ferromagnesian silicate mineral grains from different island arcs, or from different individual islands, provide important criteria for interpreting temper origins.

Feldspar grains are most typically plagioclase to the exclusion of other feldspars (PFi ≈ 100) because the latter do not occur as phenocrysts except in rhyolitic rocks that are present only along selected island arcs. Where K-feldspar mineral grains do occur as temper grains, however, they provide a special guide to temper origins. Minor volcanic quartz is present in many andesitic tempers and is abundant in some dacitic to rhyolitic tempers (variable QZi). Proportions of opaque iron oxide grains vary widely (variable OXi), and opaque grains are predominant in some placer tempers.

Bonin Temper

Two prehistoric sherds (post–0 B.C./A.D.) examined in thin section from the island of Kita Iwo Jima in the Volcano Islands of the Bonin (=Ogasawara) Group, lying between the Mariana Islands and Japan to the north (Figs. 1 and 5), contain poorly sorted volcanic sand tempers (one coarse and one fine) composed almost exclusively of andesitic (to basaltic) lithic fragments. The subangular lithic fragments include both microlitic and yellowish vitric grains, the latter composed of hydrous palagonite. Rare plagioclase grains derived from phenocrysts in volcanic rock are also present. The temper sands were probably collected from either colluvial or alluvial deposits proximal to pyroclastic sources, but the heterogeneity of the abraded lithic fragments in the temper sands implies enough sedimentary reworking of pyroclastic debris to mingle detritus derived from varied pyroclastic strata. The tempers differ in both composition and texture from all Mariana tempers, and derivation from locales on Kita Iwo Jima is most likely, although ceramic transfer from another island along the Bonin chain cannot be excluded from petrographic study.

Mariana Tempers

The inhabited Mariana islands lie within the forearc belt of the Mariana arc-trench system (Fig. 6A). They are underpinned by paleovolcanic edifices erupted along the arc axis during Late Eocene to Early Oligocene time (45–30 Ma) when the Mariana frontal arc was still attached to the Palau-Kyushu Ridge before mid-Oligocene to mid-Miocene (32–17 Ma) opening of the Parece Vela Basin by backarc seafloor spreading (Fig. 6B). The island foundations are widely capped by Miocene to Pleistocene limestone that accumulated in shallow waters on a subsiding volcanic substratum, but have since been uplifted to elevations of 50–500 m by Quaternary forearc tectonism (Fig. 13). Transient resumption of minor Middle Miocene (15–12 Ma) volcanism produced local lavas and breccias interbedded with limestone on both Guam and Saipan. The Middle Miocene volcanic assemblage was erupted before opening of the Mariana Trough by a second episode of backarc seafloor spreading while the Mariana frontal arc was still attached to the West Mariana Ridge, a remnant arc where the chief record of Miocene volcanism along the Mariana arc-trench system is preserved (Meijer et al., 1983). The opening of the Mariana Trough (Martinez et al., 1995) was initiated by intra-arc rifting 7–6 Ma (Hawkins, 2003), with seafloor spreading since 4–3 Ma (Fryer, 1995). The neovolcanic chain of the active Mariana arc was probably initiated in Pliocene time (Dickinson et al., 2001), and lies west of the paleovolcanic chain (Fig. 13).

For megascopic appraisal, archaeologists subdivide Mariana tempers into several generic groups (Dickinson et al., 2001): CST—exclusively calcareous sands; VST—generally quartz-free volcanic sands; VQT—volcanic sands in which quartz is prominent; QT—volcanic sands in which quartz is the dominant grain type; CQT (or QCT)—hybrid sands composed exclusively or predominantly of quartz and calcareous grains; MST—hybrid sands composed of volcanic sand mixed with calcareous grains; ST—grog aggregates (rare and not studied petrographically). The terrigenous detritus in all Mariana temper types was derived from andesitic and dacitic bedrock exposed on inhabited islands rather than islands of the active volcanic chain (Fig. 13).

Main Temper Distinctions

Hybrid and placer sand tempers are well-sorted aggregates derived from modern island beaches or from paleobeach deposits on emergent marine terraces. Calcareous grains in the beach sands are rounded to subrounded, whereas terrigenous grains are subrounded to subangular. Calcareous reef components in the beach sands are dominantly skeletal debris of coralgal and bivalve origin, but micrite grains derived from wave attack on consolidated reef tracts are present sparingly. Nonhybrid and nonplacer temper sands are poorly to only moderately sorted aggregates of subangular to subrounded grains, and represent stream sands derived from unknown local drainages. The stream sand tempers in some sherds from Saipan contain subordinate detrital limeclasts (6%–18%) reworked from Neogene limestone exposed within eroding uplands. Although textural contrasts in grain size between sand temper and clay paste are minimal for some stream sand tempers, the fresh and unaltered character of volcanic glass within abundant vitric grains implies that they are manually added tempers, as are all the beach sand tempers.

Many prehistoric Mariana sherds contain calcareous tempers undiagnostic of specific origins. Detrital limeclasts provide no more guide to island of origin than debris from modern reef tracts because exposed limestones are as similar in lithology from island to island as are the offshore reefs yielding calcareous debris. Both exclusively terrigenous and hybrid tempers in sherds from many sites nevertheless provide a record of interisland ceramic transfer, primarily from Saipan to several other islands but secondarily also from Guam to Rota (Dickinson et al., 2001). No terrigenous or hybrid Mariana tempers are known to derive from any islands

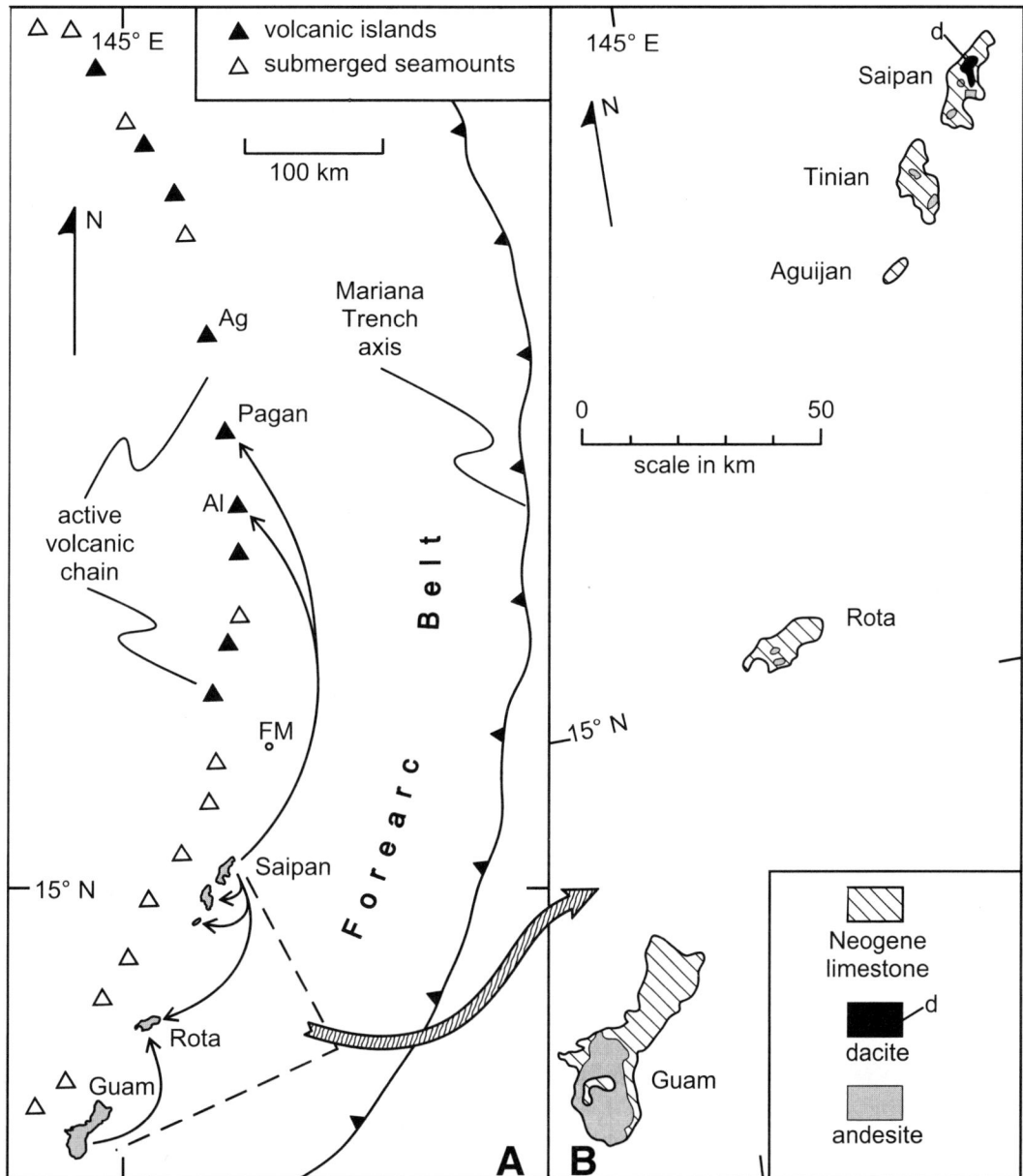

Figure 13. Documented ceramic transfer (arrows in A on left) and sources of terrigenous temper sand (lithology in B on right) in the Mariana Islands after Dickinson et al. (2001). Islands: Ag—Agrihan; Al—Alamagan; FM—Farallon de Medinilla.

other than Saipan and Guam, although restricted occurrences of potential temper sand occur on intervening Rota. In general, however, outcrops of volcanic bedrock on Rota and Tinian are too limited to provide sources for terrigenous temper sands, and no volcanic rocks are exposed on Aguijan (Fig. 13).

Volcanic temper sands from Saipan and Guam can be distinguished by several petrographic means (Dickinson et al., 2001). Table 10 is a summary of compositional contrasts among Mariana tempers derived from bedrock assemblages exposed on Saipan and Guam. Quartz-rich dacitic tempers (Fig. A16) are restricted to Saipan where Upper Eocene dacite domes, lavas, and breccias are exposed (Cloud et al., 1956; Schmidt, 1957), but are absent on other Mariana islands (Schmidt, 1957, p. 153; Stark, 1963; Tracey et al., 1964; Reagan and Meijer, 1984). Some quartz in Saipan tempers was recycled through quartz-bearing Eocene to Oligocene volcaniclastic strata derived from dacitic bedrock. Dacitic Saipan tempers include a variety of sands available in residual soils and along streambeds, or as the terrigenous fraction

TABLE 10. COMPOSITIONAL CONTRASTS OF MARIANA TEMPERS FROM SAIPAN AND GUAM

Temper type	%CAL	%VRF	QZi	OXi	OLi	PXi
Saipan hybrid CQT (n=18)	35-95	0-15	78-100	~100	-	-
Saipan dacitic VST (n=20)	0	20-85	36-90	~100	-	-
Saipan andesitic VST (n=18)	0	20-75	~0	30-85	0	76-88
Guam andesitic VST (n=36)	0	45-90	~0	25-85	0	~100
Guam placer hybrid MST (n=8)	0-18	12-30	0	4-20	3-13	97-100

Notes: n is number of sherd tempers tabulated. Megascopic temper types: CQT, calcareous-quartzose sand; VST, volcanic sand; MST, mixed volcanic-calcareous sand. %CAL is percentage of calcareous grains in total grain population. %VRF is percentage of volcanic lithic fragments in terrigenous grain population. Supplemental grain parameters (Table 6): QZi, quartz index; OXi, oxide index; OLi, olivine index; PXi, pyroxene index (PFi=VLi=PYi=100; SLi=0 for all Mariana temper types).

of hybrid sands present both on modern beaches and within the sediment cover of emergent Pleistocene marine terraces (Dickinson et al., 2001).

Ceramic transfer (Fig. 13) from Saipan to nearby Tinian and Aguijan, to more distant Rota and Guam, and to Alamagan and Pagan of the active volcanic chain north of Saipan is documented by temper analysis (Dickinson et al., 2001). Limited collections of sherds from the small volcanic islands apparently derive solely from the nearest large island, Saipan, from which the northern volcanic islands can be reached with relative ease by sailing on a single long reach with the prevailing wind on the beam and successive islands almost constantly in sight to the west. Sherds recovered from Rota represent wares from nearby Guam as well as from more distant Saipan.

Volumetric relations of interisland ceramic transfer within the Mariana Islands cannot be established from petrographic analysis of the selected sherds examined petrographically, but full archaeological appraisal of the voluminous Mariana sherd collections, which include a high proportion of sherds with undiagnostic calcareous tempers, is beyond the scope of this report. No Mariana tempers have been detected, however, in any sherds collected elsewhere in Micronesia (Intoh and Dickinson, 2002), nor are any tempers exotic to the Mariana Islands present in any Mariana sherds studied petrographically (Dickinson et al., 2001).

Saipan Dacitic Tempers

The terrigenous detritus in VQT, QT, and CQT Mariana tempers was derived from dacitic bedrock on Saipan. The dacitic tempers include both stream sands (VQT) and hybrid beach sands (CQT), in which dacitic detritus is mixed in variable proportions with abundant calcareous grains (Table 10). Volcanic lithic fragments are dominantly vitric grains composed of colorless to tan or pale brown felsic volcanic glass that is optically isotropic and variably pumiceous, with microvesicular internal structure visible in many grains. Minor felsitic grains are also present. The vitric and felsitic grains were derived from the groundmasses of dacite lava or breccia blocks, and the abundant quartz grains from phenocrysts in dacite. Some volcanic lithic fragments are vitrophyric with quartz microphenocrysts set in volcanic glass. Stretched microvesicles that are common in pumiceous grains were deformed by flowage of dacite lava after eruption. The quartz grains are equant, in part bipyramidal in habit, and display the uniform extinction characteristic of volcanic quartz, and some have smoothly embayed margins reflective of magmatic resorption before eruption. Proportions of volcanic lithic fragments and quartz mineral grains are highly variable (Figs. 14A and 14B), and minor feldspar grains that are exclusively plagioclase accompany quartz (Fig. 14D) in quite variable proportions (Figs. 14B and 14C) as well, but ferromagnesian grains are minor constituents and ferromagnesian silicate mineral grains are absent or rare (Figs. 14C and 14D).

Exotic Saipan sherds from Unai Chulu on Tinian, Mochong on Rota, and Laguas on Guam contain more ferromagnesian silicate mineral grains (Fig. 14D), chiefly clinopyroxene but also including biotite in the Mochong sherds. The quartz contents of these somewhat anomalous tempers are nevertheless indicative of origins on Saipan from primarily dacitic source rocks. The presence of biotite flakes (3%–4%) in the Mochong sherds is unique for Mariana tempers, but biotite is present as phenocrysts in some Saipan dacites (Schmidt, 1957). Some volcanic lithic fragments in the Laguas and Unai Chulu sherds are somewhat darker brown than typical for Saipan dacitic tempers, and suggest admixture of dacitic with andesitic detritus that may have contributed the ferromagnesian mineral grains.

Mariana Andesitic Tempers

The terrigenous detritus in VST and MST Mariana tempers containing little or no quartz (QZi < 5) was derived from andesitic bedrock on Saipan or Guam. The stream sand tempers of andesitic parentage are a heterogeneous array of volcanic sands in which relative proportions of plagioclase mineral grains and volcanic lithic fragments are highly variable (Fig. 14B), but contents of ferromagnesian mineral grains are generally higher than in Saipan dacitic tempers (Fig. 14D). Placer andesitic tempers from Guam are beach sands, in part hybrid, in which ferromagnesian mineral grains are concentrated (Fig. 14A). Relative proportions of opaque iron oxide grains and ferromagnesian silicate mineral grains are highly variable (OXi = 4–84) in both Saipan and Guam andesitic tempers (Fig. 14D).

Andesitic tempers from Saipan tend to be more feldspathic and less lithic than those from Guam, but there is broad overlap over much of the observed compositional range (Fig. 14B). Volcanic lithic fragments are dominantly vitric grains of brown mafic volcanic glass in Guam andesitic tempers (Fig. A8), but are more heterogeneous in Saipan andesitic tempers, with microlitic grains typically more abundant than vitric or felsic grains. Among volcanic lithic fragments, 70%–100% are vitric grains

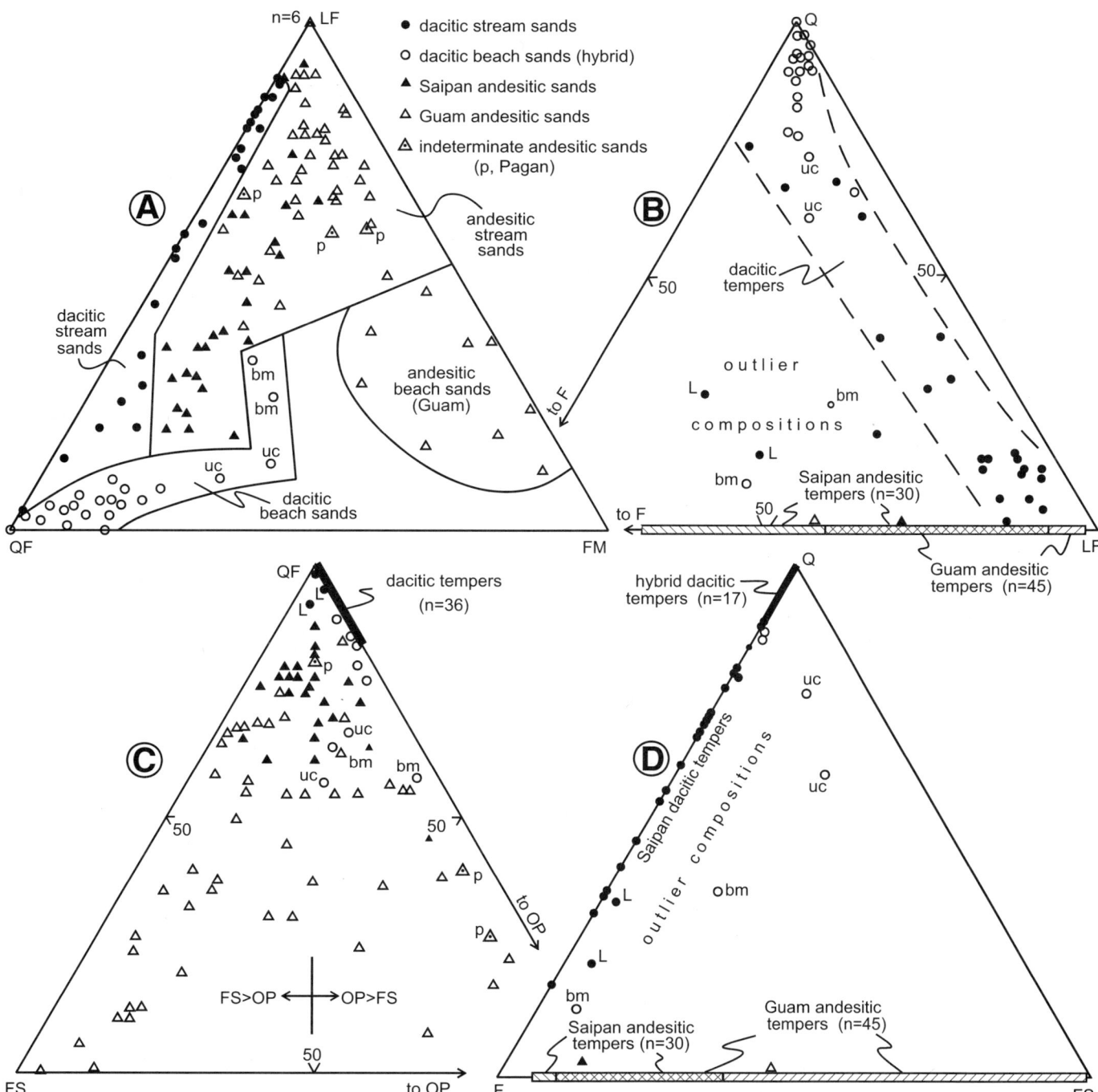

Figure 14. Compositions (grain frequency percentages) of beach and stream sand tempers in sherds from the Mariana Islands (bm—biotitic Saipan tempers in sherds from Mochong, Rota; L—dacitic Saipan tempers in sherds from Laguas, Guam; p—indeterminate but probable Saipan tempers in sherds from Sanmeima, Pagan; uc—Saipan tempers in sherds from Unai Chulu, Tinian). A: LF-QF-FM diagram. B: Q-F-LF diagram. C: QF-FS-OP diagram. D: Q-F-FS diagram. See Table 5A for poles.

in nonplacer Guam andesitic tempers, and 55%–75% in Guam placer tempers, whereas 50%–75% of the volcanic lithic fragments are microlitic in most Saipan andesitic tempers, with felsitic grains also present. Proportions of volcanic lithic types are not, however, a fully reliable criterion for distinguishing between andesitic tempers from Saipan and Guam.

Proportions of different ferromagnesian silicate mineral grains afford a basis for discrimination between Saipan and Guam andesitic tempers (Table 10). Clinopyroxene mineral grains are most abundant in both. Orthopyroxene is also present in significant amounts in Saipan tempers (PXi = 76–88) but not in Guam tempers (PXi > 95 and most typically ~100). Moreover, olivine is present in some Guam placer tempers (OLi as high as 10–15), but not in any Saipan tempers. Where contents of ferromagnesian silicate mineral grains are too low in Mariana andesitic tempers for either pyroxene (PXi) or olivine (OLi) indices to be determined with precision, alternate origin of exotic sherds from Saipan or Guam is indeterminate (Fig. 14). For nonplacer Guam andesitic tempers, however, the vitric volcanic lithic fragments are so dominant as to provide a convincing indication of probable origin on Guam for selected exotic sherds collected on nearby Rota. Exotic sherds of indeterminate pyroxene index (PXi) from Pagan most probably derive from Saipan because they are associated with sherds containing undoubted Saipan temper sands of both dacitic and andesitic character. The high olivine index (OLi > 50) of a beach sand from Agrihan suggests that none of the temper sands observed in Mariana sherds derive from the active neovolcanic islands north of Saipan.

Yap Volcanic Tempers

The tempers most diagnostic of Yap origin are quartz-free tectonic highland tempers derived from metavolcanic greenschist that forms the basement of Yap. However, half the sherds of Yap origin studied to date contain volcanic sand tempers (Table 11), most of them quartz-rich, derived from weathered remnants of younger volcanic cover (Fig. 15). The Miocene Tomil Volcanics were erupted when Yap was a continuation of the West Mariana Ridge before backarc seafloor spreading had opened the interarc Mariana Trough (Fig. 6A). A thick red lateritic soil overlies outcrops of the Tomil Volcanics (Johnson et al., 1960), typically weathered to clayey saprolite in which volcanic textures and structures are visible in roadcut outcrops and mining pits, but volcanic mineralogy is largely unpreserved. Lava (subaerial flow-breccia), tuff-breccia, lapilli-tuff, and tuff of a dissected andesitic to basaltic stratocone are all present within the volcanic assemblage, although varied pyroclastic breccias are dominant.

Typical quartzose Yap tempers are poorly sorted and generally unabraded sands composed of both monocrystalline and polycrystalline (aggregate) quartz grains, microcrystalline chalcedonic quartz grains, composite quartz-albite grains, and rare aggregate epidote grains. Textural relations between the quartzose detritus and clay paste imply that the quartzose temper sands are mostly natural temper embedded in clay bodies (Intoh and Dickinson, 2002). Because derivative quartzose detritus is not expected from the mafic character of the Tomil Volcanics, early analysis of Yap tempers suggested that the sherds containing quartzose temper might be exotic to Yap (Dickinson, 1982a). Close examination of volcanic exposures on Yap indicates, however, that the Tomil Volcanics are laced by hydrothermal quartz veins, in part gold-bearing (Rytuba and Miller, 1990). The grain types in the quartzose tempers could all be derived from hydrothermal veinlets. On the ground in Yap, quartzose particles occur both as fragments within lateritic soil over veined and altered volcanic rock, and as float scattered over eroded surfaces of laterite exposures. Quartz sand is also known to be present as a constituent of clay-rich residual soils in parts of Yap where volcanic cover is present (Intoh and Leach, 1985, p. 124). The dominant temper in a number of Yap sherds consists of outsized ferruginous particles of opaque iron oxide interpreted as iron-rich nodules from lateritic soil. These pedogenic particles occur both independently and admixed with quartzose detritus (Table 11). Fine quartzose silt is also typically dispersed within the clay pastes of Yap sherds containing grog temper (Table 11).

Chalcedonic quartz grains and quartz-albite aggregates in the quartzose tempers were derived from hydrothermal veinlets but were misinterpreted as felsitic and microgranitic fragments of felsic igneous rock in previous reports (Dickinson, 1982a; Intoh and Dickinson, 2002). This misinterpretation led to the erroneous conclusion that felsic plutonic or hypabyssal igneous rocks might be present locally within the subvolcanic bedrock of Yap (Intoh and Dickinson, 2002), but personal reconnaissance of Yap does not support that inference. A few Yap sherds contain sparse quartz-free volcanic sand tempers (Table 11) in which plagioclase and clinopyroxene mineral grains are the only prominent grain types, presumably concentrated by weathering or erosion from the andesitic to basaltic Tomil Volcanics.

TABLE 11. DISTRIBUTION OF SHERDS CONTAINING YAP VOLCANIC TEMPER VARIANTS

Temper type	Yap	Ngulu	Ulithi	Fais	Lamotrek
Quartzose Yap volcanic temper (*n*=16)	6	1	6	3	-
Mixed Yap ferruginous and quartzose temper (*n*=4)	2	1	1	-	-
Yap ferruginous temper with quartzose paste (*n*=12)	7	-	5	-	-
Non-quartzose Yap volcanic temper (*n*=3)	-	1	1	1	-
Yap grog temper with silty quartzose paste (*n*=16)	3	1	1	7	4
[total Yap tempers of andesitic arc class, *n*=51]	[18]	[4]	[14]	[11]	[4]

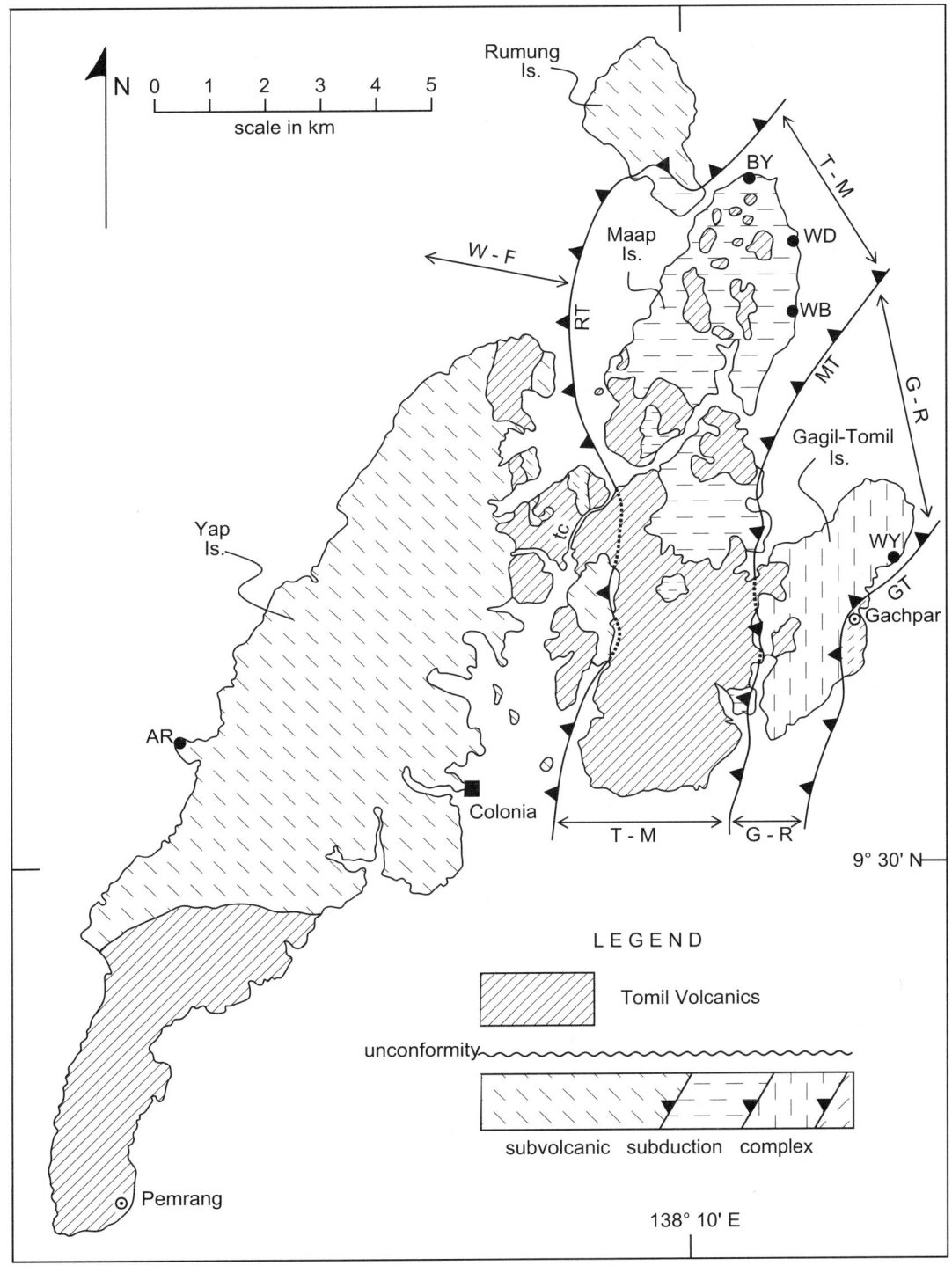

Figure 15. Geologic map of Yap (alluvium and coastal sand deposits not shown), adapted after Johnson et al. (1960) from personal reconnaissance in 2002, showing subduction complex of Oligocene to Early Miocene age overlain unconformably by Tomil Volcanics of probable Miocene age. Barbs on overriding plates of thrusts (dotted where concealed beneath younger Tomil Volcanics cover): GT—Gachpar thrust; MT—Madalai thrust; RT—Rumung thrust. Double-headed arrows denote span of thrust panels (schist and mélange): W-F—Weloy-Fanif; T-M—Tomil-Maap; G-R—Gagil-Riquen. Tageren Canal (tc) dug between Yap and Gagil-Tomil Islands indicated by double line. Collecting sites of beach sands (dots): AR—Aringel; BY—Bechiyal; WB—Wacholab; WD—Woned; WY—Waanyaan.

Palau Tempers

Babeldaob, the principal island of Palau (Fig. 16), exposes an elongate faulted stratocone of Oligocene age (38–32 Ma) erupted along the crest of the ancestral Palau-Kyushu remnant arc before backarc seafloor spreading had opened the interarc Parece Vela Basin (Fig. 6B). The eroded volcanic edifice is intruded locally by dikes of earliest Miocene (ca. 23 Ma) age (Meijer et al., 1983; Haston et al., 1988; Rytuba and Miller, 1990) that reflect waning igneous activity along the remnant arc. Local volcanic bedrock is a calc-alkalic basalt-andesite-dacite assemblage with the low strontium isotopic signature characteristic of intraoceanic island arcs (Matsuda et al., 1977). Mafic andesite and basalt dominant at lower stratigraphic horizons are succeeded by felsic andesite and dacite dominant at higher stratigraphic horizons where local dacite domes and plugs are locally prominent. Clinopyroxene is the dominant phenocrystic ferromagnesian silicate mineral throughout the stratal succession, although orthopyroxene and hornblende are also present sparingly at all stratigraphic horizons (Corwin et al., 1956). Volcanic and volcaniclastic strata of the ancient volcanic edifice are onlapped from the south by Miocene and younger limestone that forms a chain of small islands extending southward to include Peleliu and Angaur (Fig. 16).

Most sherds from Palau contain grog or composite tempers that form a gradational temper spectrum (Fitzpatrick et al., 2003), with ~60% of the sherds examined in thin section containing only grog particles as temper (Table 12). The terrigenous grains in composite Palau tempers are natural tempers of poorly sorted and little-abraded volcanic sand that was embedded within pedogenic or colluvial clay bodies available on the landscape as a residue from rock weathering. The composition of the terrigenous sand in composite Palau tempers (Fig. 17) is compatible with derivation from the bedrock of Babeldaob and neighboring offshore islets (Fig. 16), with the presence of quartz indicative of contributions from dacitic source rocks that are distinctly subordinate to andesite and basalt on Babeldaob. Conceivably, soil clays on dacite are more attractive as ceramic resources than soils on andesite or basalt, hence were selected preferentially by Palau potters for fabrication of ceramic vessels. If so, mapped occurrences of dacite on Babeldaob may provide a clue to the places where ceramic clay was collected.

The feldspar grains in the composite tempers are exclusively plagioclase (PFi = 100), volcanic lithic fragments are dominantly felsitic (72.5% of total volcanic lithic fragments, which include ~10% microphaneritic hypabyssal grains probably derived from dacitic plugs), and opaque iron oxide grains are much more abundant than ferromagnesian silicate mineral grains (OXi ≈ 90).

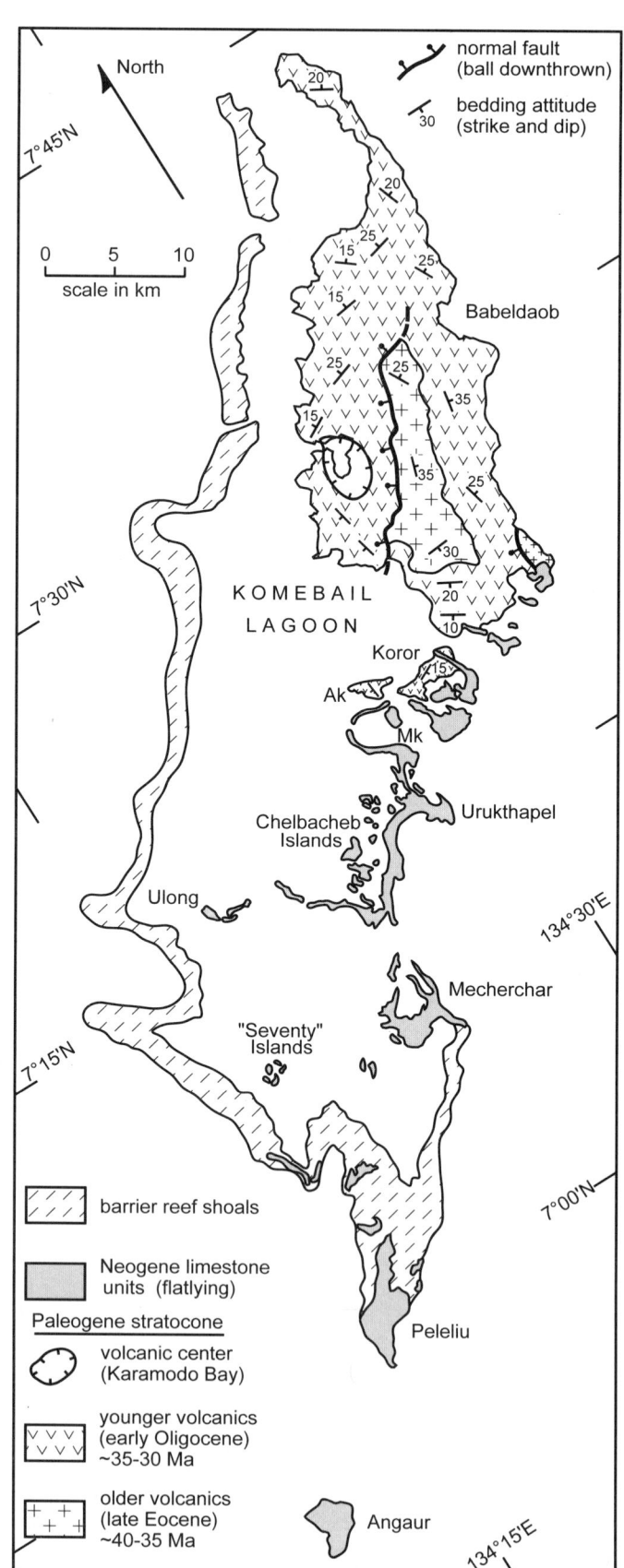

Figure 16. Geologic framework of Palau adapted after Corwin et al. (1956) with radiometric (K-Ar) ages for volcanic rocks (Haston et al., 1988; Rytuba and Miller, 1990). Small bedrock islands: Ak—Arakabesan; Mk—Malakal.

TABLE 12. GEOGRAPHIC DISTRIBUTION OF SHERDS CONTAINING BABELDAOB TEMPERS

Island	Grog and composite tempers				Volcanic and hybrid sand tempers			
	A	B	C	D	feld	lith	pla	hyb
Fana (Sonsorol Islands)	-	1	-	-	-	-	-	-
Angaur (Palau)	1	1	1	-	-	1	1	-
Peleliu (Palau)	2	2	-	-	-	-	-	-
Ulong (Palau)	6	1	-	1	1	1	-	7
Koror (offshore bedrock islet)	12	7	7	5	-	-	1	-
Orrak (offshore limestone islet)	41	8	6	-	1	6	3	1
Babeldaob (mainland sites)	35	27	4	-	2	1	-	-
Kayangel atoll (north of Palau)	2	-	2	-	-	-	-	-
Ngulu atoll (southwest of Yap)	2	-	-	-	-	-	-	-
Fais island (northeast of Yap)	6	-	-	-	-	-	-	-
[total sherds of Palau origin]	[107]	[47]	[20]	[6]	[4]	[9]	[5]	[8]

Notes: Islands listed from southwest to northeast. Grog and composite tempers (classified after Fitzpatrick et al., 2003): A, no terrigenous grains; B, rare terrigenous grains; C grog particles>terrigenous grains; D, grog particles<terrigenous grains. Volcanic and hybrid sand tempers: feld, feldspathic (F>LF); lith, lithic (LF>F); pla, placer (FM>65%), including three hybrid tempers in sherds from Orrak containing <3% calcareous grains; hyb, hybrid (16%-64% calcareous grains).

Occurrences of Palau composite tempers in sherds from limestone islands of Palau (Angaur, Orrak, Peleliu, Ulong), Fana in the Sonsorol Islands to the south, Kayangel atoll to the north, as well as Ngulu and Fais lying closer to Yap, is evidence for ceramic transfer on a large scale from Babeldaob or neighboring bedrock islets of similar geologic character (Fig. 16).

A subordinate number of Palau sherds (n = 26) from Babeldaob and nearby islets (Orrak and Koror), and from the limestone islets of Ulong and Angaur lying respectively ~25 and ~55 km to the southwest of Babeldaob (Fig. 16), contain volcanic sand tempers (Fig. 17) derived from andesitic bedrock of Babeldaob or one of its neighboring offshore islets (Table 12). Half of these sherds contain hybrid or placer tempers clearly of beach origin (Table 12), and the other volcanic temper sands are inferred to be beach sands as well from their consistent good sorting and rounded to subrounded character. No exclusively calcareous tempers are known from Palau despite the dominance of reef detritus on modern island beaches.

In Palau volcanic and hybrid sand tempers, feldspar is exclusively plagioclase (PFi = 100) and quartz is absent (QZi = 0). Volcanic lithic fragments are microlitic and vitric varieties in widely varying proportions (1:4–3:1). The dominant ferromagnesian silicate mineral grains are clinopyroxene (PYi > 95), with only trace amounts of hornblende and orthopyroxene, except for one anomalous temper dominated by hornblende and oxyhornblende (PYi = 17) in a sherd from Chelab in the interior of Babeldaob (Fig. 17). No olivine is present (OLi = 0), but the proportion of opaque iron oxide grains is highly variable (OXi = 27 ± 19, with a range of 0–65). Ratios of plagioclase mineral grains to volcanic lithic fragments are also highly variable (Fig. 17D), but both were reduced markedly in abundance by placering (Fig. 17A) that concentrated ferromagnesian mineral grains to produce pyroxenic temper sands (Figs. 17B and 17C).

In a key excavation on Ulong, all sherds from less than a meter depth contain grog or composite tempers, whereas all sherds from greater depths contain exclusively volcanic or hybrid sand tempers. Radiocarbon ages on charcoal from associated cultural horizons indicate that the transition in temper type dates to ca. 500 B.C. (Fitzpatrick et al., 2003; Clark, 2004). Adoption of grog temper as the dominant ceramic temper in Palau may reflect reduced availability of beach sand over time as alluviation in response to anthropogenic deforestation in the interior of Babeldaob choked the mouths of local streams and estuaries with sediment, converting former beach fronts to mangrove swamp. The dominance of grog temper in several millennia of Palauan ceramics may reflect this paleoenvironmental factor.

Halmahera Tempers

A limited number of sherds (n = 5) from extensive ceramic collections made on Halmahera (Fig. 18) and neighboring islets in the northern Molucca Islands were studied in a search for the origin of sherds containing exotic temper recovered from the southern Molucca and Aru Islands far to the south. All the Halmahera tempers are volcanic sands composed of volcanic lithic fragments of varied internal texture and mineral grains of plagioclase, clinopyroxene, opaque iron oxide, and hornblende, in that order of abundance. Each temper sand is somewhat different, however, in both texture and composition, and petrographic sampling to date is too restricted for an adequate appraisal of temper variation or mean compositions of tempers from different locales. The temper in a sherd from the Pulau Kumo site (Fig. 17) on northern Halmahera is discussed in a later section dealing with ceramic transfer from Banda to other islands.

Lease Temper

The small islands of Ambon, Haruku, and Saparua forming the Lease chain south of Seram (Fig. 18) at the curved northern end of the Neogene Banda arc (Fig. 6A) are formed by extinct volcanic edifices exposing a volcanic assemblage ranging from basalt through andesite and dacite to rhyolite (Hamilton, 1979,

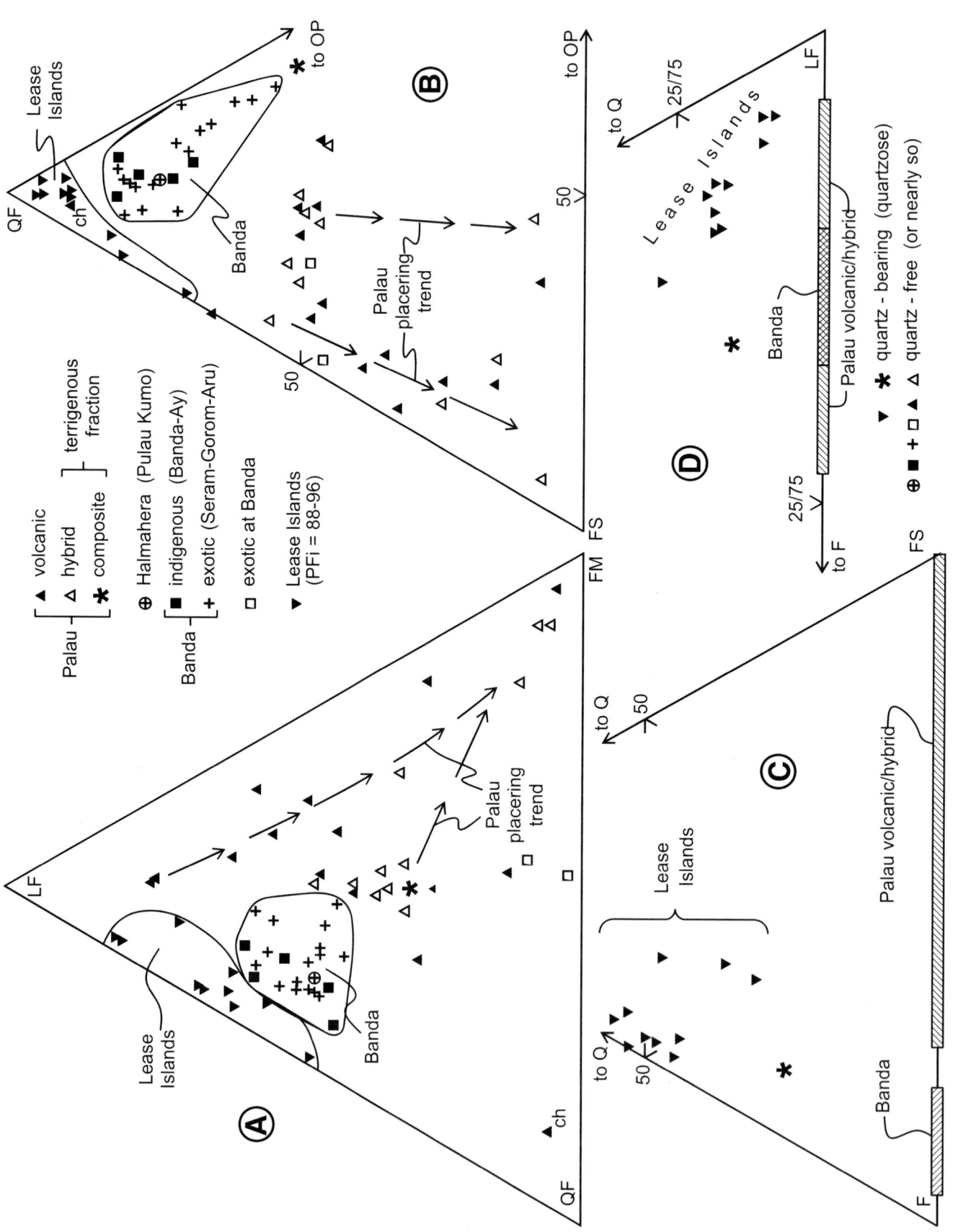

Figure 17. Compositions (grain frequency percentages) of volcanic sand tempers in sherds from the western fringe of Pacific Oceania (Palau and the Molucca Islands) where ch denotes anomalous temper in a sherd from Chelab in Ngaraard State of interior Babeldaob (Palau). A: LF-QF-FM diagram. B: QF-FS-OP diagram. C: Q-F-FS diagram. D: Q-F-LF diagram. Composition of volcanic sand in terrigenous fraction of Palau composite temper determined from a frequency count of 500 total grains in multiple selected sherds containing type C and type D composite tempers (Table 12). See Table 5A for poles.

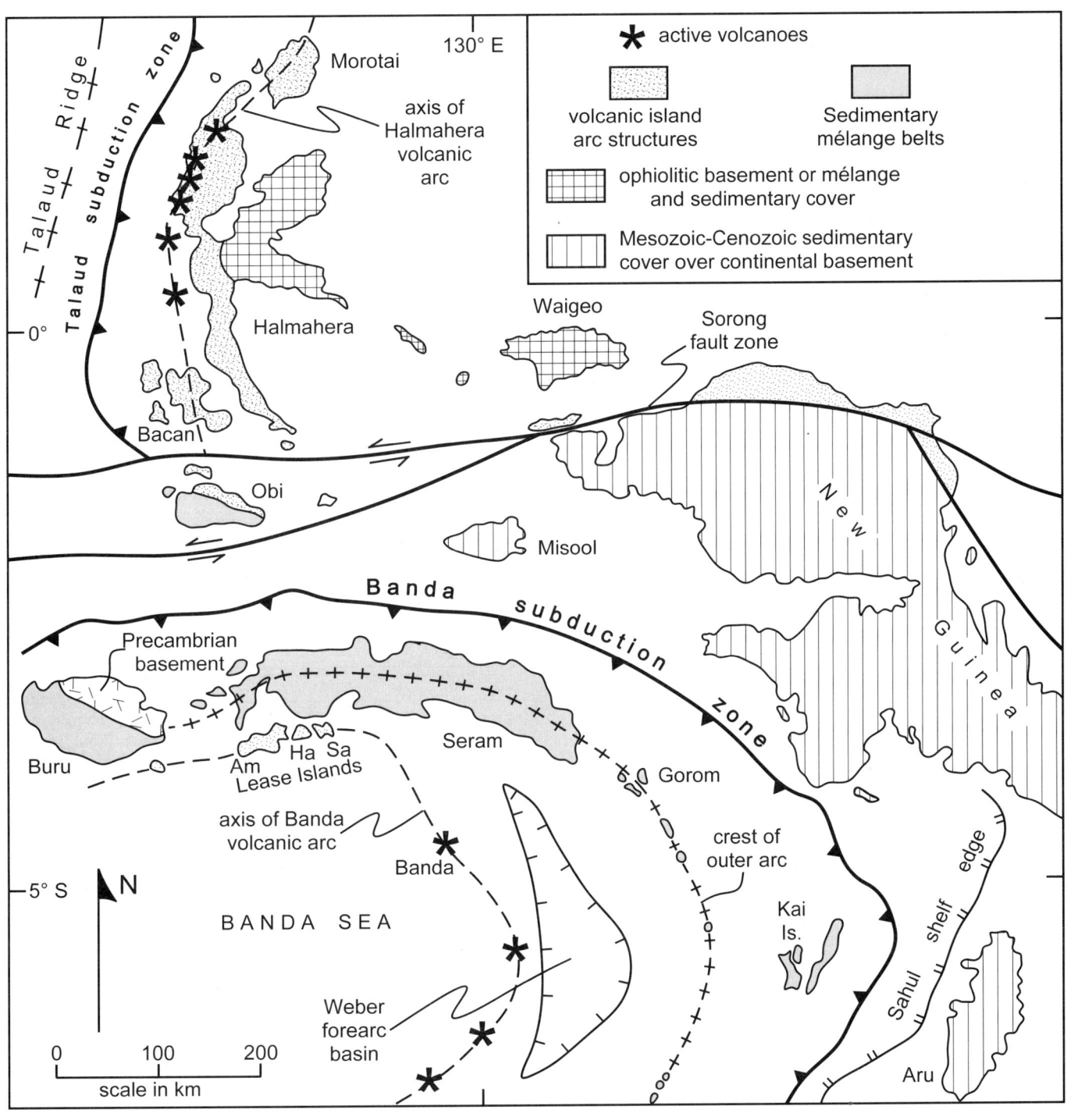

Figure 18. Geotectonic framework of the Molucca Islands in eastern Indonesia west of New Guinea, adapted after Hamilton (1979, plate 1, p. 133–138, 190–191) and Hall et al. (1988a, 1988b). Talaud subduction zone is structurally overprinted by backthrusting (not shown) of Talaud Ridge toward Halmahera arc. Lease Islands: Am—Ambon; Ha—Haruku; Sa—Saparua.

p. 118). The local volcanic rocks are both more silicic and more potassic than those of fully intraoceanic island arcs, and quartz and K-feldspar mineral grains are both present in derivative indigenous temper sands (Fig. 17). Flakes of strongly pleochroic red-brown biotite (1%–9% by grain frequency) are the only ferromagnesian silicate mineral grains present in Lease tempers. Plagioclase is largely untwinned and easy to confuse petrographically with quartz. Quartz grains include abraded bipyramidal shapes typical of euhedral quartz phenocrysts in silicic volcanic rock, and K-feldspar is largely sanidine with low optic axial angle (2V) also inviting confusion with quartz. Volcanic lithic fragments are dominantly clear isotropic volcanic glass (Fig. A10), and many either are microvesicular (pumiceous internal structure) or display perlitic cracks formed by contraction during rapid cooling of felsic lava. More mafic pale brown volcanic glass is also present as lithic fragments, but clear glass grains are much more abundant (94% ± 3% of lithic fragments, with a range of 88%–97%).

Temper sands of rhyolitic to rhyodacitic composition in eight prehistoric sherds from Ambon, one from Haruku, and an exotic Lease sherd from nearby Seram are closely similar in both texture and composition to modern stream sands, tempers in modern sherds, and sands in modern sand-clay potting mixes from Ambon and Haruku. Texturally, the temper aggregates are only moderately sorted, unplacered, and but little abraded (subangular to subrounded grains), and lack any calcareous grains of reef debris, suggesting an origin as stream sands rather than beach sands. The content of opaque iron oxide grains is widely variable (OXi range is 0–100), but both QZi (mean 51 ± 4; range 45–56) and PFi (mean 90 ± 3; range 88–96) are consistent within narrow limits, and place a diagnostic stamp of uniformity on the indigenous temper suite of the Lease Islands.

Banda Temper

A cluster of volcanic and calcareous islets form Banda at the northern end of the active Banda arc (Fig. 18). Indigenous volcanic temper sands are moderately sorted and subangular to subrounded aggregates of probable stream origin interpreted as manually added temper. The dominant grain types in Banda temper are vitric and commonly pumiceous volcanic lithic fragments, composed of felsic volcanic glass of pale hue, and plagioclase feldspar mineral grains (Fig. 17). Indigenous Banda tempers are hornblende-free (PYi = 100), reflecting the general character of eruptive assemblages (pyroxene andesite-dacite) along the Banda volcanic chain (Hamilton, 1979, p. 118; Whitford and Jezek, 1979; Bowin et al., 1980; Vroon et al., 1993). The only ferromagnesian silicate mineral grains are clinopyroxene (PXi = 100), but opaque iron oxide grains are typically more abundant (OXi = 57–71).

The clay pastes of indigenous Banda sherds contain a component of fresh volcanic glass shards of vitroclastic texture apparently derived from ash blankets that mantle Banda. The ash particles are curvilinear and forked or branching in shape, and composed of isotropic volcanic glass that is pale brown or tan to colorless in thin section. The shapes are diagnostic of glass shards formed by explosive disintegration of vesiculating felsic magma of viscous nature, and represent disrupted vesicle walls of gas bubbles expanding within erupting pumice. The shards are extremely thin (<0.05 mm), but reach lengths of ~0.25 mm. Although proportions of fine ash and clay are difficult to estimate optically, aplastic glass shards appear to form 15%–20% of Banda clay bodies but evidently did not spoil the plasticity of the clay paste in which the shards are embedded.

Two sherds recovered from Banda contain as temper hornblende-rich (PYi = 19–23) volcanic beach sands of placer and apparently hybrid character, although all calcareous grains have been leached from the sherds to leave sand-sized vacuoles. By comparison with the indigenous Banda temper type, the dominance of hornblende over pyroxene suggests that the sherds are exotic to Banda (Fig. 17). The lack of ash particles in their clay pastes supports that inference, but their origin within the Banda arc or elsewhere remains unknown.

Bismarck Arc Tempers

The Bismarck Archipelago surrounding the Bismarck Sea (Fig. 7) is part of a geotectonically complex region of intersecting subduction zones, spreading centers, and transform faults lying east of New Guinea within Near Oceania where a representative microcosm of nearly all Oceanian temper classes occur (Fig. 19). Andesitic arc tempers of the Bismarck Archipelago are known

Figure 19. Geotectonic setting of islands within and adjacent to the Bismarck and Solomon Seas east of New Guinea in Near Oceania, adapted after Dickinson (2000a, 2002b), showing distribution of known indigenous temper types (tempers on small islets offshore from New Britain plotted as indigenous even though partly exotic from nearby New Britain; similarly, tempers in sherds from Roviana Lagoon derive from multiple islands of the New Georgia Group). Key patterns of known ceramic transfer denoted by arrows. Heavy shading denotes dismembered and eroded Paleogene arc assemblage of Bismarck Archipelago dispersed by backarc seafloor spreading to open the Manus Basin. Arc tectonic affinity of Witu Islands (Wi) after Johnson and Arculus (1978). Barbs on overriding plates at subduction zones; extension directions assumed perpendicular to trends of spreading centers; half-arrows denote lateral motions across transform faults. Other geotectonic features: IRZ—intracontinental rift zone of D'Entrecasteaux Islands on strike with oceanic Woodlark Basin spreading system; RMS—Ramu-Markham suture where bedrock of the Huon Peninsula was accreted to New Guinea by Neogene crustal collision with an ancestral Schouten–New Britain island arc (Fig. 9). Selected islands and archaeological sites: Aw—Arawe Islands; Ba—Baluan; BG—Boduna-Garua; DY—Duke of York Islands; Fe—Fergusson; Fi—Fissoa; Go—Goodenough; Ke—Kreslo; Ki—Kieta (-Pidia-Asio); Ko—Kolombangara; La—Lasigi; Lo—Lossu (Lesu); Lu—Lou; Mb—M'Buke; Me—Meinakapa; Mu—Muyua (Woodlark); NB—New Britain; NG—New Georgia; NH—New Hanover (Lavongai); NI—New Ireland; No—Normanby; Pa—Paubake (Buin); Re—Rendova; Ro—Roviana Lagoon; Si—Siassi Islands; Sh—Shortland Islands; Ti—Trobriand Islands; Ve—Vella Lavella; Wa—Watom; Wi—Witu Islands; WP—Willaumez Peninsula. TLTF—Tabar-Lihir-Tanga-Feni chain. Goodenough (Go), Fergusson (Fe), and Normanby (No) form the D'Entrecasteaux Islands in the Solomon Sea.

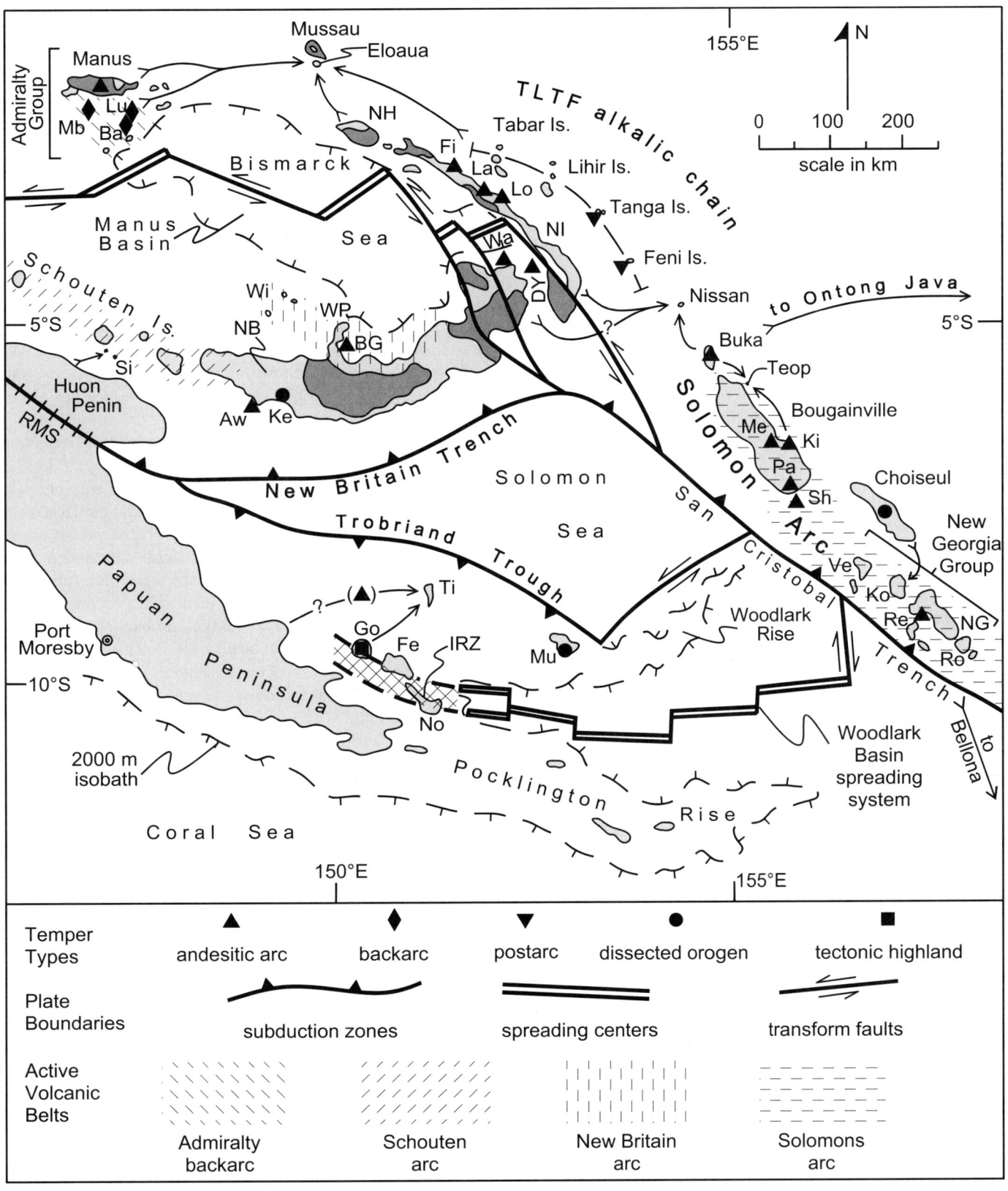

from both indigenous and exotic contexts at various archaeological sites on multiple islands. The degree of placering of Bismarck andesitic arc tempers is highly variable (Fig. 20), and no reliable distinctions can be drawn between different temper types from megascopic estimates of relative proportions of light and heavy mineral grains and volcanic lithic fragments. The Bismarck andesitic arc temper sands were derived, however, from four different volcanogenic assemblages of contrasting age and petrology that yield temper sands of differing mineralogy (Table 13A).

Vitiaz Arc Tempers

Paleogene volcanogenic assemblages of the ancestral Vitiaz island arc (Fig. 9) are exposed on all the major islands of the Bismarck Archipelago (Fig. 19) to which they have been dispersed by Neogene seafloor spreading that opened the intervening Manus Basin of the Bismarck Sea (Johnson, 1979; Martínez and Taylor, 1996; Dickinson, 2000a). Although buried over wide areas beneath younger volcanic and sedimentary strata, the Paleogene arc assemblages have yielded detritus incorporated into local temper sands (Fig. A22) that cannot be distinguished with confidence from island to island. Eocene to Oligocene volcanic assemblages forming the basement complexes of the different major islands (Manus in the Admiralty Group, New Hanover or Lavongai, New Ireland, New Britain) are given varied local names, but are petrogenetically similar, and are overlain by Miocene volcanic and limestone formations of diverse stratigraphic nomenclature but similarly analogous lithology (Hine and Mason, 1978; Hohnen, 1978; Whalen and McDougall, 1985; Exon and Marlow, 1988; Francis, 1988; Stewart and Sandy, 1988; Dickinson, 2000a). Paleogene volcanic strata contain both hornblende and clinopyroxene in widely varying proportions, but generally lack orthopyroxene or olivine in any appreciable amounts. Minor quartz present in some local temper sands was derived from cogenetic granitic intrusions that were emplaced locally into the Paleogene volcanic assemblage on each island, and reflect a gradational relationship to Bismarck dissected orogen tempers to be described in a later section of this report. Nonvolcanic igneous lithic fragments are also present in subordinate amounts in selected Vitiaz arc tempers (VLi > 77 in Lasigi-Lossu tempers of Table 13A7).

Bismarck tempers derived from the Paleogene Vitiaz arc assemblage include indigenous tempers in sherds from Manus (Table 13A1) and New Ireland (Table 13A7–13A8), Manus tempers in exotic sherds from smaller islands lying farther south within the Admiralty Group (Table 13A2), tempers of probable Manus origin (Dickinson, 2000a) in sparse exotic sherds from the tiny islet of Watom off the north tip of New Britain (Table 13A12), apparently exotic Lapita sherds of probable New Britain origin from the Duke of York Islands lying in the narrow strait between New Britain and New Ireland (Table 13A11), exotic Lapita sherds from Nissan atoll (Spriggs, 1991) east of New Ireland (Table 13A9–13A10), and several disparate temper types in exotic Lapita sherds from the outlying Mussau (Saint Matthias) Group at the northern fringe of the Bismarck Archipelago (Table 13A3–13A6).

Occurrences of prehistoric sherds are widespread on both Manus (Kennedy, 1981, 1982) and New Ireland (White and Downie, 1980; Golson, 1991, 1992; White and Murray-Wallace, 1996), but available sherd collections (Dickinson, 1980a) are still too limited to provide an adequate census of local temper types (Clay, 1974; Ambrose, 1991; White, 1997). The exotic Mussau sherds containing andesitic arc tempers were probably derived from Manus to the west (Fig. 19) because they occur in close association with other exotic sherds containing quartz-bearing backarc felsitic tempers from the Admiralty Group (Dickinson, 2006b). On strictly geologic grounds, however, alternate derivation of the exotic Mussau sherds containing andesitic arc tempers from New Hanover (Lavongai) to the south, or from New Ireland via New Hanover (Fig. 19), cannot be precluded. Most exotic Nissan sherds (Fig. 20; Table 13A9) containing Vitiaz arc tempers were probably derived from nearby New Ireland (Fig. 19), but rare olivine-bearing placer tempers nearly lacking in hornblende (Fig. 20; Table 13A10) may reflect derivation of some Nissan Lapita sherds from New Britain (Fig. 19).

Vitiaz arc temper sands (Fig. 20) are sedimentologically varied, ranging from only moderately sorted and unplacered or weakly placered subangular variants of probable stream origin to well-sorted and more strongly placered subrounded variants, in part hybrid, of beach origin. Volcanic lithic fragments are dominantly microlitic (most typically hyalopilitic) or vitric (brownish volcanic glass), with both probably derived from mafic to basaltic andesite, but also include felsitic grains probably derived from felsic andesite or dacite. Both hornblende and clinopyroxene mineral grains are commonly present, though in widely variable subequal proportions (PYi of Table 13A), and minor biotite (<5%) is also present in some Vitiaz tempers. Pyroxene is absent from some exotic tempers of unknown exact derivation (Table 13A4), and hornblende is absent in placer tempers of sherds from Lasigi on New Ireland (Table 13A7) and from exotic tempers in Duke of York sherds (Table 13A11) probably derived from nearby New Britain (Thomson and White, 2000). The pyroxene-free and hornblende-free temper variants are regarded as end members of a heterogeneous Vitiaz temper suite. The variability of temper sands in sherds from Lasigi and Lossu on New Ireland suggests that both local and nonlocal indigenous tempers are apt to be present in sherds from some archaeological sites on the larger Bismarck islands. Both orthopyroxene and olivine mineral grains are uniformly lacking, however, in any detectable amounts from all Vitiaz temper aggregates (OLi = 0; PXi = 100) except for exotic Nissan sherds possibly derived from New Britain (Table 13A10). The absence of orthopyroxene serves to discriminate between temper sands derived from the ancestral Vitiaz arc assemblage of Paleogene age and temper sands derived from Neogene volcanic assemblages related to active subduction at the New Britain Trench (Table 13A13–13A15).

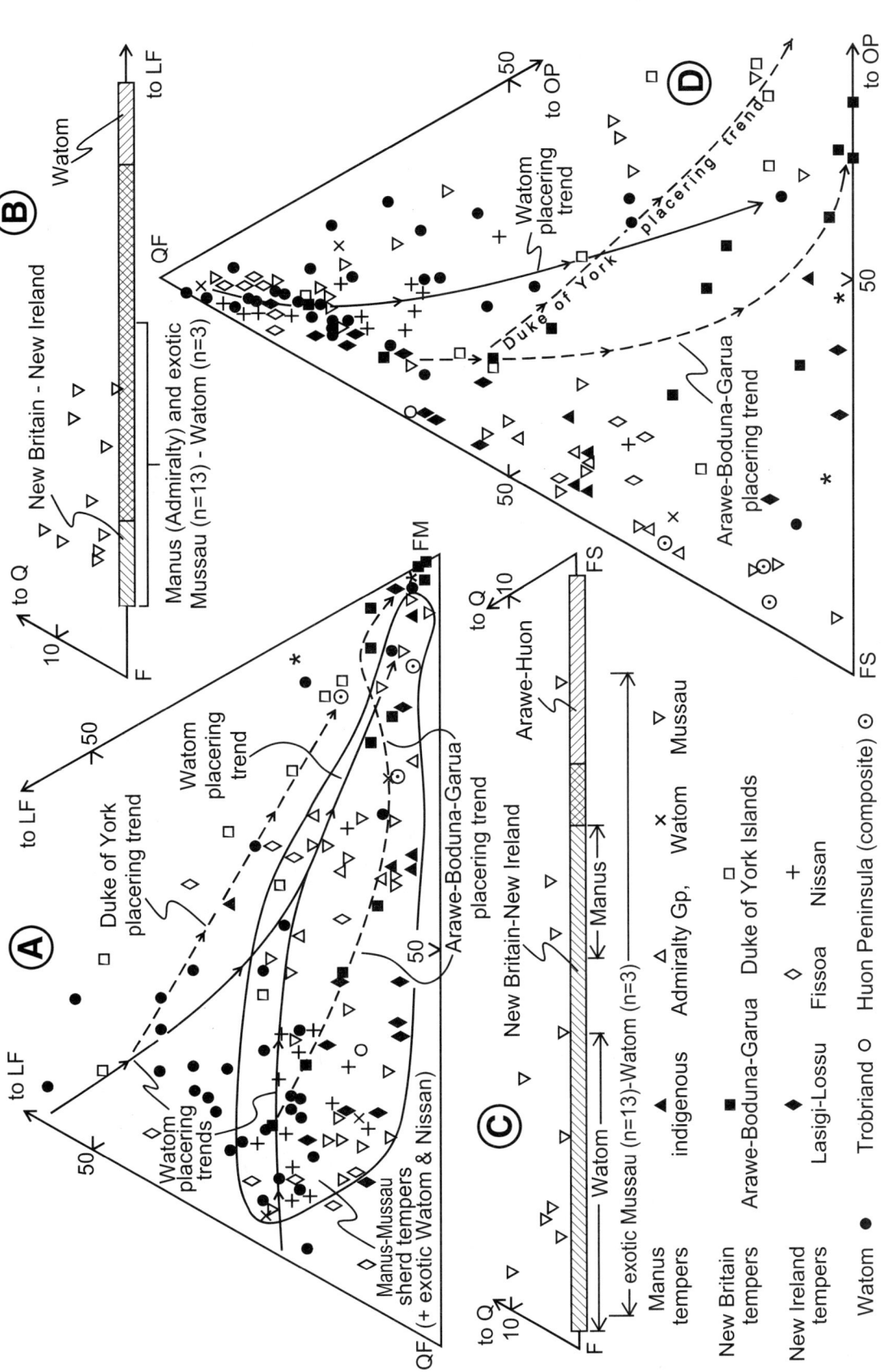

Figure 20. Compositions (grain frequency percentages) of andesitic arc tempers in sherds from the Bismarck Archipelago. A: LF-QF-FM diagram. B: Q-F-LF diagram for nonplacer tempers (FM <90%). C: Q-F-FS diagram. D: QF-FS-OP diagram. Solid symbols—tempers of indigenous sherds including suites from tiny Arawe and Boduna-Garua islets derived from nearby New Britain; open symbols—tempers of exotic sherds (+ and x also exotic sherds). For B and C, New Britain–New Ireland fields denote all sherd suites (of legend at lower left) not otherwise identified. For A and D, asterisks denote placer tempers of uncertain derivation (from within the Bismarck Archipelago) in Lapita sherds from Nissan. See Table 5A for poles.

TABLE 13. KEY SUPPLEMENTAL GRAIN PARAMETERS FOR ANDESITIC ARC TEMPERS FROM THE BISMARCK ARCHIPELAGO, SOLOMON ISLANDS, VANUATU, FIJI, LAU, AND TONGA

Temper type	N	QZi	OXi	OLi	PYi	PXi
A. Bismarck Archipelago						
1 - Manus (Admiralty Group) indigenous sherds	5	0	7-20	0	32-88	100
2 - Manus (exotic sherds within Admiralty Group)	5	0	4-13	0	32-87	100
3 - Mussau Type H (exotic from Manus or Lavongai)	2	0	5-14	0	44-53	100
4 - Mussau Type E (exotic from Manus or Lavongai)	4	0	26-65	0	0	-
5 - Mussau Type F (exotic from Manus or Lavongai)	6	2-6	30-78	0	30-45	100
6 - Mussau Types GJ (exotic Manus or Lavongai)	11	2-17	6-82	0	70-94	100
7 - Lasigi-Lossu (New Ireland) indigenous sherds	10	0	19-41	0	52-99	100
8 – Fissoa (New Ireland) indigenous sherds	10	0	12-50	0	50-91	100
9 - Nissan (exotic from nearby New Ireland)	13	0	20-47	0	24-99	100
10 - Nissan (exotic, possibly from New Britain)	2	0	25-47	6-9	98-99	100
11 - Duke of York (exotic from nearby New Britain)	8	0	19-86	0	100	100
12 - Watom (exotic sherds probably from Manus)	3	0	9-65	0	0-48	100
13 - Watom (island off New Britain) indigenous	31	0	16-67	tr	100	88-94
14 - Boduna-Garua (exotic from nearby New Britain)	4	0	19-39	3-9	90-94	62-78
15 - Arawe Islands (exotic from nearby New Britain)	8	0	30-87	0-3	95-99	89-94
16 - Huon Peninsula (Papua) indigenous sherds	3	0	2-8	0	98-99	98-99
B. Solomon Islands						
1 - Meinakapa (natural temper, east Bougainville)	3	0	9-22	0	3-30	100
2 - Pidia (natural temper, east Bougainville)	4	0	29-45	0	3-7	100
3 - Pidia (manually added temper, east Bougainville)	2	0	15-39	0	4-35	100
4 - Kieta (east Bougainville, exotic sherds on Teop)	6	0	3-62	0	3-12	100
5 - Paubake (natural temper, south Bougainville)	11	0	8-27	0	17-37	100
6 - Paubake (manually added temper, Bougainville)	9	0	4-45	0	21-57	100
7 - Buin (south Bougainville, exotic Shortland sherds)	2	0	~0	0	32-68	100
8 - Shortland Islands (south of Bougainville)	15	0-1	0-21	0	5-25	100
9 - Roviana Lagoon feldspathic (New Georgia Group)	6	0	32-44	0	7-40	100
10 - Roviana Lagoon lithic (New Georgia Group)	3	0	13-26	12-63	86-99	100
11 - Roviana Lagoon placer (New Georgia Group)	8	0	4-96	20-30	91-99	100
C. Vanuatu						
1 - Taumako (Duff Islands) non-placer temper	11	4-15	30-90	0	100	100
2 - Taumako (Duff Islands) placer temper	9	0	4-23	0	96-99	100
3 - Nendö quartz-pyroxene temper (exotic)	1	58	10	0	96	100
4 - Nendö tempers (indigenous Lapita wares)	20	0	1-16	2-11	>98	100
5 - Reef Islands Lapita (exotic from Nendö)	12	0	1-14	4-14	>98	100
6 - Reef Islands post-Lapita (exotic from Nendö)	18	0	1-11	5-12	>98	100
7 - Reef Islands hornblende-rich (exotic)	2	0	42-45	0	8-14	100
8 - Reef Islands hornblende-pyroxene (exotic)	2	0	14	0	75-78	100
9 - Tikopia (exotic Sinapupu wares)	7	0	42-85	0	100	100
10 - Vanikolo (indigenous and/or exotic from Banks)	7	0	22-88	0	100	100
11 - Pakea (indigenous Banks Islands ware)	4	0	36-64	0	100	100
12 - Torres Islands (probably exotic from Santo)	5	0	31-67	0	67-83	100
13 - Santo (island also called Espiritu Santo)	12	0	25-55	0	55-83	100
14 - Malo (exotic from neighboring Santo)	15	0	19-47	0	72-94	100
15 - Exotic Santo (?) tempers (on other islands)	9	0	12-41	0	64-94	100
16 - Malakula (including neighboring islets)	25	0	17-67	0	12-96	100
17 - Shepherd Islands (Tongoa and Emae)	3	0	45-67	0	100	100
18 - Efate (including neighboring islets)	39	0	18-83	0	100	100
19 - Erromango (Ponamla and Ifo non-placer tempers)	8	0	8-36	0-1	100	100
20 - Erromango (Ponamla and Ifo placer tempers)	2	0	13-19	5-8	100	100

(*continued*)

TABLE 13. KEY SUPPLEMENTAL GRAIN PARAMETERS FOR ANDESITIC ARC TEMPERS FROM
THE BISMARCK ARCHIPELAGO, SOLOMON ISLANDS, VANUATU, FIJI, LAU, AND TONGA (continued)

Temper type	N	QZi	OXi	OLi	PYi	PXi
D. Fiji and Lau						
1 - Udu Peninsula, Vanua Levu (exotic on Cikobia)	5	12-23	44-86	0	100	~90
2 - Yasawa Islands, west of Viti Levu	9	0	6-19	0	97-99	96-99
3 - Navua Delta, Viti Levu, central Fiji	9	0	15-40	1-2	100	100
4 - Kadavu (-Bulia-Ono), south of Viti Levu	8	0	22-82	0	0-43	100
5 - Exotic Kadavu tempers on other islands	9	0	18-71	0	0-33	100
6 - Aiwa-Lakeba-Nayau (southern Lau)	33	0	3-67	0	100	100
E. Tonga						
1 - Niuatoputapu (northernmost Tonga)	16	0	3-22	0	100	83-88
2 - Vava'u Group (northern Tonga)	12	0	5-35	0	100	91-95
3 - Ha'apai Group (central Tonga)	25	0	2-16	0	100	81-90
4 - Tongatapu (southern Tonga)	15	0-12	2-15	0	100	82-88
5 - 'Ata (southernmost Tonga)	2	0	0	5-11	100	88-89
6 - 'Eua gabbroic temper (VLi = 50-80)	3	0	54-56	0	100	100
7 - Exotic temper in Nukuleka (Tongatapu) sherds	4	11-61	6-13	0	97-99	100
8 - exotic Tonga sherd on Mauke (Cook Islands)	1	10	11	0	100	96

Notes: N is number of sherds for which petrographic frequency counts of grain types in thin section are tabulated. See Table 6 for definitions of supplemental grain parameters (or indices). Large ranges in OXi values for many temper types reflect variable intensity of placering. Anomalous temper compositions represented by only one sherd not tabulated. Possible Lavongai origin of exotic tempers in Mussau sherds (Bismarck Archipelago) includes potential transit of wares northward from New Ireland through Lavongai (New Hanover). Indigenous Watom (A13) and Huon Peninsula (A16) tempers tabulated as terrigenous fractions of largely hybrid and exclusively composite tempers, respectively. Tempers of individual indigenous sherds from Aoba, Pentecost, and Tanna in Vanuatu not tabulated. Exotic sherds derived from Santo (line C15) have been collected on Efate, Malakula, Pakea (Banks Islands), Tikopia, and Vanikolo (Fig. 23). Navua Delta tempers (line D3) include both local sherds on the Navua Delta and non-local sherds from nearby Karobo (Fig. 26). Exotic sherds derived from Kadavu (line D5) have been collected on Lakeba (Fig. 28) and Moturiki (Fig. 26), and from the Rewa Delta (Nasilai) and the Navua Delta of Viti Levu (Fig. 26). Tempers C3 and E7 correlated as evidence for Lapita ceramic transfer from Melanesia to Tonga.

Schouten Arc Temper

A dormant segment of the active Schouten island arc (Johnson, 1976, 1982), lying offshore from the Huon Peninsula of New Guinea, extends into the Shrader and Andrews (or Andewa) massifs of westernmost New Britain (Fig. 19). Schouten volcanic suites are mainly pyroxene-bearing basalt and basaltic andesite in which orthopyroxene is rare and hornblende is typically absent (Palfreyman and Cooke, 1976; McKee et al., 1976; Johnson et al., 1985). The mineralogy of temper sands in sherds from Apalo (Gosden, 1989, 1991) and Adwe (Summerhayes, 2001b) in the Arawe Islands (Table 13A15), offshore from western New Britain (Fig. 19), is compatible with derivation from the Andewa massif inland from the adjacent coast of New Britain from which the sherds on the tiny offshore islets are inferred to derive (Dickinson, 2000a). The temper sands in all sherds studied to date from the Arawe Islands are well-sorted and subrounded hybrid (Apalo) or placer (Adwe) aggregates clearly of beach origin (Fig. A27). Sparse hornblende (PYi > 95) and orthopyroxene (PXi > 90) mineral grains in Arawe tempers (Table 13A15) are unexpected, but not prohibitive of the derivation inferred.

New Britain Arc Temper

An active island arc extends along the north coast of New Britain, from the offshore Witu Islands in the Bismarck Sea on the west, and past the Willaumez Peninsula projecting northward into the Bismarck Sea, to near the complex of transform faults that disrupts the northeastern end of New Britain (Fig. 19). Along the arc trend, pyroxene andesite is dominant, together with associated basalt and rhyodacite, and commonly contains both pyroxenes (clinopyroxene and orthopyroxene) in significant amounts (Lowder and Carmichael, 1970; Johnson et al., 1972, 1983; Blake and Ewart, 1974). The mineralogy of temper sands in sherds from the tiny islets of Boduna (Specht, 1974; Ambrose and Gosden, 1991; Specht et al., 1991) and Garua (Torrence and Stevenson, 2000; Torrence, 2001), lying off the Willaumez Peninsula (Fig. 19), is compatible with derivation from the arc assemblage of the adjacent New Britain coast from which the sherds on the offshore islets are inferred to derive (Dickinson, 2000a). PXi values (62–78) for the related tempers in Boduna-Garua sherds (Table 13A14) are the lowest in any Oceanian sherd suites and reflect the high ratio of orthopyroxene to clinopyroxene in volcanic rocks of the north New Britain magmatic arc. Low hornblende contents (PYi > 90) are also expected from that source of sand.

Watom Temper

The sherd assemblage from Watom (Fig. 19), a small island only 4–5 km in diameter off the Gazelle Peninsula at the northeastern tip of New Britain, has been studied with care (Dickinson,

2000a) because Lapita pottery was first discovered by a German missionary working on Watom in 1909 (Green, 2000a). Much of the Watom sherd suite is post-Lapita (Green and Anson, 2000b), but the excavated Lapita horizon (Specht, 2003) was fortuitously sealed beneath an ash-fall deposit (Specht, 1968; Green and Anson, 1991) now reliably dated to the interval A.D. 850–650 (Anson, 2000a). The characteristic tempers in plain (largely or partly post-Lapita) and decorated Lapita sherds from Watom are indistinguishable petrographically and are thought to be indigenous to Watom for three reasons (Dickinson, 2000a). First, they represent by far the most abundant temper type in Watom sherds, and derivation from elsewhere would require wholesale importation, on a massive scale, of temper sands or finished wares for which there is no independent archaeological evidence. Second, the grain types present in the characteristic Watom tempers are exactly those expected from Watom bedrock. Third, microprobe analysis has established that the elemental compositions of clay pastes in typical Watom sherds are indistinguishable from the compositions of locally sampled clay deposits (Green and Anson, 2000a). Rare exotic sherds from Watom containing Bismarck Vitiaz arc and Bismarck dissected orogen tempers are described elsewhere in this report.

Watom bedrock, flanked locally by coastal exposures of emergent Quaternary reef limestone, is entirely basalt and basaltic andesite lava and scoriaceous breccia on the northwest flank of the Rabaul volcanic edifice ~15 km from the rim of the Rabaul caldera (Walker et al., 1981). Phenocrysts in the Rabaul volcanic suite include plagioclase, clinopyroxene, subordinate orthopyroxene, and rare olivine (Heming, 1974, 1977), but hydrous ferromagnesian silicates (hornblende, biotite) are absent (Wood et al., 1995). Eruption of the Rabaul volcanic edifice is linked to incipient rupture of New Britain by strands of a transform fault system (Fig. 19) along which New Ireland has translated northward with respect to New Britain (Lindley, 1988; Madsen and Lindley, 1994), and the Rabaul volcanic assemblage is not contiguous with the New Britain arc assemblage erupted farther west along the north coast of New Britain (Fig. 19). Although some orthopyroxene is present in the Rabaul eruptive suite, it is much less abundant in the unique geotectonic setting of the Gazelle Peninsula than along the main New Britain magmatic arc to the west.

Indigenous Watom tempers (Table 13A13) include a spectrum of variably placered volcanic and hybrid sands that are dominantly well-sorted and subrounded aggregates of beach origin, although more poorly sorted terrigenous sands of probable terrestrial origin are also represented in the sherd collection. Megascopic distinction between exclusively terrigenous and hybrid temper sands is difficult because calcareous grains have been removed from many sherds by postdepositional leaching (Green and Anson, 2000a; Dickinson, 2000a). Volcanic lithic fragments in Watom tempers include gradationally related vitric and microlitic varieties in subequal proportions (50% ± 10% each). Typical pale brown to red vitric fragments of mafic glass are partly devitrified to cryptocrystalline alteration products imparting a faint birefringence to otherwise isotropic volcanic glass. The microlitic lithic fragments, dominantly of hyalopilitic internal texture, contain variable proportions of plagioclase microlites embedded in similar volcanic glass. All the lithic fragments are inferred to derive from local island bedrock (Dickinson, 2000a).

The intensity of placering reflected by Watom temper compositions is commonly less than for temper sands in sherds from Boduna, Garua, and the Arawe and Duke of York Islands, but there is wide overlap in the content of ferromagnesian mineral grains in the tempers from all these small islets off New Britain (Fig. 20). Watom tempers differ, however, from nearly all other Bismarck andesitic arc tempers by containing no hornblende, in keeping with the petrology of the Rabaul volcanic edifice, and differ from Bismarck andesitic arc tempers derived from the Paleogene Vitiaz arc assemblage by containing significant amounts of orthopyroxene (Table 13A). As expected from regional geologic relations, the orthopyroxene content ($PXi = 88–94$) of Watom tempers is distinctly less than in tempers of Boduna-Garua sherds ($PXi = 62–78$) derived from the magmatic arc along the north coast of New Britain to the west, but comparable to the orthopyroxene content of tempers derived from the eastern segment of the Schouten arc in sherds from the Arawe Islands ($PXi = 89–94$) off southwest New Britain.

Solomon Arc Tempers

The modern Solomon island arc (Coulson, 1985) at the easternmost extremity of Near Oceania (Fig. 7) includes three geotectonic belts of contrasting geology (Fig. 21A). On the northeast, the Pacific province is composed of deformed and uplifted basaltic lavas and overlying pelagic sedimentary rocks derived from the edge of the otherwise submerged Ontong Java Plateau (Hughes and Turner, 1977; Tejada et al., 1996, 2002; Neal et al., 1997; Petterson et al., 1997), the largest oceanic plateau in the

Figure 21. Geologic sketch maps of Solomon island arc (A), Bougainville and neighboring islands (B), and the New Georgia Group (C), showing distribution of principle kinds of local geologic sources in relation to key archaeological sites at Paubake-Buin, Kieta, Meinakapa, and Pidia on Bougainville; on the islets of Alu, Buka, Mono, and Teop off Bougainville; and within Roviana Lagoon in the New Georgia Group. Volcanic rocks in the New Georgia Group are exclusively or predominantly of Quaternary age; multiple Quaternary stratocones on Bougainville are subdivided into northern (N), central (C), and southern (S) clusters of differing petrology (see text). Local subvolcanic stocks (shown in black) include diorite-monzonite on Bougainville and diorite-gabbro (with minor tonalite) on New Georgia. Areal geology (B, C) adapted after Blake and Miezitis (1967), Stanton and Bell (1969), Turner and Ridgway (1982), Ridgway and Coulson (1987), Hilyard and Rogerson (1989), and Rogerson et al. (1989). Tectonic provinces of Solomon island arc (A) modified after Coulson (1985), Coulson and Vedder (1986), Petterson et al. (1999), and Petterson (2004). Principal islands (A): Bou—Bougainville; Cho—Choiseul; Gua—Guadalcanal; Mak—Makira (San Cristobal); Mal—Malaita; NGG—New Georgia Group; SA—Santa Ana; SI—Santa Isabel.

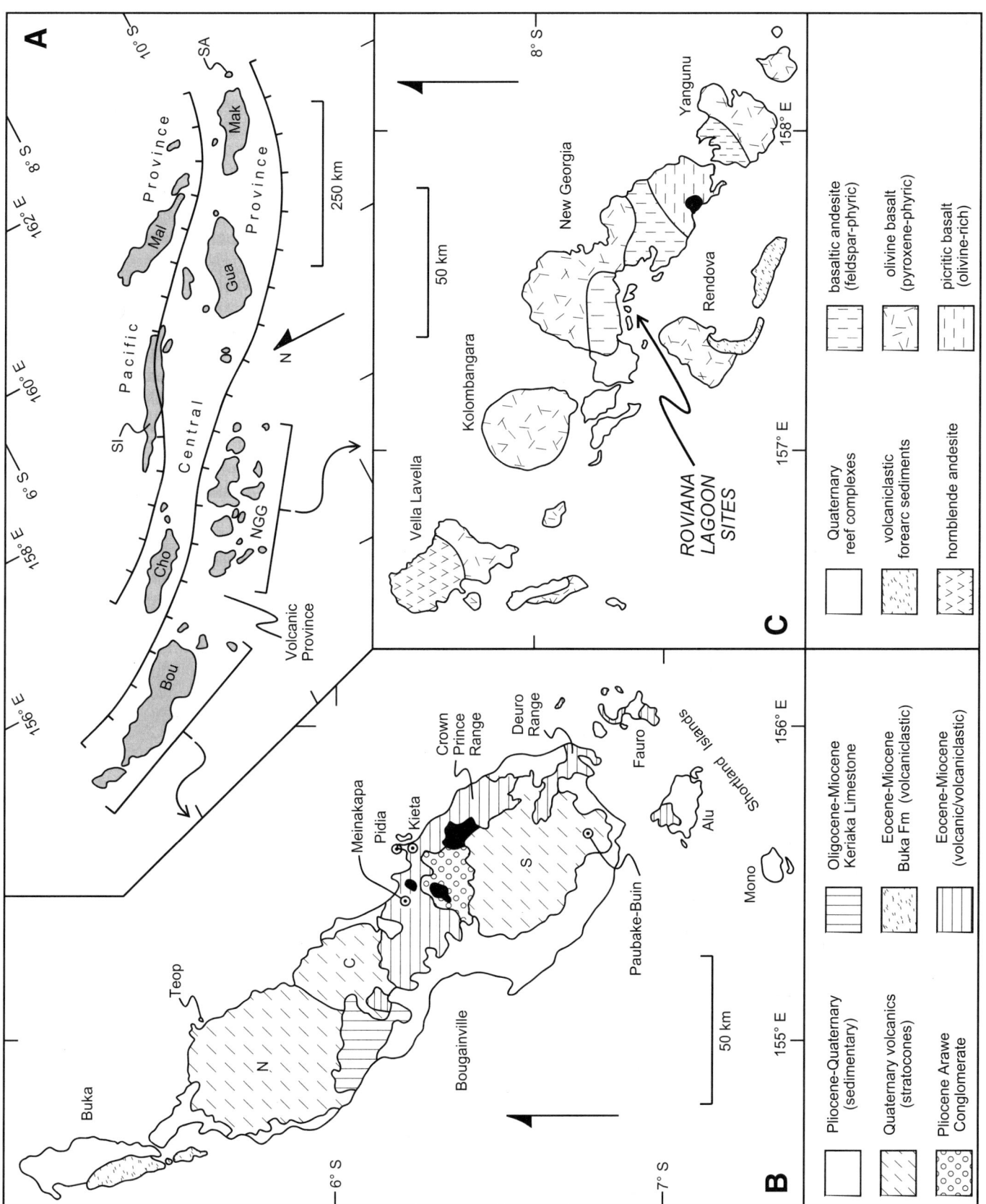

world with an overthickened crustal profile of 36–42 km (Petterson et al., 1999). Accretion of the Ontong Java Plateau to the central province late in Miocene time (ca. 10 Ma) induced reversal of arc polarity that terminated the Solomon segment of the ancestral Vitiaz island arc (Fig. 9) and initiated the modern Solomon island arc (Dunkley, 1983; Kroenke, 1984; Wells, 1989a, 1989b; Musgrave, 1990). Although the principal subduction zone then shifted from the Vitiaz Trench along the northeast flank of the Solomon arc to the San Cristobal Trench on its southwest flank (Fig. 7), limited thrusting continues along the trend of the Vitiaz paleotrench (Cooper and Taylor, 1985; Ridgway, 1987; Petterson et al., 1999). The central province is composed of dissected and partly metamorphosed volcanogenic assemblages of the ancestral Vitiaz arc overlain locally by postreversal arc volcanics of latest Miocene (ca. 7.5 Ma) or younger age (Petterson et al., 1997, 1999). On the southwest, the volcanic province is blanketed by Pliocene and younger volcanic cover of the active Solomon magmatic arc.

Solomon ceramic assemblages have been studied only from selected islands of the volcanic province where varied andesitic arc tempers are known (Fig. 19). The absence or paucity of ceramic collections from other Solomon islands (Reeve, 1989) is a continuing puzzle (Spriggs, 2000) that may reflect only lack of adequate archaeological exploration, or may indicate that pottery-making peoples moving outward from the Bismarck Archipelago largely bypassed the Solomon chain on their way to island groups lying farther to the southeast. Late-Lapita ceramic assemblages in the New Georgia Group of the volcanic province may reflect back-migration of Lapita peoples who initially leapfrogged the Solomon chain (Summerhayes and Scales, 2005).

Indigenous Solomon andesitic arc tempers are dominantly feldspathic, although placer tempers also occur more sparingly (Fig. 22). Tempers from Bougainville and the nearby Shortland Islands were derived almost exclusively from andesitic source rocks, but tempers from the New Georgia Group include important contributions from basaltic source rocks prominent in the local arc assemblage. Along the trenchward fringe of the Solomon arc axis, volcaniclastic forearc strata composed of detritus dispersed from erupting and eroding arc stratocones are exposed on Buka northwest of Bougainville (Fig. 21B) and on islets along the southern fringe of the New Georgia Group (Fig. 21C).

All tempers from Buka, an important Lapita and post-Lapita ceramic center (Specht, 1972; Wickler, 2001a) from which ceramic wares were widely dispersed to other islands, incorporate a component of nonvolcanic hypabyssal detritus. Buka-derived tempers are discussed in a later section of this report dealing with ceramic transfer, as are the temper sands in exotic sherds from Roviana Lagoon (Fig. 21C), the principal ceramic site of the New Georgia Group (Sheppard et al., 1999; Felgate, 2001; Felgate and Dickinson, 2001).

Bougainville-Shortland Tempers

Sherds studied from Bougainville include (1) indigenous post-Lapita wares from sites near Kieta on the east coast (Fig. 21B) and from the Paubake survey area near Buin close to the south coast (Dickinson, 1973c, 1975b) and (2) exotic post-Lapita wares recovered from the tiny islet of Teop off the northeast coast but derived from near Kieta on the east coast (Dickinson, 1975c). Post-Lapita sherds studied from the Shortland Islands to the south (Dickinson, 1972b) represent mainly indigenous wares but also exotic wares from nearby Buin on Bougainville.

Bougainville volcanic assemblages include a range of dominantly andesitic rocks of varying age (Fig. 21B). Eocene to Miocene volcanic and derivative volcaniclastic strata, erupted along the ancestral Vitiaz arc and exposed in both the Crown Prince Range and the areally more restricted Deuro Range of southeast Bougainville, have counterparts on Fauro and Alu in the Shortland Islands to the southeast. Middle Tertiary limestone, deposited before Solomon arc reversal, forms the Keriaka Plateau in central Bougainville where the calcareous strata locally overlie the Vitiaz arc assemblage. Subsequent eruptions formed high Quaternary stratocones capping Tertiary strata over extensive areas of both northern and southern Bougainville. Quaternary reef complexes or Pliocene and younger sedimentary aprons of terrigenous volcaniclastic debris mask volcanogenic bedrock along coastal tracts of Bougainville, as well as over large areas on Buka and in the Shortland Islands. Quaternary volcanic and volcaniclastic strata of Bougainville locally interfinger with the Pleistocene Sohano Limestone widespread on Buka.

All Bougainville and Shortland temper sands are mineralogically similar, with plagioclase the only feldspar present and the predominant grain type except in placer aggregates, quartz absent or nearly so, and heterogeneous populations of volcanic lithic fragments that include microlitic, vitric, vitrophyric, and felsitic varieties in the varying proportions typical of andesitic-dacitic detritus. Only contrasts in the ferromagnesian mineral grains in different temper types (Table 13B1–13B7) allow the probable geologic sources of each to be inferred from the varying petrologic character of eruptive volcanic assemblages on Bougainville (Blake and Miezitis, 1967; Hilyard and Rogerson, 1989; Rogerson et al., 1989).

All volcanic assemblages of Bougainville are composed chiefly of pyroxene or hornblende andesites in which orthopyroxene is sparse and olivine is rare. Quaternary volcanic edifices are composed on northern Bougainville (N of Fig. 21B) of pyroxene andesite lacking hornblende but containing biotite that is locally abundant, on central Bougainville (C of Fig. 21B) of hornblende andesite with subordinate pyroxene but no biotite, and on southern Bougainville (S of Fig. 21B) of varied andesites with subequal proportions of hornblende and clinopyroxene but lacking either orthopyroxene or biotite. Pre-Quaternary volcanic and derivative volcaniclastic strata of the older Atamo (or Kieta) Volcanics on southeast Bougainville and in coeval strata exposed in the Shortland Islands (Turner and Ridgway, 1982; Ridgway and Coulson, 1987) form a heterogeneous basalt-andesite-dacite assemblage containing pyroxene and hornblende in variable proportions.

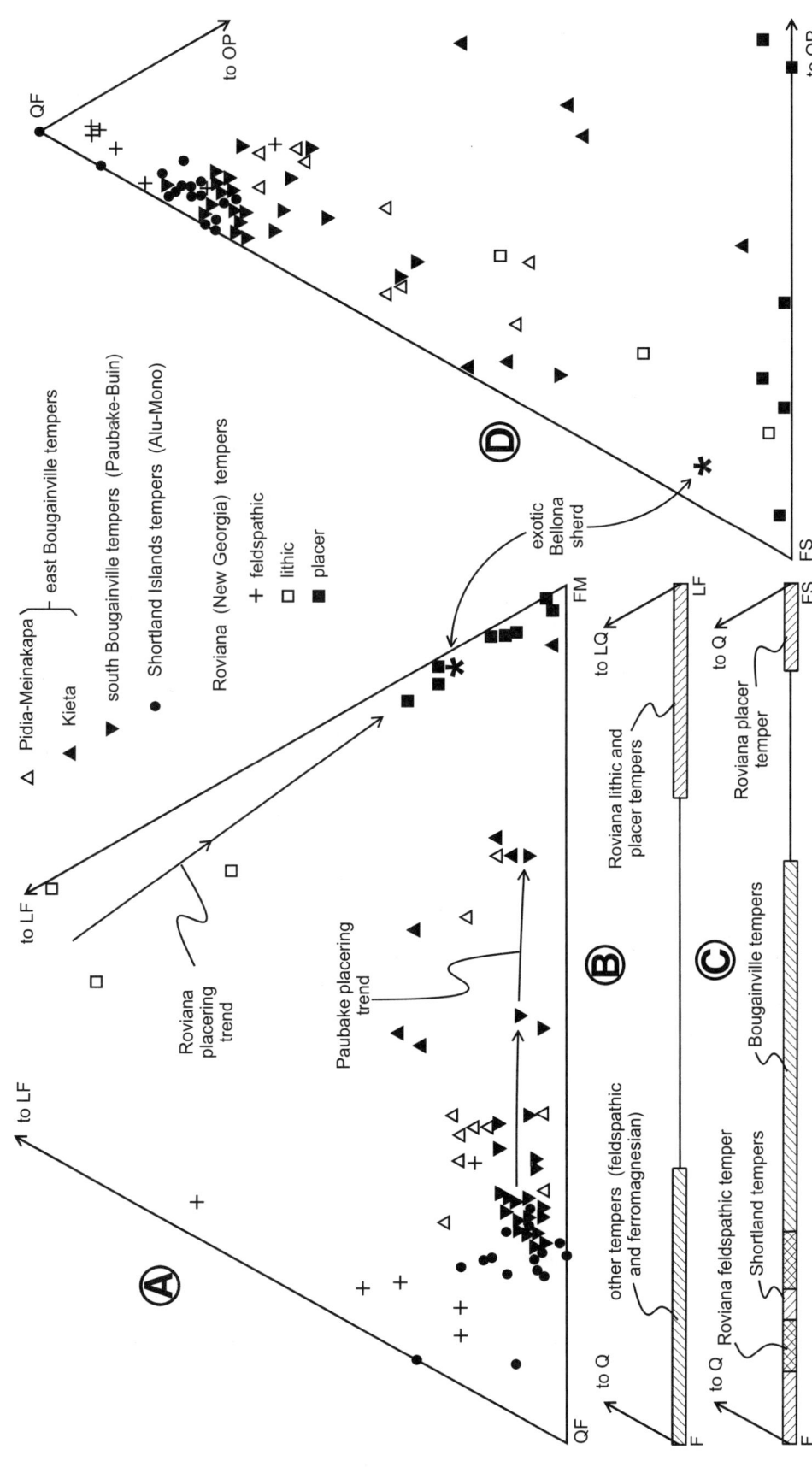

Figure 22. Compositions (grain frequency percentages) of andesitic arc tempers in sherds from Bougainville, the Shortland Islands, and the New Georgia Group within the volcanic arc province of the Solomon Islands. A: LF-QF-FM diagram. B: Q-F-LF diagram. C: Q-F-FS diagram. D: QF-FS-OP diagram (two Roviana placer temper sands that plot closer to OP pole not shown). See Table 5A for poles.

The lack of appreciable biotite in Bougainville andesitic arc tempers argues against derivation of any known temper sands from northern Bougainville. The dominance of hornblende over clinopyroxene (low PYi) in sherds from sites on eastern Bougainville (Table 13B1–13B4) suggests derivation either from nearby Quaternary volcanic sources in central Bougainville or from pre-Quaternary volcanic rocks widely exposed in the vicinity of the sites (Fig. 21B). Greater abundance of pyroxene (higher PYi) in sherds from the Paubake-Buin area (Table 13B5–13B7) near the southern tip of Bougainville implies derivation from the flanks of nearby Quaternary stratocones (Fig. A21).

Shortland sherds contain temper sands (Table 13B8) with hornblende contents (PYi = 12 ± 6 for n = 10) comparable to Meinakapa-Pidia-Kieta tempers from eastern Bougainville (PYi = 9 ± 3 for n = 15) but not to Paubake-Buin sherds from southern Bougainville (PYi = 35 ± 15 for n = 22). The similarity suggests that Shortland and Meinakapa-Pidia-Kieta temper suites are sands derived jointly from pre-Quaternary volcanic source rocks exposed both on southeast Bougainville and in the Shortland Islands (Fig. 21B), whereas Paubake-Buin temper sands are detritus from the Quaternary stratocones of southern Bougainville. The mineralogical similarity is not evidence for ceramic transfer from eastern Bougainville to the Shortland Islands because Meinakapa-Pidia-Kieta tempers are less feldspathic than all Shortland tempers, with no overlap in gross composition (Figs. 22A and 22D).

Hornblende grains of Meinakapa-Pidia-Kieta tempers and indigenous Shortland tempers are uniformly green hornblende, pleochroic in exclusively green tints (pale to dark), whereas hornblende grains of indigenous Paubake-Buin sherds are more variably pleochroic in combined green or yellowish and brown tints, as are hornblende grains in exotic Shortland sherds derived from Buin (Table 13B7). This contrast in the character of hornblende is supportive of consistent petrologic differences between pre-Quaternary and Quaternary volcanic sources on Bougainville and neighboring islands as reflected by the derivative temper sands.

Bougainville and Shortland temper sands are texturally varied, but contrasting sedimentological origins are not in general reflected by any prominent variations in mineralogical composition. There is a systematic tendency, however, for higher ratios of clinopyroxene grains to hornblende grains in manually added Bougainville tempers as opposed to natural Bougainville tempers (Table 13B2–3, 13B5–6), perhaps because more cleavable hornblende was preferentially removed from sand aggregates by reduction in grain size during sedimentary reworking of volcanic detritus. Sherds from the Paubake survey area near Buin in southern Bougainville and from Pidia on the east coast include both naturally tempered clay bodies in which sparse and poorly sorted temper grains are dispersed in clay pastes (Table 13B2, 13B5), and sherds in which moderately well sorted stream sands composed of subangular to subrounded grains were added manually to clay bodies (Table 13B3, 13B6). Natural Paubake temper is compositionally indistinguishable from the sand embedded within fired samples of impure clay from the Paubake survey area (Dickinson, 1973c). Indigenous sherds from the Shortland Islands (Table 13B8) probably also include both natural and manually added tempers, but textural relations are equivocal in most instances. Meinakapa sherds contain only natural temper (Table 13B1), but well-sorted and partially placered stream sands in exotic Teop sherds (Figs. 22A and 22D) derived from Kieta (Table 13B4) were added manually to clay bodies.

New Georgia Tempers

Multiple intertidal ceramic sites (Sheppard et al., 1999; Felgate, 2001), the record of stilt villages built over shallow waters, are aligned along a string of coral islets formed by emergent Quaternary reef complexes within Roviana Lagoon in the central part of the New Georgia Group (Fig. 21C). The local ceramic assemblage includes late-Lapita as well as post-Lapita elements, making Roviana Lagoon the best-known Lapita site in the Solomon chain south of Buka, and consequently a regionally important locality for understanding Oceanian prehistory. Local beach sands in Roviana Lagoon are exclusively calcareous, but many Roviana sherds contain terrigenous temper sands inferred to derive both from elsewhere within the New Georgia Group and from farther afield. Sherds containing exclusively volcanic sand as temper are interpreted as indigenous to the New Georgia Group (Table 13B9–13B11), but the volcanic sand tempers are nonlocal with respect to the coral islets within Roviana Lagoon.

The Neogene volcanic suite of the New Georgia Group has been erupted along an anomalous segment of the Solomon magmatic arc where the Woodlark spreading system is being subducted at the San Cristobal Trench (Fig. 19), and arc volcanism occurs <50 km from the trench axis (Crook and Taylor, 1994; Petterson et al., 1997, 1999). The local eruptive suite is more picritic (magnesian) and richer in olivine than most Oceanian arc assemblages (Stanton and Bell, 1969; Ramsay et al., 1984; Schuth et al., 2004). The unusual petrology is reflected in derivative temper sands by higher olivine contents than any other andesitic arc tempers, with OLi ≈ 25 in placer aggregates (Table 13B11). The abundance of olivine in New Georgia tempers invites confusion with tempers of the oceanic basalt class, but New Georgia tempers also include hornblende, which is absent from oceanic basalt tempers except for trachytic variants. Moreover, volcanic lithic fragments in the olivine-bearing New Georgia tempers are typically vitric to microlitic varieties unlike the lathwork grains prominent in oceanic basalt tempers.

Islands of the New Georgia Group (Fig. 21C) expose pyroxene-phyric olivine basalt, less picritic feldspar-phyric basaltic andesite, and hornblende-phyric andesite (and minor dacite). Roviana tempers (Fig. 22; Table 13B10–13B11) containing olivine mineral grains (OLi = 12–63) were apparently derived from widespread olivine basalt exposures present on multiple islands surrounding Roviana Lagoon (Fig. 21C), but the exact island or islands of origin cannot be specified with present information. Well-sorted and subrounded Roviana placer tempers (Fig. 22; Table 13B11) are similar to modern beach sand collected on Rendova (Fig. 21C), lying immediately south of Roviana Lagoon, but other beaches within the New Georgia Group are not precluded as potential sources for the olivine-bearing Roviana temper sands. Highly variable ratios of ferromagnesian oxide grains to ferromagnesian silicate mineral grains (OXi = 4–96) imply that the placer sands were derived from multiple locales within the New Georgia Group. Roviana lithic tempers (Fig. 22; Table 13B10) are only moderately sorted aggregates of subangular grains and may be interior stream sands rather than beach sands. It is unknown from archaeological studies whether nonlocal indigenous tempers in Roviana sherds reflect ceramic transfer of wares from nearby bedrock islands to Roviana Lagoon, or transport of loose temper by canoe from the nearby islands to islets within Roviana Lagoon.

Feldspathic Roviana temper sands (Fig. 22; Table 13B9) are only moderately to poorly sorted aggregates of subangular to subrounded grains and are probably stream sands rather than beach sands. Ferromagnesian mineral grains include clinopyroxene, green hornblende, green-brown hornblende, and oxyhornblende, but not olivine, with hornblende (predominantly green) dominant over pyroxene (PYi = 7–40). Volcanic lithic fragments are dominantly felsitic rather than microlitic varieties. The feldspathic Roviana tempers were probably derived from olivine-free but hornblende-rich andesitic bedrock forming Vella Lavella, which lies nearly 100 km northwest of Roviana Lagoon (Fig. 21C). As transport of temper over such a long distance seems unlikely, the presence of feldspathic Roviana temper in selected sherds suggests that all or most Roviana wares containing terrigenous volcanic sand as temper were made on multiple islands within the New Georgia Group and brought to the intertidal sites in Roviana Lagoon as finished vessels. As both clay resources and wood for ceramic firing are more abundant on the bedrock islands of the New Georgia Group than on the coral islets of Roviana Lagoon, a consistent pattern of ceramic transfer from multiple islands to Roviana Lagoon seems a logical inference given the variety of Roviana temper types observed. Late-Lapita surface sherds recently recovered from Kolombangara (Summerhayes and Scales, 2005) between Vella Lavella and Roviana Lagoon (Fig. 21C) have not yet been studied petrographically, but include both placer and nonplacer tempers that should eventually contribute further insight into New Georgia temper relations.

Vanuatu Tempers

All igneous assemblages of the New Hebrides island arc (Fig. 23), including the main Vanuatu island chain and eastern outliers of the Solomon Islands (as a political entity) to the north, are products of arc magmatism (Mitchell and Warden, 1971; Colley and Warden, 1974; Dupuy et al., 1982). Volcanic assemblages were erupted both before and after Late Miocene polarity reversal, which gave birth to the modern southwest-facing New Hebrides arc-trench system from the precursor northeast-facing Vitiaz arc-trench system (Figs. 8 and 9). Andesitic rocks are dominant, but the full eruptive suite ranges from basalt to dacite.

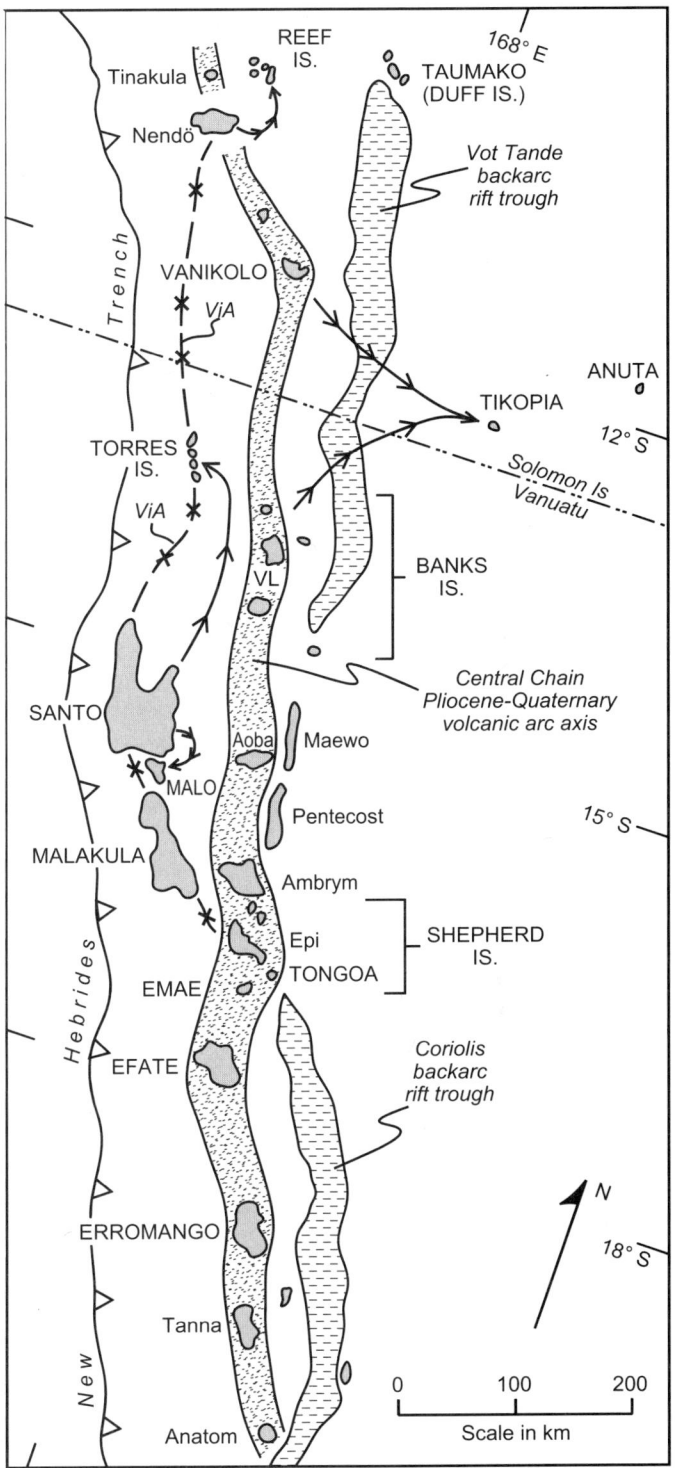

Figure 23. Distribution of principal islands along New Hebrides island arc and backarc. All-capitals denote islands from which sherd tempers are known. In the Banks Islands, Pakea (Table 13C11) is a tiny cay ~1 km off the east coast of Vanua Lava (VL). Arrows denote inferred systematic ceramic transfer (isolated cases of sherds on Tikopia, Vanikolo, Pakea, Malakula, and Efate derived from Santo not shown). ViA—ancestral (late Oligocene to early Miocene) Vitiaz arc axis.

plateau caps overlying older volcanic and volcaniclastic strata on many islands of Vanuatu, and locally contribute detritus to temper sands.

Upper Oligocene to Lower Miocene volcanic and volcaniclastic strata and associated subvolcanic intrusions inherited from the ancestral Vitiaz island arc are exposed along the western flank of the present New Hebrides island arc (Fig. 23) from Nendö on the north to Santo and Malakula on the south (Carney et al., 1985; MacFarlane et al., 1988; Greene et al., 1994). The youngest geologic elements (14–13 Ma) of the Vitiaz eruptive and intrusive suites are locally overlain by superposed Vitiaz intra-arc basins filled with Middle Miocene (16–11 Ma) volcaniclastic strata underlying an unconformity of Late Miocene age (11–8 Ma). The oldest volcanic rocks (8–7 Ma) erupted along the modern New Hebrides island arc at the time of polarity reversal are exposed on Maewo and Pentecost along the rear side of the modern arc where eruptions continued until at least the middle of Pliocene time. The active to dormant volcanic edifices lying along the axis of the modern volcanic chain, built partly within sedimented intra-arc summit basins aligned along the crest of the arc structure, are largely latest Pliocene and younger (younger than 2.5 Ma), but include Miocene-Pliocene rocks (5.3–3.4 Ma) on Erromango (Carney and MacFarlane, 1982; Bellon et al., 1984). Backarc rift troughs (Fig. 23) have evolved since ca. 7 Ma during the time span of New Hebrides arc volcanism to the west (Monjaret et al., 1991; Maillet et al., 1995).

Indigenous Vanuatu tempers are all volcanic sands that plot in partly overlapping fields on triangular compositional diagrams (Figs. 24 and 25). Much of their variability reflects different degrees of placering, but textural variations in rounding and sorting, supplemental grain parameters (Table 13C), and contrasts in the internal textures of volcanic lithic fragments (Table 14) allow the differentiation of temper suites from different islands (Dickinson, 2001a). Although the differences among Vanuatu arc tempers are strictly fortuitous in a regional sense, they can be related systematically to local island geology. Ceramic assemblages have been available for petrographic study from more than half the major islands and island groups (Fig. 23), but not from the central volcanic islands of Aoba and Ambrym, the eastern islands of Maewo and Pentecost (Mallick and Neef, 1974; Carney, 1986), or the southern islands of Tanna and Anatom (Carney and MacFarlane, 1979).

Taumako Tempers

Sherds from Taumako (Fig. 23), also known as the Duff Islands, contain two distinct temper types, respectively nonplacer and placer in character (Fig. 24; Table 13C1–13C2). Although the placer aggregates are better sorted and more abraded, both temper types are interpreted provisionally as beach sands. No gradations between the two temper types are present in the set of prehistoric sherds examined in thin section, but sands with compositions closely similar to both types of temper occur on modern beach faces of Taumako, and both temper types are interpreted as indigenous to Taumako.

Bedrock exposures on Taumako include andesitic to basaltic lava, breccia, and tuff of uncertain Late Miocene to Pliocene age erupted by largely phreatomagmatic activity in a submarine setting, and cut locally by basaltic dikes and sills (Hughes et al., 1981, p. 1–16). Taumako lies near the northern extremity of the Vot Tande backarc rift trough (Fig. 23), and Taumako volcanism may have accompanied extensional arc rifting. Islands along the rear flank of the New Hebrides island arc constructed during initial stages of backarc rifting carry the petrochemical signature of arc volcanism (Hughes, 1978; Marcelot et al., 1983).

The dominant mineral grains in Taumako tempers are plagioclase and clinopyroxene, as expected for derivation from the Taumako bedrock assemblage. Lithic fragments include sparse microgranular varieties (Table 14) derived from local hypabyssal intrusions (mean VLi = 97). Taumako nonplacer tempers also contain grains of quartz, largely polycrystalline and partly microcrystalline of chalcedonic character. Though minor ($QZi < 15$ from Table 13C1), the quartzose grains are consistently present in proportions not seen in any other Vanuatu temper suites (Figs. 24C and 24D). As quartz phenocrysts are not reported from Taumako volcanic rocks, the quartzose grains are inferred to derive from veinlets or amygdules in the altered Taumako volcanic assemblage (Dickinson, 2001a). Green-brown hornblende, which is present in minor amounts in most Taumako placer tempers ($PYi > 95$ from Table 13C2), also occurs sparingly in Taumako volcanic rocks. The general absence of quartz from Taumako placer tempers and of hornblende from Taumako nonplacer tempers can be attributed to placering effects that reduced the ratio of plagioclase to clinopyroxene (Fig. 24B). Even though the much more prevalent plagioclase and clinopyroxene mineral grains are prominent constituents of both temper types, quartz grains were essentially removed from placer aggregates, whereas hornblende was not sufficiently concentrated to be detectable in thin sections of the nonplacer aggregates.

Nendö-Reef Tempers

The most abundant tempers in Lapita sherds (Green, 1976; McCoy and Cleghorn, 1988) from the large bedrock island of Nendö and from the exclusively calcareous Reef Islands in the northern Santa Cruz Group are petrographically indistinguishable (Dickinson, 1978) (Table 13C4–13C5). The temper match documents systematic transfer of ceramic products on a large scale (Green, 1996) over the intervening 50 km of open water (Fig. 23). The paucity of clay deposits in the Reef Islands implies that finished wares, and not just temper sand, were carried from origins on Nendö to the Reef Islands (Dickinson, 2001a). Recent petrographic study has indicated that post-Lapita wares of the Reef Islands (Table 13C6) were also derived from Nendö.

Nendö temper aggregates are variably placered beach sands of well-sorted and subrounded character (Fig. A23), some are hybrid sands with admixed reef debris, and there is broad compositional overlap in the temper spectra observed for indigenous Nendö and exotic Reef Islands sherds (Figs. 24A and 24B). The most strongly placered sands occur as temper in selected post-

Lapita sherds (Fig. 24) and may reflect longer duration of placering action on the beach faces of Nendö by the time the younger wares were made. Clinopyroxene and plagioclase are the most abundant mineral grains in Nendö tempers, but the presence of olivine (OLi = 9 ± 3 for 36 characteristic sherds) distinguishes Nendö tempers from all other temper types known along the New Hebrides island arc except for Erromango (Table 13C20) nearly 1000 km to the south. All grain types in the related Nendö-Reef tempers were derived from volcanic and volcaniclastic Lower Miocene basaltic bedrock of Nendö (Hughes et al., 1981, p. 34–63), a remnant of the Vitiaz island arc (Fig. 23). Olivine is present, in part as microphenocrysts, in half the specimens of volcanic rock studied microscopically from Nendö (Hughes et al., 1981).

Four Lapita sherds from the Reef Islands contain two hornblende-bearing temper types compositionally dissimilar to Nendö tempers (Table 13C7–13C8). Although one of the anomalous exotic temper types plots within the Nendö LF-QF-FM compositional field (Fig. 24A), both plot outside the Nendö QF-FS-OP compositional field (Fig. 24B). The prominence of hornblende mineral grains in both implies non-Nendö origin because only trace amounts of hornblende are present in Nendö tempers (Table 13C4–13C6).

Both temper types are well-sorted sands of probable beach origin, although only subrounded to subangular, but the more FM-rich (Fig. 24A) and FS-rich (Fig. 24B) temper type is distinctly coarser-grained, and the two may not derive from the same locale. The sparse lithic fragments in both anomalous temper types are exclusively volcanic (dominantly vitric to felsitic). The seemingly attractive possibility (Green, 1976, p. 260) that the two anomalous tempers might derive from nearby Tinakula (Fig. 23) is precluded by knowledge that the Tinakula eruptive suite (pyroxene andesite and basaltic andesite) lacks any reported hornblende (Hughes et al., 1981, p. 25–33). Tempers in sherds from islands in central Vanuatu all contain hornblende in varying proportions (Table 13C12–13C16), but no known temper types from central Vanuatu closely resemble the tempers in exotic Reef Islands sherds, which may conceivably derive instead from either the Solomon Islands or the Bismarck Archipelago far to the northwest (Dickinson, 2001a). Their origin will remain uncertain until a future temper match can be made with some presently unknown ware or wares.

A single Lapita sherd from Nendö also contains an exotic temper (Figs. 24A and 24B; Table 13C3), well-sorted but subangular beach sand of equally unknown derivation, in which clinopyroxene is predominant over subordinate hornblende, but quartz is more abundant than plagioclase. The quartz grains are of volcanic derivation, as indicated by straight extinction, inclusions of volcanic glass in some, and shreds of volcanic glass adhering to the margins of others. The unusual combination of clinopyroxene and quartz as dominant mineral grains in a volcanic sand is otherwise unknown for Oceanian tempers except in exotic Lapita sherds from Tonga thought to derive from the same unknown origin as the exotic Nendö sherd (Burley and Dickinson, 2001).

The source rock for the temper was probably pyroxene dacite from some island within Vanuatu (Hughes, 1978) that cannot be specified with present information.

Vanikolo-Banks Tempers

Understanding temper relationships between sherds from Vanikolo in the southern Santa Cruz Group and from the Banks Islands farther south (Fig. 23) is a severe challenge because the islands are quite similar geologically (Ash et al., 1980; Hughes et al., 1981, p. 67–74). Exotic sherds of the Sinapupu ceramic tradition on Tikopia in the backarc region farther east represent wares probably derived from either Vanikolo or the Banks Islands (Dickinson, 2001a). Typical temper sands (Fig. 24) of all three sherd suites are similar poorly to moderately sorted and generally subangular aggregates of probable stream origin, and some could be natural temper.

The pyribole mineral grains in tempers of the Vanikolo-Banks-Sinapupu sherd suite are predominantly clinopyroxene (Table 13C9–13C11), as are the pyriboles in Taumako and Nendö tempers from islands farther north (Table 13C1–13C6), but hornblende as well as clinopyroxene is prominent in Santo and Malakula tempers from farther south in central Vanuatu (Table 13C12–13C16). The absence of hornblende in Vanikolo-Banks tempers accords with the nature of the local volcanic assemblages, which are predominantly basalt and basaltic andesite containing no hornblende (Ash et al., 1980, p. 27–36; Hughes et al., 1981, p. 70–71). Proportions of different textural varieties of volcanic lithic fragments in Vanikolo-Banks tempers are also similar to those in tempers from islands to the north but different from those in tempers from islands farther south in central Vanuatu (Table 14).

Empirically, the tempers in sherds from the tiny cay of Pakea (Ward, 1979) off the east coast of Vanua Lava in the Banks Islands (Fig. 23) are more lithic-rich than most tempers in Vanikolo sherds (Fig. 24A). If this distinction between Banks and Vanikolo tempers is reliable, Sinapupu wares on Tikopia may well derive from both Vanikolo and the Banks Islands (Figs. 23 and 24A). Temper relationships in sherds from Vanikolo, the Banks Islands, and Tikopia are further complicated, however, by the interpretation that some Vanikolo sherds may be exotic (Kirch, 1982a, 1983), and by the observation that individual sherds from all three locales contain anomalous hornblende-bearing temper sands. The latter are apparently exotic to the local settings and were probably derived from Santo as discussed in the next section.

Santo-Malo Tempers

Post-Lapita sherds from Santo (Fig. 23), and Lapita sherds from the nearby calcareous islet of Malo (Hedrick, 1971), contain volcanic sand tempers (Fig. 25) in which the dominant ferromagnesian silicate mineral grains are clinopyroxene (Fig. A15), but in which subordinate hornblende mineral grains are consistently present in significant proportions (Table 13C13–13C14). The temper sands are inferred to derive from western Santo because

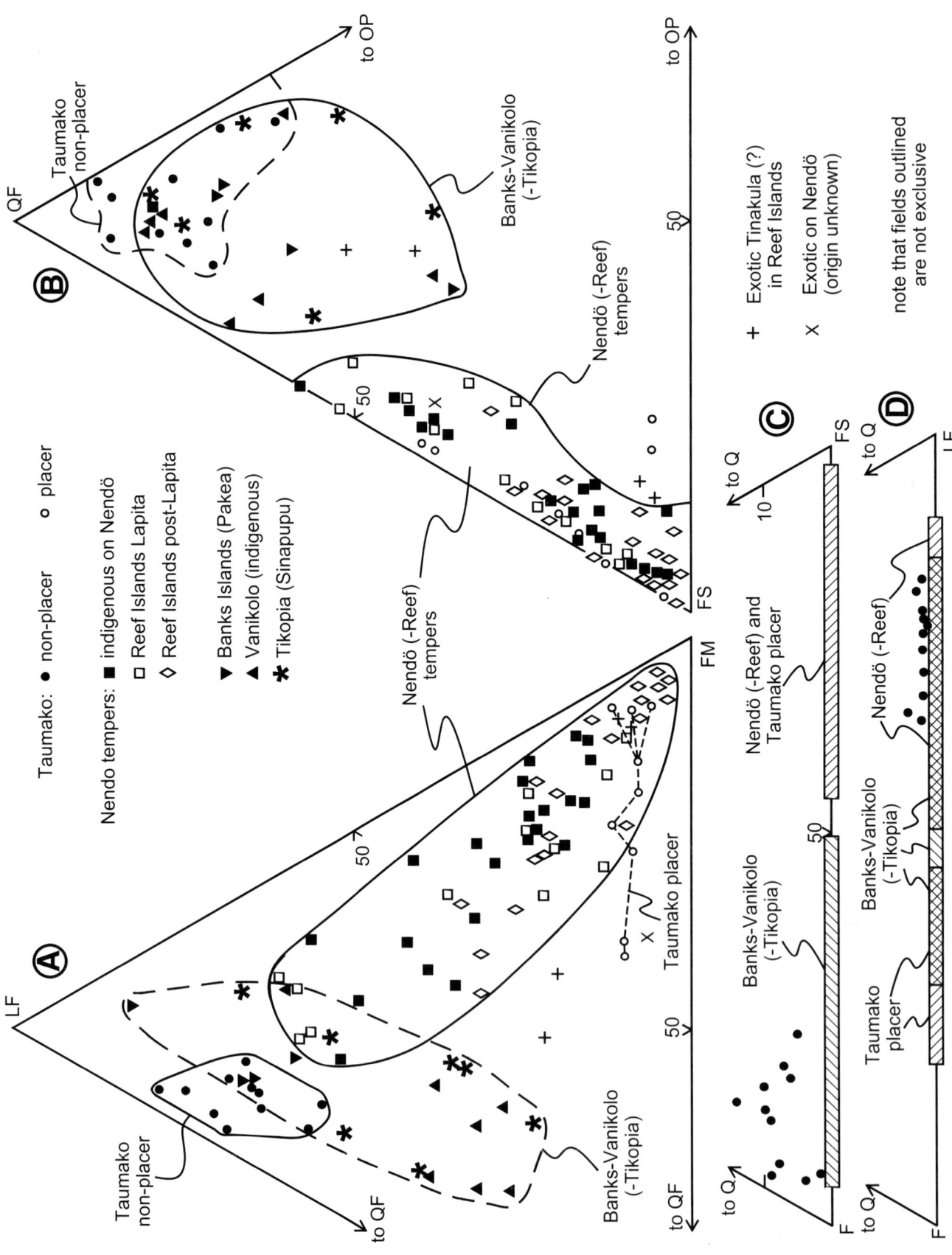

Figure 24. Compositions (grain frequency percentages) of andesitic arc tempers in sherds from the northern New Hebrides island arc (northern Vanuatu and eastern outliers of the Solomon Islands). A: LF-QF-FM diagram. B: QF-FS-OP diagram. C: Q-F-FS diagram. D: Q-F-LF diagram. Exotic Nendö-Reef tempers omitted from plots C and D. See Table 5A for poles.

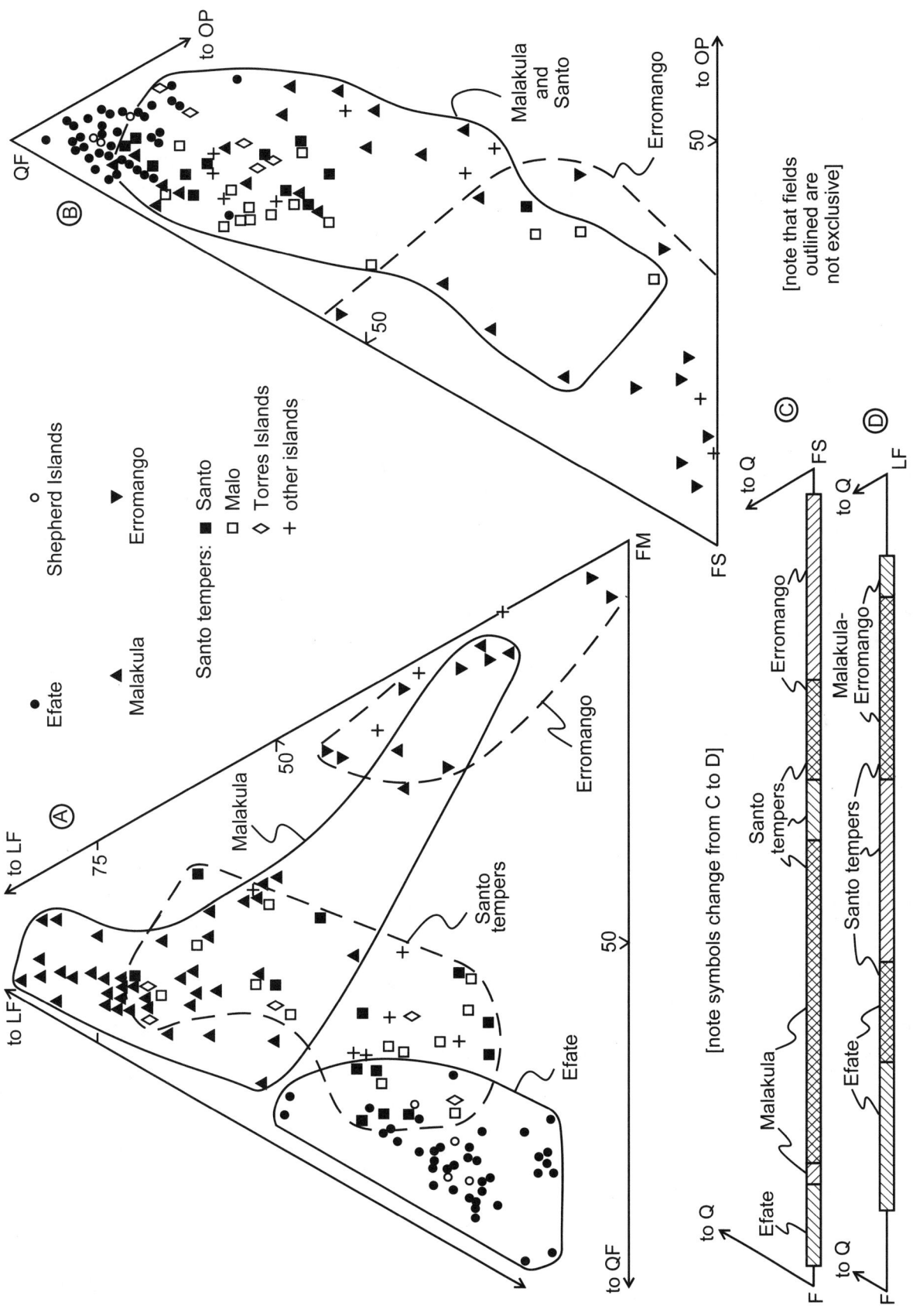

Figure 25. Compositions (grain frequency percentages) of andesitic arc tempers in sherds from central and southern Vanuatu (southern New Hebrides island arc). A: LF-QF-FM diagram. B: QF-FS-OP diagram. C: Q-F-FS diagram. D: Q-F-LF diagram. Erromango tempers and selected Malakula-Malo tempers also contain up to ~15% of detrital limeclasts reworked from uplifted limestone terraces (plotted compositions recalculated free of calcareous grains). Excessively lithic (LF > 70%) and strongly placered (FM > 75%) Malakula tempers omitted from plots B and C. See Table 5A for poles.

TABLE 14. MEAN PROPORTIONS OF IGNEOUS LITHIC FRAGMENTS
WITH DIFFERENT INTERNAL TEXTURES IN VANUATU TEMPER SANDS

Island(s) and temper type	N	Lathwork	Microlitic	Felsitic	Vitric	Microgranular	VLi
Taumako (non-placer)	8	-	62	9	23	3	97
Nendö (-Reef)	42	4	62	-	34	tr	>99
Vanikolo-Banks	10	-	58	-	42	-	100
Santo (-Malo)	18	-	68	14	18	tr	>99
Malakula (+ offshore islets)	18	-	6	78	12	4	96
Efate (+ offshore islets)	36	-	tr	-	100	-	100
Erromango (Ifo/Ponamla)	8	32	24	-	44	-	100

Notes: Figures tabulated are frequency percentages of grain types as fractions of total populations of volcanic lithic fragments. Placer temper sands containing less than ~15% volcanic lithic fragments omitted from compilation (N is number of sherds used to calculate mean). VLi (Table 6) denotes the relative proportion of extrusive volcanic lithic fragments (lathwork, microlitic, felsitic, vitric) where hypabyssal microgranular fragments of intrusive igneous rock (microphanerite) are treated as non-volcanic. Vitric volcanic lithic fragments are entirely felsic for Efate but dominantly mafic for other tempers listed.

no volcanic bedrock is exposed on either Malo or the limestone plateau of adjacent eastern Santo (Robinson, 1969; Mallick and Greenbaum, 1977). Western Santo lies along the axis of the ancestral Vitiaz island arc (Fig. 23) and is largely underlain by a heterogeneous assemblage of Lower to Middle Miocene andesitic to basaltic volcaniclastic rocks and associated sedimentary strata expected to yield derivative sands containing both clinopyroxene and hornblende (Mallick and Greenbaum, 1977, p. 68–69). Santo-Malo temper sands include an array of variably placered aggregates (Figs. 25A and 25B), which range from poorly sorted and little-abraded stream sands, some of which may be natural temper, to well-sorted and subrounded beach sands, which are dominant in sherds from Malo (Dickinson, 1971b). Volcanic lithic fragments are heterogeneous (Table 14), including microlitic, felsitic, and vitric varieties.

Sherds from the Torres Islands (Fig. 23) contain temper sands indistinguishable from Santo-Malo tempers (Fig. 25; Table 13C12) and are thought to derive from Santo, in part because exposures of Oligocene-Miocene volcanic bedrock in the Torres Islands are severely limited beneath Neogene limestone cover (Greenbaum et al., 1975) and in part because archaeological considerations point to ceramic transfer from Santo (Galipaud, 1998). Individual sherds (Fig. 25) apparently containing Santo temper sands (Table 13C15) have also been recovered from several other islands in Vanuatu (Table 13 notes), and provisionally document widespread dispersal of post-Lapita wares from Santo over a total distance of ~550 km along the New Hebrides island arc (Fig. 23). The exotic Santo sherds can be identified on Pakea (Banks Islands), Tikopia and Vanikolo to the north by the presence of hornblende as well as clinopyroxene mineral grains, on nearby Malakula by a paucity of felsitic volcanic lithic fragments (Table 14), and on Efate to the south both by the presence of hornblende in addition to clinopyroxene and by the abundance of nonvitric volcanic lithic fragments (Table 14).

Malakula Tempers

Abundant post-Lapita and rare Lapita sherds (Bedford, 2001) from Malakula (Fig. 23) and multiple offshore islets (Atchin, Rano, Uri, Uripiv, Vao, Wala) lying within a kilometer of the northeast coast contain volcanic sand tempers that are typically rich in lithic fragments (Fig. 25A). Selected sherds contain placer tempers in which either clinopyroxene or hornblende is the dominant grain type (Table 13C16). The placer tempers are well-sorted and subrounded beach sands, but the more abundant nonplacer tempers are moderately to poorly sorted stream sands, some of which may be natural temper. As the offshore archaeological sites are all on small calcareous islets, sherds recovered from them are inferred to represent wares brought from Malakula, or in some cases possibly local wares incorporating temper sands brought from Malakula.

The characteristic hallmark of Malakula temper sands is the dominance of felsitic varieties (Fig. A7) among volcanic lithic fragments (Table 14), with microgranular fragments of hypabyssal igneous rock also present in minor amounts. Moreover, the subordinate microlitic lithic fragments present in Malakula tempers commonly display pilotaxitic internal fabrics composed of oriented and densely packed plagioclase microlites, and contain as little interstitial volcanic glass as the quartz-feldspar mosaics of the more abundant felsitic grains. The contrast in proportions of different textural varieties of volcanic lithic fragments in Malakula and Santo tempers (Table 14) is entirely empirical, for descriptions of the Miocene volcanogenic assemblages of Malakula (Mitchell, 1966, 1971) and western Santo (Robinson, 1969; Mallick and Greenbaum, 1977) as nearby segments of the ancestral Vitiaz island arc (Fig. 23) give no indication of systematic differences in fundamental character. Multiple small subvolcanic intrusions of hypabyssal microdiorite on Malakula are the inferred source, however, of the microgranular lithic fragments that are present in most Malakula tempers, and are predominant among the lithic fragments of one anomalous temper sand in a sherd from Port Stanley on the east coast. Scattered subvolcanic intrusions are also exposed on Santo but did not contribute any detritus to known Santo temper sands.

Efate-Shepherd Tempers

Efate ceramic assemblages include only a limited sampling of Lapita ware but abundant post-Lapita (Mangaasi-style) collections (Garanger, 1966, 1971, 1972; Bedford and Spriggs, 2000;

Spriggs and Bedford, 2001). Sherds from Efate (Fig. 23) and neighboring offshore islets (Lelepa, Eretoka) contain a characteristic temper type (Dickinson, 1972a, 2001a; Dickinson et al., 1999a) with a consistently high proportion of subangular to quite angular plagioclase mineral grains (Fig. 25) and subordinate volcanic lithic fragments that are exclusively pale tan and commonly pumiceous particles of felsic volcanic glass (Table 14). Some of the vitric grains display concavely arcuate margins and filamentous extensions typical of airborne glass shards of volcanic ash. Variable proportions of minor clinopyroxene ($PYi = 100$; $PXi = 100$) and opaque iron oxides are the only other grain types present in Efate temper sands (Table 13C18). Moderately to poorly sorted textures imply origins of the volcanic detritus as stream sand or reworked tuff derived from the erosion of widespread Pliocene and younger pyroclastic deposits erupted locally along the active volcanic axis of central Vanuatu (Ash et al., 1978; Carney and MacFarlane, 1982). Sand concentrated from soil covering the alluviated Mele Plain on Efate contains the same grain types as Efate sherd tempers (Dickinson et al., 1999a). Uplifted reef limestone capping coastal terraces and interior plateaus on Efate contributed no detritus to known Efate temper sands.

Similar temper sands occur in sherds from Emae and Tongoa (Figs. 25A and 25B) in the Shepherd Islands (Fig. 23) north of Efate where the Neogene eruptive assemblage is analogous. Shepherd tempers (Table 13C17), however, contain trace amounts of basaltic (microlitic) volcanic lithic fragments not present in Efate tempers. The occurrence of mafic volcanic detritus reflects the existence of local basaltic as well as more widespread felsic eruptive centers within the Shepherd Islands (Warden, 1967). By contrast, exposures of basaltic rocks on Efate are restricted to isolated interior outcrops and selected offshore islands (Ash et al., 1978; Coulon et al., 1979; Raos and Crawford, 2004), the latter as yet untested for ceramic yield.

Erromango Tempers

Sherds from the Ifo and Ponamla sites (Spriggs, 1999; Bedford, 1999, 2003) on Erromango (Fig. 23) contain volcanic sand tempers in which lathwork grains of comparatively coarse grain size are more abundant relative to other textural types of volcanic lithic fragments than in other Vanuatu temper suites (Table 14). The relative abundance of lathwork volcanic lithic fragments reflects the dominantly mafic character of the Erromango eruptive assemblage, more than half of which is basaltic with the remainder basaltic andesite (Colley and Ash, 1971). Erromango placer tempers (Table 13C20) also contain subordinate but significant amounts of olivine mineral grains ($OLi = 5-10$), and olivine also occurs as microphenocrysts in Erromango basalts.

Because Nendö Lapita tempers from nearly 1000 km to the north contain comparable proportions of olivine mineral grains, and Lapita sherds are rare on Erromango (Spriggs and Bickler, 1989), a direct visual comparison of Nendö and Erromango placer tempers was made to test for possible ceramic transfer from Nendö to Erromango during Lapita occupation of the two islands. The gross compositions of Erromango and Nendö tempers suites are similar (Figs. 24A and 25A), although the ratio of opaque iron oxide grains to pyroxene mineral grains is somewhat higher in many Erromango tempers (Figs. 24B and 25B). Several petrographic observations indicate, however, that all Erromango tempers studied to date are indigenous and not derived from Nendö. Nendö tempers contain trace amounts of hornblende mineral grains (Table 13C4–13C6), whereas hornblende is absent from Erromango tempers (Table 13C19–13C20). Clinopyroxene in Erromango tempers has a distinct greenish cast, whereas clinopyroxene in Nendö tempers is almost colorless. Volcanic lithic fragments in Nendö tempers include prominent brownish to reddish vitric to microlitic (hyalopilitic) grains that are not present in Erromango tempers, whereas the lathwork volcanic lithic fragments so common in Erromango tempers are sparse in Nendö tempers. A Lapita sherd from Ifo contains a temper similar to tempers in non-Lapita sherds from the same site, and a Lapita sherd from Ponamla contains a strongly placered temper sand similar to the temper in a non-Lapita sherd from Ifo. Finally, Erromango tempers contain sparse detrital limeclasts (<15%), not present in Nendö tempers, which were reworked from uplifted limestone terraces.

Fiji Platform Arc Tempers

Erosional remnants of four Late Miocene to Pliocene magmatic arc assemblages (Fig. 26) are exposed on or adjacent to the Fiji platform (Colley and Greenbaum, 1980; Colley and Hindle, 1984; Gill et al., 1984; Rodda and Kroenke, 1984; Rodda and Lum, 1990; Rodda, 1994), which is underpinned by thick crust representing a rotated segment of the Vitiaz island arc (Fig. 9). Each of the four inactive arc assemblages has contributed derivative detritus to local temper sands. Three Upper Miocene assemblages were erupted along the ancestral Vitiaz island arc (Fig. 9) near the time of arc reversal (Fig. 26): (1) a dacitic assemblage (ca. 7 Ma) exposed on the Udu Peninsula of northern Vanua Levu; (2) an andesitic-dacitic assemblage (8–6 Ma) forming the Mamanuca-Yasawa island chain west of Viti Levu; and (3) an eroded capping of Upper Miocene andesitic strata (ca. 6 Ma) resting unconformably on more deformed Vitiaz basement rocks of the Wainimala orogen (Eocene-Miocene) in the interior of Viti Levu inland from the Navua Delta. The fourth Fijian arc assemblage is a younger suite of Pliocene volcanogenic strata (3.0–2.4 Ma) forming the Kadavu island arc (Fig. 26), which was active along the southern edge of the Fiji platform during its sinistral rotation following reversal of Vitiaz arc polarity (Dickinson, 2001a). Tempers in Udu Peninsula and Kadavu sherds are rich in plagioclase mineral grains, whereas tempers in sherds from the Yasawa Islands and Navua Delta are more placered sands rich in clinopyroxene mineral grains (Fig. 27).

Udu Peninsula

No indigenous sherds have been studied from the Udu Peninsula of northernmost Vanua Levu (Fig. 26), but temper sands derived from the distinctive Udu eruptive assemblage of silicic

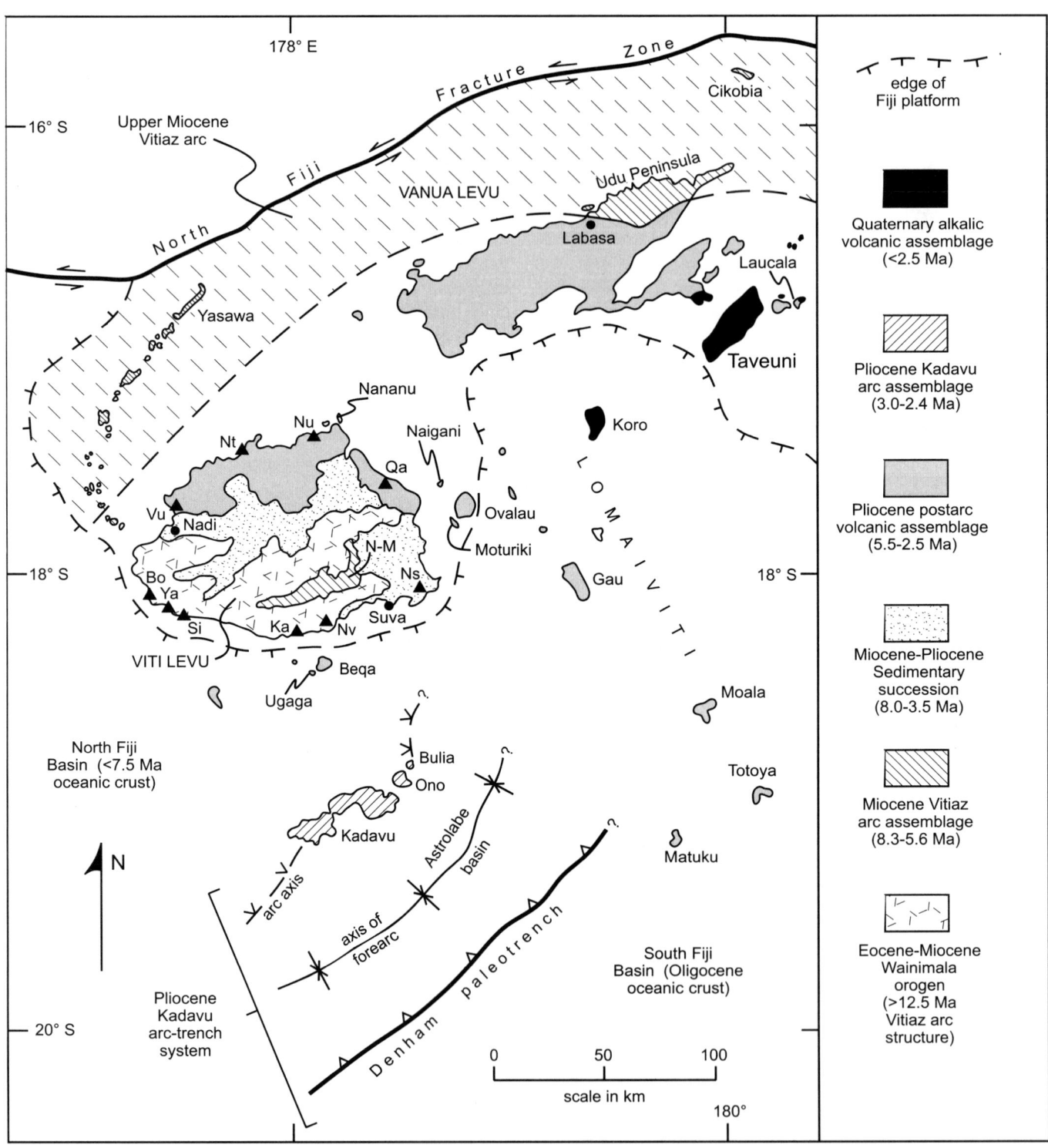

Figure 26. Geotectonic setting of islands on the Fiji platform and adjoining intraoceanic realms of Lomaiviti and Kadavu (Fig. 28 for Lau adjoins east edge on right). Ceramic sites (filled triangles) on Viti Levu: Bo—Bourewa; Ka—Karobo; Ns—Nasilai (on Rewa Delta); Nt—Natunuku; Nu—Navatu; Nv—Navua Delta; Qa—Qaqaruku; Si—Sigatoka; Vu—Vuda; Ya—Yanuca (and nearby Yadua). N-M—Upper Miocene (6.0–5.5 Ma) Namosi-Medrausucu volcanic and volcaniclastic rocks of ancestral Vitiaz arc on Viti Levu. Adapted after Dickinson (2001a).

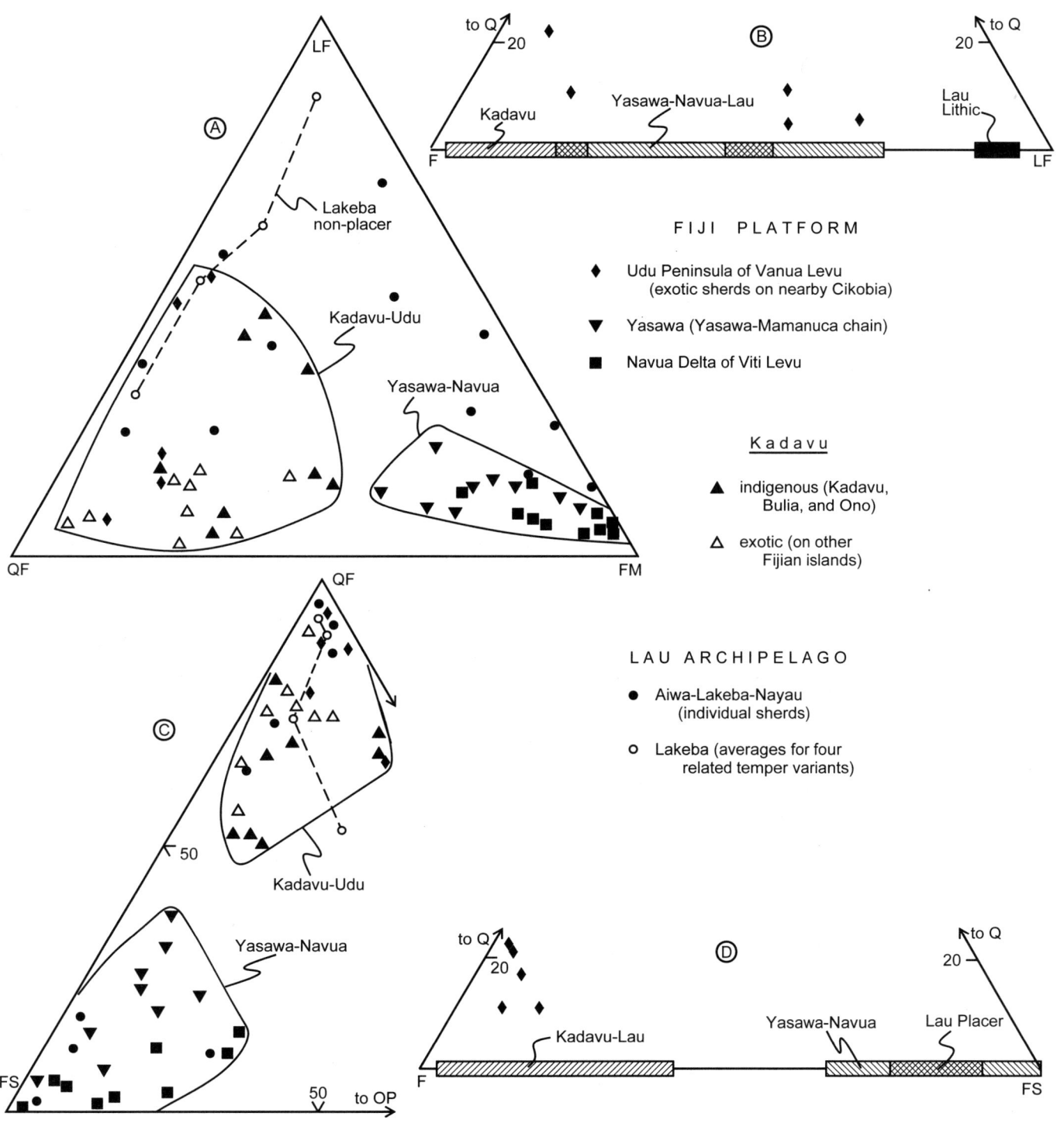

Figure 27. Compositions (grain frequency percentages) of andesitic arc tempers in sherds from Fiji and Lau. A: LF-QF-FM diagram. B: Q-F-LF diagram. C: QF-FS-OP diagram. D: Q-F-FS diagram. Four opaque-rich placer tempers omitted from plot C. See Table 5A for poles.

volcanic rocks occur in exotic sherds from nearby Cikobia (Fig. 26). The Udu suite of felsic andesitic and quartz-rich dacitic to rhyodacitic lavas and breccias (Colley and Greenbaum, 1980; Colley and Hindle, 1984) is petrologically unlike other arc assemblages of Vanuatu, Fiji, Lau, or Tonga, and the derivative quartz-bearing volcanic temper sands are compositionally unique for Fiji or Lau (Fig. 27). Phenocrysts of sand size in the Udu Volcanics include subordinate quartz as well as more abundant plagioclase (Rickard, 1966; Gill and Stork, 1979; Gill et al., 1984).

Tempers in five sherds studied from Cikobia (Fig. 26), where most sherds contain more mafic volcanic sands as temper, are quartz-bearing feldspathic sands (Fig. 27; Table 13D1) that document ceramic transfer from the Udu Peninsula lying 40–60 km to the south (Fig. 26). Quartz grains are clear volcanic quartz with straight extinction. Most (~90%) of the volcanic lithic fragments are either vitric grains of colorless to pale tan and partly pumiceous volcanic glass, typical of dacitic-rhyodacitic rock, or felsitic grains formed by advanced devitrification of felsic volcanic glass. Aggregate epidote grains present sparingly in some of the exotic tempers reflect hydrothermal alteration of parts of the Udu volcanic assemblage near sulfide ore deposits (Colley and Rice, 1975, 1978). Both well-sorted aggregates of probable beach origin and moderately sorted aggregates of possible stream origin are present in Cikobia sherds from the Udu Peninsula, but no exact geographic origin can be specified for either temper type.

Yasawa Islands

Sherds from Waya (Cochrane, 2002) at the southern end of the Yasawa island chain were not available for this study, but surface sherds from Yasawa (Fig. 27) at its northern end contain indigenous andesitic temper sands of beach origin (Bentley, 2000). All the tempers in the sherds studied are hybrid placer sands, but frequency percentages of calcareous grains are consistently <10%. The dominant grain type is clinopyroxene, but minor hornblende (PYi > 95) and orthopyroxene (PXi > 95) are also present (Table 13D2). Among volcanic lithic fragments, microlitic grains (60%–90% of total lithic fragments) are dominant over vitric grains of mafic brown volcanic glass.

An independent study (Aronson, 1999) of the sand tempers in 25 Lapita and post-Lapita sherds from the Qaranicagi rockshelter on Waya showed that most are quartz-free (n = 14) or quartz-poor (QZi < 5) feldspathic and pyroxenic (placer) volcanic sands with mineralogical compositions comparable to beach sands collected from shorelines of Waya. This dominant component of the Qaranicagi sherd suite is interpreted to represent indigenous Yasawa ware. Seven of the studied sherds recovered from the shallower levels of the excavation contain quartzose temper sands (QZi = 36–41) of exotic origin, and may have been derived from nearby Viti Levu (Aronson, 1999).

Navua Delta

The tempers of indigenous sherds from the Navua Delta on the south coast of Viti Levu (Fig. 26) contain well-sorted beach placer sands of volcanic detritus brought down the Navua River from inland exposures of the Namosi Andesite and the Medrausucu Group of associated volcaniclastic strata (Table 13D3). Bold exposures of volcanogenic strata in the rugged Korobasabasaga Range rise above the middle reaches of the Navua River as a proximate source for the volcanic detritus. Clinopyroxene is the dominant grain type in the placer sands (Fig. 27). Navua tempers lack either hornblende or orthopyroxene, but contain minor biotite (~1%) and olivine (OLi = 1–2), as expected from the known mineralogy of the volcanogenic source rocks.

Kadavu Arc

Few indigenous sherds are available from the Kadavu island arc (Fig. 26), which is virtually unexplored archaeologically, but surface sherds from northern Kadavu and the nearby islands of Ono and Bulia (Table 13D4) contain moderately sorted and subangular to subrounded volcanic sand tempers of probable stream origin. Within the Fiji-Lau-Tonga region, Kadavu temper sands are uniquely hornblende-rich (PYi = 0–43) feldspathic sands (Fig. 27), with oxyhornblende as well as common hornblende present, and a number also contain significant biotite (1%–10%). The temper compositions reflect the geology of Kadavu where hornblende andesites are dominant over pyroxene andesites and oxyhornblende is a common phenocrystic mineral in the hornblende andesites (Woodrow, 1980). The unusual joint abundance of hornblende, oxyhornblende, and biotite mineral grains in Kadavu tempers provides a template for detection of ceramic transfer from Kadavu to multiple islands within Fiji and Lau (Table 13D5). Clinopyroxene is the dominant if not the exclusive ferromagnesian silicate mineral in all other arc, postarc, and backarc Fiji-Lau-Tonga temper suites. Volcanic lithic fragments in Kadavu tempers tend to be vitric or felsitic varieties derived from felsic hornblende andesite, rather than microlitic varieties (<25% of total volcanic lithic fragments) typical of more mafic andesitic source rocks.

Lau Remnant Arc Temper

Along the Lau Ridge (Fig. 28), the remnant arc of the Tongan arc-trench system in eastern Fiji, a number of islands expose a Miocene (14–6 Ma) andesitic arc assemblage (Gill, 1976; Woodhall, 1985a; Cole et al., 1985) that was erupted along the ancestral Tongan island arc before opening of the Lau Basin by seafloor spreading separated the Lau Ridge from the active Tongan magmatic arc (Fig. 9). Intra-arc rifting began ca. 6 Ma, with seafloor spreading under way by ca. 5 Ma (Hawkins, 1995, 2003), and younger Lau volcanic assemblages (younger than 5 Ma) are postarc igneous suites (Fig. 8).

Lau andesitic arc tempers are known only from central Lau where Best (1984) studied ~1000 sherds from Lakeba in thin section. My own study of Lakeba tempers has been limited to ~100 of his thin sections thought to be representative of the sherd suite as a whole. My study also included, however, examination of the tempers in an additional 21 sherds thought to reflect ceramic transfer from Lakeba to the nearby limestone islands of Nayau and Aiwa (Fig. 28). Indigenous Lakeba sherds and exotic Nayau-Aiwa sherds (O'Day et al., 2004) contain variably placered but mineralogically simple volcanic sand tempers (Fig. 27), with quartz absent and clinopyroxene the only ferromagnesian mineral present (Table 13D6). Varied volcanic lithic fragments include subequal proportions of vitric grains, composed of tan or brown to red volcanic glass, and grains with internal microlitic textures, both pilotaxitic and hyalopilitic including dark tachylitic lithic fragments that are commonly microporphyritic. The absence of orthopyroxene distinguishes Lau andesitic arc tempers (PXi = 100) from all Tongan andesitic arc tempers (PXi < 95).

The textural relations of temper sands and clay pastes in the Nayau-Aiwa sherds indicate that all the tempers were manually added to clay bodies, but include poorly sorted and lithic-rich aggregates of colluvial or alluvial sand, moderately sorted and generally subangular feldspar-rich aggregates of probable stream sand, pyroxene-rich placer sands of stream or beach origin (or both), opaque-rich hybrid beach sands, and largely calcareous sands containing only minor terrigenous detritus. The relative percentages of studied sherds containing the various temper variants are almost identical for Lakeba and Nayau-Aiwa sherd suites (Table 15), as expected if the Nayau-Aiwa sherds were derived from Lakeba. The textural and compositional variety of Lakeba temper sands is taken to reflect the richness of the Lakeba ceramic heritage at multiple locales on the island (Best, 1984).

Tongan Tempers

The active Tongan (Tofua) island arc (Fig. 29A) extends >750 km parallel to the Tonga Trench (Bryan et al., 1972; Ewart and Bryan, 1973; Ewart et al., 1973, 1977; Ewart, 1976; Ewart and Hawkesworth, 1987). The Tonga volcanic assemblage is dominantly andesite, largely mafic or basaltic but also felsic, accompanied by minor basalt and dacite. Typical eruptive products, whether lavas or pyroclastic particles, are porphyritic with phenocrysts of plagioclase, both pyroxenes (clinopyroxene and orthopyroxene), and minor opaque titanomagnetite set in vitric to microlitic groundmasses. Rare olivine is also present in some volcanic rocks, but hydrous ferromagnesian minerals (hornblende, biotite) are wholly lacking.

A chain of widely spaced Quaternary volcanic islands, and many more submerged seamounts, is flanked to the east in central Tonga by a chain of emergent forearc islands (Fig. 29B) composed dominantly of uplifted Neogene limestone capping a thick Cenozoic forearc basin of buried sedimentary strata (Herzer and Exon, 1985; Cunningham and Anscombe, 1985; Scholl et al., 1985; Austin et al., 1989; Scholl and Herzer, 1992; Tappin, 1993).

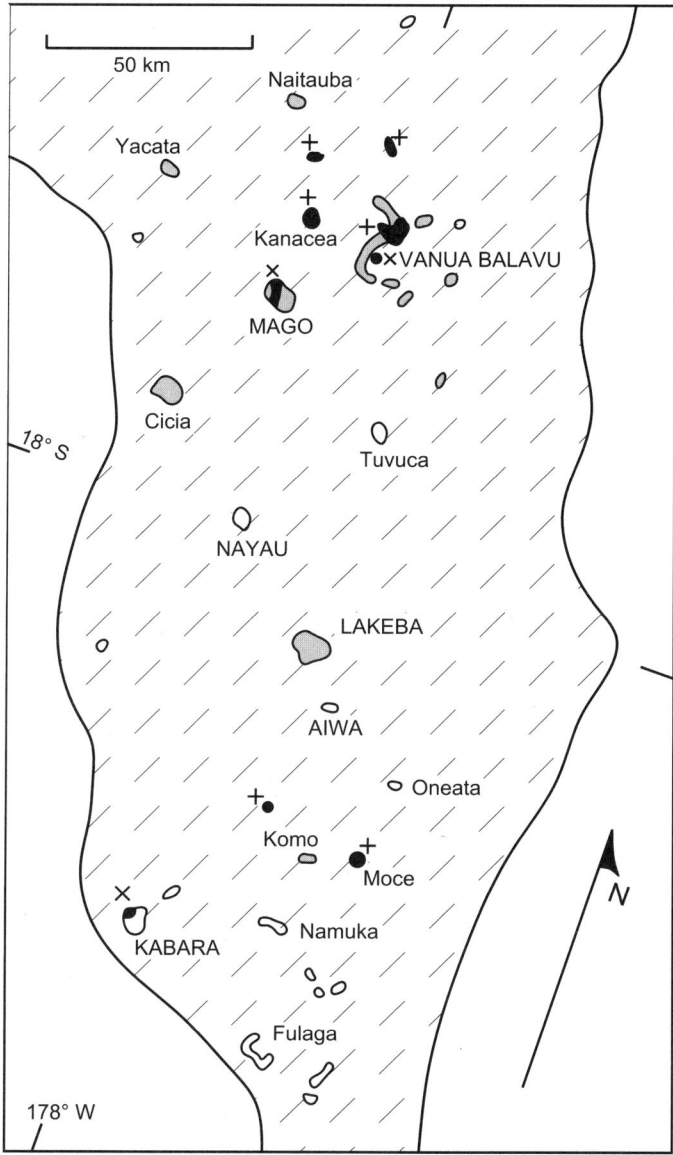

Figure 28. Geology of islands within the central and northern Lau Archipelago (eastern Fiji) capping the Lau Ridge remnant arc (diagonal dashes) of the Tongan arc-trench system, adapted after Woodhall (1985a) and Cole et al. (1985). All-capitals denote islands from which sherd tempers are known. Island symbols: *shaded*—islands exposing Miocene (14–6 Ma) Lau calc-alkalic arc assemblage of ancestral Tongan volcanic arc; *black*—islands exposing postarc Pliocene (4.5–2.5 Ma) Korobasaga eruptive assemblage (+) of transitional subalkalic petrology or Quaternary (younger than 2.5 Ma) alkalic eruptive assemblage (x); *blank*—islands exposing only or mainly uplifted Neogene limestone capping arc volcanic substratum. Subalkalic (+) and alkalic (x) basalt symbols also denote islands where postarc volcanics overlie Lau arc volcanics.

TABLE 15. TEMPER TYPES IN LAKEBA AND NAYAU-AIWA SHERDS OF CENTRAL LAU

Sherd assemblage	Lithic	Feldspathic	Pyroxenic	Hybrid	Calcareous
Lakeba	29	24	10	19	19
Nayau-Aiwa	24	26	13	17	20

Note: figures tabulated are percentages of temper types (n=168) for Lakeba (Best, 1984) and percentages of individual sherds (n=21) for Nayau-Aiwa (Fig. 27, Table 13D6).

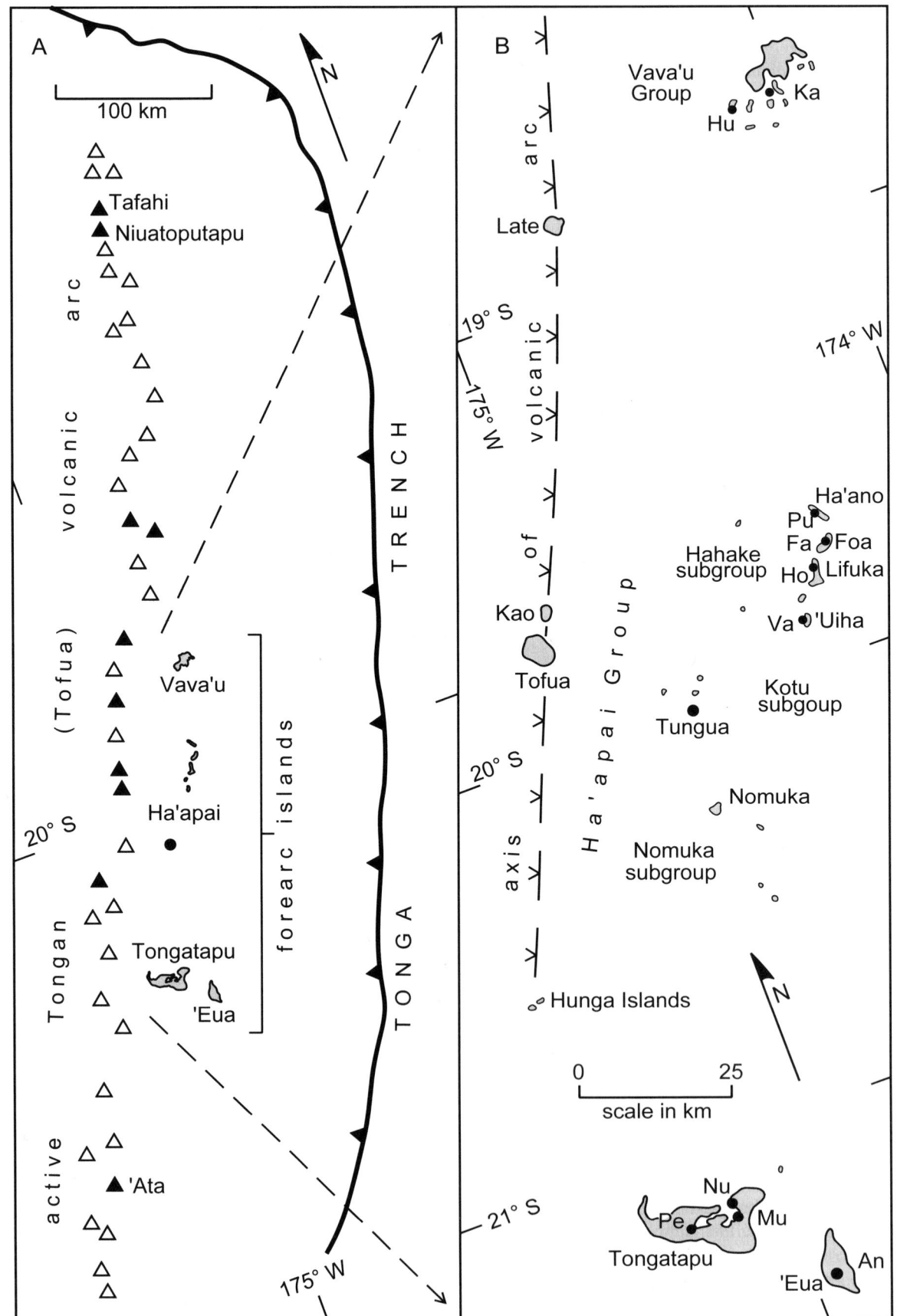

Figure 29. Location of key ceramic sites in Tonga. A: Triangles denote active or dormant to extinct volcanic edifices (black—islands; blank—seamounts). B: Filled circles denote archaeological sites: An—'Anatu-Anokula; Fa—Faleloa; Ho—Holopeka; Hu—Hunga; Ka—Kapa; Mu—Mu'a; Nu—Nukuleka; Pe—Pea; Pu—Pukotala; Tu—Tungua; Va—Vaipuna.

Restricted exposures of Miocene volcaniclastic strata underlying Quaternary limestone form island cores in the Nomuka subgroup (Fig. 29B) within the forearc belt of central Tonga, but are insignificant for temper analysis. On the other hand, altered volcanogenic bedrock of Middle Eocene age exposed unconformably beneath the forearc sedimentary succession on 'Eua (Figs. 29A and 29B) serves as a local source of temper sand that is otherwise distinctly anomalous for Tonga.

The ceramic record of the volcanic islands is restricted to voluminous collections from multiple sites on Niuatoputapu and related sparse sherds from adjacent Tafahi on the north (Rogers, 1974; Kirch, 1978, 1988c), and a few sherds from 'Ata on the south (Burley et al., 2004; Dickinson, 2004a). Most of the ceramic record of Tonga stems instead from multiple archaeological sites among the populous limestone islands of the forearc belt (Davidson, 1971; Shutler et al., 1994; Burley, 1994; Burley et al., 2001). The volcanic sand tempers characteristic of Tongan sherds from the forearc islands are coastal beach sands (Fig. A24), in part hybrid, reworked from thick tephra blankets that were deposited across the surfaces of the limestone islands from airborne ash clouds generated by explosive eruptions along the volcanic chain to the west (Dye and Dickinson, 1996). The preference of ancient Tongan potters for terrigenous sand as temper presumably reflected experience with the deleterious effects produced by calcining of calcareous tempers during ceramic firing. Even though noncalcareous black sand occurs only as local patches and pockets on island beaches that are composed dominantly of reef-derived calcareous detritus, the white calcareous sands were disfavored for use as ceramic temper. Close textural and compositional similarities between volcanic temper sands in central Tongan sherds and modern placer black sands of the Ha'apai Group leave no doubt of the origin of the tempers as beach sands on the limestone islands (Dye, 1987; Dye and Dickinson, 1996; Dickinson et al., 1996).

Indigenous tempers in sherds from Tongatapu, the Ha'apai Group, and the Vava'u Group of central Tonga (Figs. 29A and 29B) are variably placered volcanic sands that plot along the same placering trends in compositional space on ternary plots (Figs. 30A and 30B). The volcanic detritus is composed of phenocrystic mineral grains (plagioclase, pyroxenes, opaques) and volcanic lithic fragments derived from the island tephra blankets. The total variability of central Tongan tempers is somewhat greater than depicted by Figure 30 when point counts from other petrographic studies are taken into consideration (Key, 1987; Dye, 1987), but overall trends of temper variation are not substantially altered (Dye and Dickinson, 1996). Strongly placered Tongan temper sands are rich in pyroxene mineral grains that include both clinopyroxene and orthopyroxene (Table 13E2–13E4), although the former is consistently more abundant (PXi in the range of 80–95). Internal textures of volcanic lithic fragments in central Tongan tempers are varied (lathwork, microlitic, felsitic, vitric), but the most abundant are microlitic and vitric grains, with the latter including tachylitic grains in some temper sands.

Because the forearc tephra cover from which the volcanic temper sands were derived is lithologically similar on all the limestone islands of central Tonga, distinctions among Tongatapu, Ha'apai, and Vava'u temper suites are subtle and perhaps equivocal. For example, no systematic differences in proportions of microlitic and vitric volcanic lithic fragments have been detected. Tongatapu tempers display somewhat higher ratios of plagioclase feldspar mineral grains to volcanic lithic fragments than Ha'apai-Vava'u tempers (Fig. 30C). Vava'u tempers (Table 13E2) display slightly higher ratios of clinopyroxene to orthopyroxene than Ha'apai-Tongatapu tempers (Table 13E3–13E4), with PXi > 90 for Vava'u tempers but PXi < 91 for Ha'apai-Tongatapu tempers. The contrast in pyroxene index for Ha'apai and Vava'u tempers is inferred to reflect comparable differences in the ratio of clinopyroxene to orthopyroxene within the ash clouds that spread eastward across the forearc limestone islands from volcanic centers (Fig. 29B) lying west of Ha'apai (Tofua) and west of Vava'u (Late). The frequency percentage of rare quartz grains (~1%) in Tongatapu tempers (Table 13E4) is generally higher than in nearly quartz-free Ha'apai-Vava'u tempers (Table 13E2–13E3), and suggests derivation from dacitic as well as andesitic tephra deposits.

Niuatoputapu Temper

Lapita sherds from Niuatoputapu and neighboring Tafahi, lying well to the north of central Tonga (Fig. 29A), contain tempers (Table 13E1) that lie along Tongan placering trends, but are typically more lithic sands than typical central Tonga tempers (Figs. 30A and 30B). Their pyroxene index (PXi = 83–88 from Table 13E1) is similar, however, to values for Ha'apai and Tongatapu tempers and is characteristic of the typical ratio of clinopyroxene to orthopyroxene in detritus derived from most Tongan volcanic sources. The volcanic island of Tafahi, and the volcanic core of neighboring Niuatoputapu only 10–12 km distant, are likely places for accumulations of unplacered volcanic sand, and the tempers are interpreted as indigenous to the two adjacent islands (Dye and Dickinson, 1996). The volcanic lithic fragments in Niuatoputapu-Tafahi tempers are almost all characteristically pumiceous vitric grains (Fig. A9) of pale tan silicic volcanic glass not observed in other Tongan tempers (Dickinson, 1974b, 1988). The vitric lithic fragments reflect derivation of the temper sands from local pyroclastic deposits, rather than from the reworking of distal tephra layers that form the source for volcanic sands on the forearc islands. The Niuatoputapu tempers richest in pumiceous vitric grains may have been collected from unconsolidated ash blankets on Niuatoputapu or nearby Tafahi, whereas temper sands containing higher proportions of plagioclase and pyroxene mineral grains may represent locally reworked or winnowed volcanic ash deposits.

'Ata Temper

The temper sands in two surface sherds of post-Lapita Polynesian Plainware collected from the central plateau of 'Ata (Fig. 29A) in southern Tonga are poorly to moderately sorted volcanic

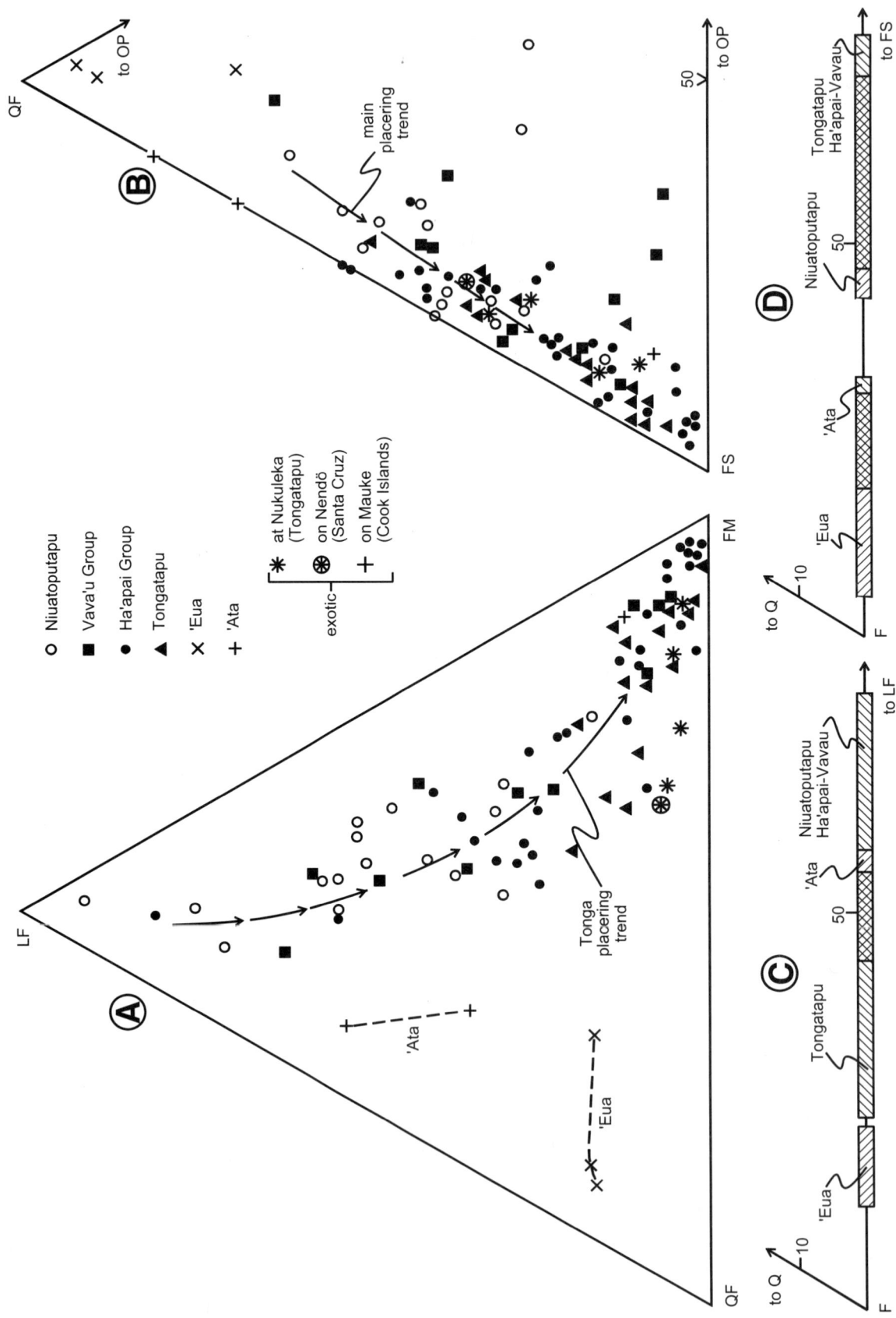

Figure 30. Compositions (grain frequency percentages) of Tongan andesitic arc tempers, four non-Tongan andesitic arc tempers in exotic Lapita sherds from Nukuleka site on Tongatapu (comparable temper sand in exotic sherd from Nendö in Santa Cruz Group also plotted for comparison), and the temper in an exotic Mauke sherd (Cook Islands) derived from Tonga. A: LF-QF-FM diagram. B: QF-FS-OP diagram. C: Q-F-LF diagram. D: Q-F-FS diagram. Two opaque-rich placer tempers (one each from Ha'apai and Vava'u) and one excessively lithic temper (from Niuatoputapu) omitted from plot B. Exotic tempers omitted from plots C and D, and 25 placer tempers (FM > 75%) omitted from plot D. See Table 5A for poles.

sands (Dickinson, 2004a) derived from pyroclastic strata exposed over wide areas in the interior of 'Ata (Johnstone, 1978). The tempers are more feldspathic than most Tongan tempers (Fig. 30A), and contain a higher content of olivine mineral grains than other Tongan tempers (Table 13E5), but the ratio of clinopyroxene to orthopyroxene (PXi) lies within the range of most other Tongan tempers. The average proportions of clinopyroxene, orthopyroxene, and olivine in 'Ata temper sands (82:10:8) are statistically indistinguishable from the proportions in modern beach sand (83:11:6) collected from the north coast of 'Ata.

The high ratio of plagioclase to ferromagnesian mineral grains in 'Ata tempers mimics proportions of phenocrysts present in 'Ata volcanic rocks (Vallier et al., 1985), the prominence of olivine in derivative temper sands reflects the presence of significant phenocrystic olivine in the 'Ata eruptive assemblage (Vallier et al., 1985), an abundance of tan to brown vitric grains among the volcanic lithic fragments of 'Ata tempers is characteristic of mafic pyroclastic strata widely exposed on 'Ata (Johnstone, 1978), and rare grains (~1%) of cumulus microgabbro in 'Ata tempers resemble xenoliths reported from pyroclastic breccias on 'Ata (Vallier et al., 1985).

'Eua Temper

Tempers in sherds from 'Eua, lying to the southeast of Tongatapu (Figs. 29A and 29B), are more feldspathic than any other Tongan tempers (Figs. 30A and 30B), lack any orthopyroxene (Table 13E6), and contain only minor volcanic lithic fragments (~10%) with varied internal textures, but contain distinctive microphanerite lithic fragments (~5%) of altered microgabbro. The angular and weathered nature of the temper grains suggests that the temper sand is natural temper embedded within impure colluvial or alluvial clay, material not known to have been used for temper elsewhere in Tonga. The distinctive microphanerite lithic fragments are composite grains composed of plagioclase and clinopyroxene, with the latter largely altered to fibrous amphibole (uralite). Many of the sparse clinopyroxene mineral grains in 'Eua tempers are also partly or completely altered to uralitic hornblende.

The composition of 'Eua temper reflects derivation from restricted exposures of Middle Eocene volcanogenic bedrock cropping out at intervals beneath younger sedimentary cover on the steep eastern slope of 'Eua (Stearns, 1971; Ewart and Bryan, 1972; Cunningham and Anscombe, 1985), but unexposed elsewhere in Tonga. The volcanogenic succession of altered basaltic to andesitic lava, volcanic agglomerate and conglomerate, and dacitic tuff has yielded Middle Eocene radiometric ages (47–46 Ma) from gabbro boulders within conglomerate beds, and is overlain depositionally by Middle Eocene to Lower Oligocene limestone forming the highest elevations on 'Eua. Younger strata exposed on the gentle western slope of 'Eua include deeply weathered Miocene volcaniclastic beds in local exposures, but are mainly Pliocene to Quaternary limestone deposited within the Tonga forearc belt during eruptive activity along the active volcanic chain to the west.

Plagioclase feldspar is the dominant phenocrystic mineral in volcanic rocks and reworked volcanic clasts of the 'Eua basement succession, thereby accounting for the exceptionally feldspathic character of 'Eua tempers. The gabbro boulders in the 'Eua volcanogenic succession are altered uralitic rocks, as are the microphanerite lithic fragments in 'Eua tempers. The geochemistry of the igneous components of the 'Eua basement implies an origin either as the upper level of ophiolitic oceanic crust, uplifted by subduction beneath the inner wall of the Tonga Trench, or as the oldest component of the Tongan island-arc edifice (Fleck and Vallier, 1985), with the latter interpretation preferred (Hawkins and Falvey, 1985).

POSTARC AND BACKARC TEMPERS

Along some island arcs, volcanism continued or was resumed after subduction ended along associated oceanic trenches, to erupt postarc volcanic assemblages that are commonly more alkalic than subjacent calc-alkalic arc assemblages. In other cases, backarc spreading has separated dormant remnant arcs from still-active frontal arcs, and postarc volcanism on remnant arcs or backarc volcanism within interarc basins between active and remnant arcs has also produced nonarc volcanic assemblages. The eruption of backarc volcanic islands through oceanic lithosphere has led to the affinity of some backarc temper suites with oceanic basalt tempers. Postarc and backarc temper suites contain grain populations similar to andesitic arc tempers, and were initially grouped with the latter (Dickinson and Shutler, 1968, 1971, 1979; Dickinson, 1998a). Because of their petrologic differences from arc tempers, however, there are advantages to treating the postarc-backarc tempers separately for regional temper analysis (Dickinson and Shutler, 2000; Dickinson, 2001a).

Postarc tempers are best known from the Fiji platform (Fig. A17) where postarc volcanism succeeded arc magmatism along the trend of the ancestral Vitiaz island arc after reversal in arc polarity and backarc opening of the North Fiji Basin to the west (Figs. 8 and 9). Postarc tempers are also common along the Lau Ridge remnant arc of the Tongan arc-trench system nearby to the east. Postarc temper sands of distinctly alkalic affinity are known from the TLTF (Tabar-Lihir-Tanga-Feni) island chain northeast of New Ireland in the Bismarck Archipelago (Fig. 19) where Neogene transform disruption of the Paleogene Vitiaz arc assemblage (Figs. 7 and 9) has produced an igneous assemblage of unique petrology for the Melanesian region. Backarc tempers are known from isolated basaltic islands in the North Fiji Basin and along the northern Melanesian borderland between Fiji and Samoa (Fig. 8), and from a cluster of islands exposing a rift-related bimodal igneous assemblage within the Bismarck Sea (Fig. 19).

Backarc Admiralty Tempers

The tempers in a limited sampling of sherds from Lou, Baluan, and M'Buke in the Admiralty Group south of Manus

(Fig. 19) are quartzose volcanic sands (QZi = 11–32) derived from exposures of rhyolitic to rhyodacitic rocks forming the silicic end members of a locally erupted bimodal igneous assemblage (Johnson and Smith, 1974; Johnson et al., 1978; Jaques, 1981; Smith and Johnson, 1981). The bimodal suite, present on the southwestern peninsula of Manus and southward within the Admiralty Group, developed during crustal rifting associated with continuing backarc seafloor spreading to open the oceanic Manus Basin within the Bismarck Sea north of New Britain (Fig. 7). No tempers derived from mafic end members of the bimodal assemblage are known (Dickinson, 2000a), possibly because deep tropical weathering has prevented derivation of sand attractive for temper from basaltic rocks, while allowing sand derivation from associated silicic volcanic rocks more resistant to surficial weathering.

Volcanic lithic fragments in the Admiralty backarc tempers (Fig. 31), among which textures suggest origins as both beach and stream sands, are predominantly felsitic, composed of microcrystalline anhedral mosaics of intimately intergrown feldspar and quartz. Feldspar mineral grains are exclusively plagioclase, derived from phenocrysts in felsic volcanic rocks. Ferromagnesian mineral grains include varying proportions of clinopyroxene and hornblende or oxyhornblende (Table 16A1) accompanied by minor biotite.

Several related hybrid or opaque-rich temper types of beach origin in exotic sherds from the Mussau Group (Fig. 19) are inferred to derive also from within the cluster of small islands in the southern Admiralty Group where silicic volcanic rocks of the bimodal backarc rift assemblage are exposed (temper types B-C-D of Dickinson, 2006a). The terrigenous fractions of these tempers are more placered than any known temper sands from indigenous Admiralty sherds (Fig. 31), but are nevertheless inferred to derive from the Admiralty Group because of their comparatively high quartz contents and the prominence of volcanic lithic fragments with felsitic internal textures. Detritus of a comparable petrologic character cannot be expected elsewhere within the Bismarck Archipelago. The oxide index of the hybrid and placer tempers (Table 16A2) at Mussau is consistently high (OXi = 64–94), but pyribole indices (PYi = 25–91) are comparable to those of indigenous Admiralty tempers (PYi = 31–67), and compatible with sources in the rhyodacitic lavas and breccias of the Admiralty Group.

Postarc TLTF Tempers

The TLTF chain formed by the Tabar, Lihir, Tanga, and Feni island groups off the northeast coast of New Ireland (Fig. 19) is composed of Neogene alkalic volcanic rocks (basanite, tephrite, trachybasalt) petrologically unlike the calc-alkalic arc assemblages of the larger islands farther west in the Bismarck Archipelago (Johnson et al., 1976; Wallace et al., 1983). Parallelism of the TLTF chain to multiple Neogene strands of the transform fault system that has displaced New Ireland to the northwest with respect to New Britain (Fig. 19) suggests that eruptions along the TLTF chain were related to incipient transform rupture of the flank of the ancestral Paleogene Vitiaz island arc (Fig. 9). The mineralogy of temper sands in Lapita sherds from the Feni Islands (Table 16A4–16A7) and the Tanga Islands (Table 16A8–16A11) in the southern TLTF chain imply derivation from volcanic sources within the TLTF alkalic assemblage. A few exotic sherds (Table 16A3) from the Mussau Group farther northwest (temper type K of Dickinson, 2006a) contain analogous though not compositionally identical tempers (Fig. 31) suggestive of ceramic transfer from some locale in the northern part of the TLTF chain (Kennedy et al., 1990; Rytuba et al., 1993), possibly via New Ireland or New Hanover (Fig. 19).

The most distinctive petrographic signal of TLTF tempers is the nature of their clinopyroxene mineral grains, many of which have the greenish cast and faint yellowish pleochroism typical of aegirine-augite, a sodic variety of augite characteristic of alkalic volcanic suites. The identification of aegirine-augite is not conclusive from coloration alone, but interference figures consistently document high optic axial angles (2V > 75°) atypical of normal augite (2V ≈ 60°) derived from calk-alkalic arc assemblages. Although the reported soda content of selected clinopyroxenes from the Feni Islands is too low for aegirine-augite (Heming, 1979), clinopyroxene phenocrysts in all TLTF volcanic suites are described as distinctly greenish (Wallace et al., 1983) and comparable in color to the aegirine-augite mineral grains observed in temper sands of TLTF sherds. Vitric volcanic lithic fragments of mafic brown volcanic glass are common in TLTF tempers. Most other volcanic lithic fragments have the lathwork internal textures (intergranular to intersertal) typical for lithic fragments in oceanic basalt tempers, rather than the microlitic internal textures (hyalopilitic to pilotaxitic) typical for lithic fragments in andesitic arc tempers. Felsitic volcanic lithic fragments are also present in some TLTF tempers.

In the Feni Islands (Fig. 19), Lapita sites occur on both Anir (=Ambitle) and Babase (White and Specht, 1971; Summerhayes, 2000c). Feni temper suites are heterogeneous (Fig. 31), including both well-sorted and subrounded aggregates of inferred beach origin and only moderately sorted subangular aggregates of inferred stream origin. Temper sands in sherds from Kamgot on Babase (Table 16A4) are typically more placered than temper sands in sherds from Malekolon on Anir (Table 16A7), although the two sites are only ~2.5 km apart across Salat Strait, which is <500 m wide at its narrowest. Sherds from Balbalankin not far from Malekolon on Anir (Table 16A5–16A6) contain both placer and nonplacer tempers. In Babase sherds (Table 16A4) and some Balbalankin sherds (Table 16A5), ferromagnesian silicate mineral grains are predominantly clinopyroxene (PYi > 95) and are consistently more abundant than opaque grains (OXi < 50). In other Balbalankin sherds (Table 16A6) and Malekolon sherds (Table 16A7), ferromagnesian silicate mineral grains also include prominent green-brown to red-brown hastingsitic hornblende (PYi < 77), with opaque grains subequal in abundance (OXi ≈ 50). Quartz is absent from any Feni tempers, as expected for derivation from the local alkalic volcanic assemblage. Because

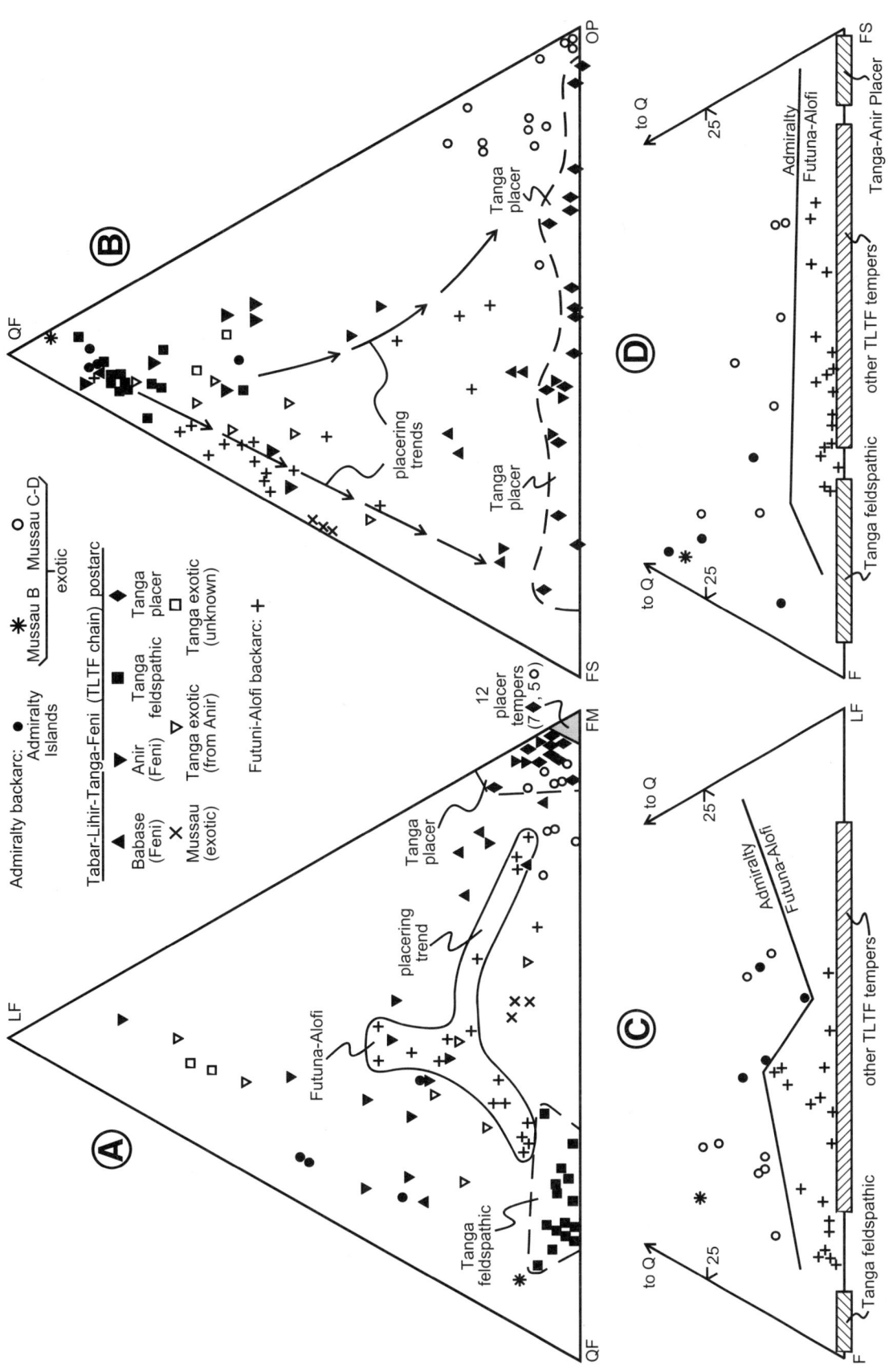

Figure 31. Compositions (grain frequency percentages) of postarc and backarc tempers in prehistoric sherds from the Bismarck Archipelago and the northern Melanesian borderland (Futuna-Alofi). A: LF-QF-FM diagram. B: QF-FS-OP diagram. C: Q-F-LF diagram. D: Q-F-FS diagram. Opaque-rich exotic tempers in Mussau sherds omitted from plots C and D. See Table 5A for poles.

TABLE 16. KEY SUPPLEMENTAL GRAIN PARAMETERS FOR BACKARC AND POSTARC TEMPERS OF THE SOUTHWEST PACIFIC REGION

Temper type	N	QZi	OXi	OLi	PYi	PXi
A. Bismarck Archipelago and northern Melanesian borderland						
1 - Admiralty Islands (Baluan-Lou-M'Buke sherds)	4	11–32	40–71	0	31–67	100
2 - Mussau Types BCD (exotic Admiralty sherds)	8	14–33	64–94	0	25–91	100
3 - Mussau Type K (exotic sherds from TLTF chain)	3	0	2	0	100	100
4 - Kamgot on Babase in Feni Islands (TLTF chain)	6	0	27–47	0	95–98	100
5 - Balbalankin pyroxenic on Anir in Feni Islands	5	0	38–68	0	93–98	100
6 - Balbalankin hornblendic on Anir in Feni Islands	3	0	39–71	0	10–17	100
7 - Malekolon on Anir in Feni Islands (TLTF chain)	6	0	12–62	0	40–77	100
8 - Tanga Islands exotic hornblendic (from Anir)	6	0	21–43	0	38–77	100
9 - Tanga Islands (TLTF chain) indigenous feldspathic	15	0	21–37	0	100	100
10 - Tanga Islands (TLTF chain) indigenous placer	16	0	21–94	0	100	100
11 - Tanga Islands exotic hybrid (TLTF chain)	2	0	38–58	0	100	100
12 – Futuna-Alofi (Horne Islands) non-placer tempers	14	2–10	6–26	0	100	100
13 - Futuna (Horne Islands) placer tempers	4	12–20	43–58	0	100	100
B. Northwestern Fiji (Vanua Levu, Viti Levu, and associated islands)						
1 - Cikobia A (indigenous or from Vanua Levu)	7	0	3–11	1–2	100	98–99
2 - Cikobia B (exotic from nearby Vanua Levu)	6	1–3	30–60	0	100	94–98
3 - Cikobia C (from Vanua Levu or Viti Levu)	1	22	21	0	95	95
4 - Vanua Levu (near Labasa on northwestern coast)	2	0	6–58	3	100	91–99
5 - Rotuma exotic (probably from Vanua Levu)	1	0	4	2	98	86
6 - Nananu-i-ra (islet off northeastern Viti Levu)	2	0	11–27	2	100	97–98
7 - Qaqaruku rockshelter, northeastern Viti Levu	1	0	6	2	100	100
8 - Navatu indigenous, northeastern Viti Levu	25	0–4	4–18	5–15	65–98	100
9 - Tuvalu exotic (from Navatu on Viti Levu)	4	0	11–16	5–11	76–90	100
10 - Tuvalu exotic (from unknown Fiji site or sites)	6	0	5–36	1–3	100	100
11 – Tokelau exotic (from unknown Fiji or Lau site)	1	0	10	0	100	100
12 - Natunuku indigenous (Tavua), north Viti Levu	16	0	5–29	2–12	>95	100
13 - Natunuku indigenous (Vatia), north Viti Levu	6	0	8–15	5–11	75–89	100
14 - Yanuca non-local (from western Viti Levu)	6	0	2–36	0	100	100
15 - Vuda indigenous (local), western Viti Levu	3	1–6	34–39	0	76–96	100
C. Southeastern Fiji (Beqa, Lomaiviti, Lau, and associated islands)						
1 - Laucala (small islet off Taveuni)	1	0	10	13	100	100
2 - Taveuni (large island east of Vanua Levu)	2	0	3–5	0	100	100
3 - Naigani (exotic from uncertain locale)	2	0	2–4	2–4	100	100
4 - Naigani (exotic from nearby Ovalau or Moturiki)	9	0	2–10	0	73–98	100
5 - Moturiki (indigenous or exotic from Ovalau)	23	0	1–13	0	72–90	100
6 - Moturiki (exotic from uncertain locale)	1	0	3	3	97	100
7 - Beqa indigenous (multiple sites)	10	0	35–67	0–6	93–99	100
8 - Ugaga exotic (from Beqa)	8	0	14–35	0	68–98	100
9 - Beqa indigenous (Kulu Bay vicinity)	4	9–18	18–44	0	40–46	100
10 - Ugaga exotic (from Kulu Bay on Beqa)	5	11–19	21–46	0	45–72	100
11 - Totoya indigenous (hybrid temper)	2	0	14–36	7–16	100	100
12- Totoya exotic (probably from Moala or Matuku)	2	0	9–16	4–8	86–91	100
13 - Vanua Balavu indigenous (northern Lau)	2	0	95–98	0	100	100
14 - Mago indigenous (northern Lau)	3	0	5–6	3–4	100	100
15 - Kedekede temper (exotic on multiple islands)	12	0	2–6	2–9	100	100

Notes: N is number of sherds for which petrographic frequency counts of grain types in thin section are tabulated. See Table 6 for definitions of supplemental grain parameters (or indices). Large ranges in OXi values for many temper types reflect highly variable degrees of placering. Anomalous temper compositions represented by only one sherd not tabulated. Kedekede placer temper (line C15), thought probably to derive from Kanacea in northern Lau (see text), occurs in exotic sherds from Kabara, Lakeba, Moturiki, Totoya, and Vanua Balavu (Figs. 26, 28). See Table 17 for other supplemental grain parameters of selected Fijian postarc tempers containing detritus from subvolcanic intrusions.

bedrock exposures on both Anir and Babase are eroded remnants of mafic Pliocene to Pleistocene stratocones of similar petrology (Wallace et al., 1983), tracing Feni temper types to specific bedrock sources on either of the two islands is not feasible with present information.

In the Tanga Islands (Fig. 19), all the main islets are remnants of the flanks of a single collapsed Pleistocene stratocone separated by a submerged caldera (Wallace et al., 1983). Lapita sherds from Tanga contain two indigenous temper types (Fig. 31A), one feldspathic (Table 16A9) and the other a placer sand (Table 16A10). Feldspathic Tanga tempers are moderately sorted and generally subangular aggregates of probable stream origin, whereas Tanga placer tempers are well-sorted aggregates of beach origin (Fig. A26). The only ferromagnesian silicate mineral grains in both types of indigenous Tanga temper are clinopyroxene (aegirine-augite), although the ratio of clinopyroxene mineral grains to opaque iron oxide grains varies widely for the placer tempers (Fig. 31B). In the Tanga sherd suite, sparse hybrid beach sand tempers of lithic character (Fig. 31A), with clinopyroxene the only ferromagnesian silicate mineral grains (Table 16A11), may be either indigenous or exotic. Subordinate numbers of Tanga sherds contain hornblende-bearing beach sand tempers, in part hybrid, with compositions inferred to document ceramic transfer from Anir in the Feni Islands (Table 16A8).

Backarc Vanuatu Tempers

Indigenous Lapitoid sherds from Anuta (Dickinson, 1973d) and Tikopia (Dickinson, 1982b) in the Vanuatu backarc region (Fig. 23) contain volcanic temper sands derived from the limited exposures of basalt and basaltic andesite present on those two tiny islets (Fryer, 1974). Tikopia is the remnant of a single Pleistocene volcano, and outcrops of volcanic rock on Anuta occupy less than a square kilometer (Hughes et al., 1981, p. 75–90). Indigenous Anuta and Tikopia temper types have broad affinity with the oceanic basalt temper class, and reflect derivation from volcanic edifices erupted within the oceanic region behind the intraoceanic New Hebrides island arc (Dickinson, 2001a). Archaeological horizons that have yielded indigenous sherds on Anuta (Kirch and Rosendahl, 1973; Kirch, 1976a) and Tikopia (Kirch and Yen, 1982) date from before 0 B.C./A.D. (Kirch, 1982b, 1984) and are unrelated to the modern aceramic culture of the current Polynesian outlier populations.

Anuta tempers include closely related volcanic and hybrid sands in which the only terrigenous grains are glass-rich volcanic lithic fragments, with internal textures ranging from vitric to hyalopilitic (microlitic) and probably derived from local ash deposits. Lithic fragments commonly contain microphenocrysts of plagioclase, as do the volcanic rocks exposed on Anuta. Subordinate clinopyroxene, rare olivine, and sparse oxyhornblende are also present as microphenocrysts in volcanic lithic fragments of the temper sands, as tiny individual crystals isolated within the clay pastes of some Anuta sherds, and within volcanic rocks of Anuta.

Hybrid sand tempers (75%–90% calcareous grains) in Lapitoid sherds of the Kiki ceramic phase on Tikopia contain subordinate terrigenous grain types indicative of local derivation. Plagioclase feldspar and dominantly microlitic (pilotaxitic) volcanic lithic fragments are most abundant, with subordinate clinopyroxene and olivine mineral grains and minor opaque iron oxide grains also present. Subequal proportions of olivine and clinopyroxene in Tikopia tempers (OLi ≈ 50) contrast with other Vanuatu sherd suites in which olivine is absent or minor (Dickinson, 2001a). Younger sherds representative of the Sinapupu ceramic phase on Tikopia contain olivine-free temper sands that document ceramic transfer of Mangaasi-style wares from islands along the main Vanuatu island chain lying ~200 km to the west (Kirch, 1982a, 1986), and were discussed previously with Vanuatu arc tempers.

Horne Islands Tempers

Lapita and Lapitoid sherds from Futuna and Alofi (Kirch, 1976b, 1981; Sand, 1990, 1993) in the Horne Islands along the northern Melanesian borderland northeast of Fiji (Fig. 8) contain slightly different but related tempers (Dickinson, 1976a) derived from sources in local island bedrock. The Horne Islands occupy a fault block uplifted along a constraining bend in the North Fiji fracture zone (Pelletier et al., 2000, 2001), and expose three superposed volcanic assemblages (Grzeszyk et al., 1987, 1991). The island foundations are subalkalic Pliocene pillow lavas and pillow breccias erupted within an interarc or backarc basin by submarine basaltic volcanism following reversal of Vitiaz arc polarity that initiated rotational migration of the New Hebrides island arc (Figs. 8 and 9). Overlying subaerial volcanic assemblages, also Pliocene in age, include (1) andesitic lavas and breccias erupted near the northern end of the Tongan island arc after opening of the interarc Lau Basin and (2) younger basaltic flow-breccias of marginally alkalic affinity representing postarc volcanism during rotation of the Fiji platform away from the ancestral Vitiaz arc trend.

Futuna-Alofi tempers include a spectrum of lithic, feldspathic, and ferromagnesian placer sands (Fig. 31) of varied sedimentological origin, as well as unstudied calcareous tempers. In Futuna sherds, rounded and well-sorted placer tempers are interpreted as beach sands, rounded but only moderately sorted feldspathic tempers as stream sands, and partly weathered lithic tempers as residual volcanic detritus forming natural temper within colluvial or alluvial clays. Dominantly felsitic internal textures of volcanic lithic fragments in Alofi tempers, as opposed to dominantly microlitic and lathwork internal textures of volcanic lithic fragments in Futuna tempers, imply different origins, as do abundant microphanerite grains derived from hypabyssal igneous rocks in the Alofi tempers. Consistent minor proportions of quartz (QZi < 20) in Futuna-Alofi temper sands were apparently derived from quartz veinlets in altered volcanic bedrock, for the local volcanic assemblages lack any phenocrystic quartz. Feldspar grains are exclusively plagioclase (PFi = 100), and ferromagnesian silicate mineral grains are predominantly clinopyroxene (OLi ≈ 0; PYi ≈ 100; PXi ≈ 100), although rare grains of olivine, hornblende, and orthopyroxene are also present. Opaque iron oxide grains are subordinate (OXi < 26) in nonplacer temper sands (Table 16A12), but more concentrated (OXi = 43–58) in placer temper sands (Table 16A13).

Modern Futuna beach sands collected in 2000 are hybrid aggregates (generally <10% calcareous grains) that do not provide a close compositional match for the sherd tempers, although sparse quartz grains confirm the presence of quartz in detritus derived from local bedrock. The terrigenous fractions of modern beach sands either contain distinctly less plagioclase feldspar (<10% by grain frequency) in placer sands, or are more lithic (>90% volcanic lithic fragments) than any of the Futuna-Alofi temper sands. As thin sections of Futuna and Alofi sherds studied more than a quarter century ago have been misplaced over the years, direct petrographic comparison of modern beach sands with the sherd tempers is not possible at present.

Postarc Fiji-Lau Tempers

Large areas of the emergent Fiji platform and offshore basins (Fig. 26) are covered by mafic volcanic assemblages erupted after reversal of Vitiaz arc polarity during sinistral rotation of the Fiji platform westward from the northern extremity of the Tongan arc-trench system (Dickinson, 2001a). A latest Miocene (ca. 5.5 Ma) tectonic and volcanic transition marked the shift from intraoceanic arc activity (Gill, 1987) to postarc magmatism triggered by incipient structural rupture of the Fiji platform during tectonic rotation (Gill, 1976; Gill and Whelan, 1989a, 1989b).

To the east, a tectonic and volcanic transition with the same timing (ca. 5.5 Ma) on the Lau remnant arc (Fig. 28) marked the initiation of seafloor spreading within the interarc Lau Basin farther east (Hawkins, 1995) and concomitant structural rupture of the Lau Ridge. Postarc Lau eruptive assemblages include the largely Pliocene (5.5–2.5 Ma) Korobasaga assemblage of subalkalic basalt and the largely Quaternary (younger than 2.5 Ma) Mago assemblage of alkalic basalt (Woodhall, 1985a; Cole et al., 1985).

The two successive phases of Lau postarc volcanism, subalkalic and then alkalic, have counterparts on the Fiji platform to the west (Whelan et al., 1985). Largely Pliocene (5.5–2.5 Ma) volcanic edifices of subalkalic to shoshonitic (K-rich) affinity are widely distributed along the northern flank of Viti Levu, over much of Vanua Levu to the north, along the main Lomaiviti island chain oriented northwest-southeast in the Koro Sea (Fig. 26), and on both Beqa and Vatulele south of Viti Levu (Band, 1968; Coulson, 1976; Rodda, 1976, 1994; Hindle and Colley, 1981). Largely Quaternary (younger than 2.5 Ma) alkalic lavas were later erupted along a younger northeast-southwest trend to form the large island of Taveuni (Cronin and Neall, 2001), smaller Koro in northern Lomaiviti, and a small islet off the coast of Kadavu (Dickinson, 2001a).

The postarc volcanic assemblages of Fiji and Lau were the sources of temper sands at multiple archaeological sites spread across a broad span of diverse geographic settings. Fijian postarc temper suites are best known from Viti Levu and nearby offshore islands (Dickinson, 1971a) because archaeological exploration of Vanua Levu has been severely limited and has been restricted to selected islands within Lomaiviti and Lau. Fiji-Lau postarc tempers are highly varied in both composition and texture, but many are placer beach sands rich in ferromagnesian mineral grains. A majority contain minor olivine, and half contain hornblende as well as clinopyroxene, but few contain orthopyroxene (Table 16B–16C). The temper sands can be related to local sources of postarc volcanic detritus with varying degrees of confidence from site to site.

Navatu Temper

At Navatu on the northeast coast of Viti Levu (Fig. 26), human occupation has spanned the past two millennia (Gifford, 1951; Clark, 1999). Temper sands in post-Lapita Navatu sherds include a spectrum of placered and unplacered aggregates (Figs. 32A and 32B) composed of detritus derived mainly from basaltic lavas and breccias of the Pliocene Rakiraki volcano (Fig. 33). Heterogeneous volcanic lithic fragments in Navatu temper were derived from groundmasses of mafic lavas or breccia blocks with internal textures ranging from lathwork (intergranular and intersertal) through microlitic (pilotaxitic and hyalopilitic) to vitric (mafic brown volcanic glass) or vitrophyric (microphenocrysts set in mafic brown volcanic glass). Mineral grains are dominantly plagioclase and clinopyroxene, although subordinate hornblende ($PYi > 65$) and minor olivine ($OLi = 5–15$) are present as well (Table 16B8). Most Navatu tempers are moderately sorted and subangular to subrounded aggregates of probable stream origin, but better-sorted and better-rounded placer variants of the Navatu temper suite are beach sands, in part hybrid with minor admixtures of calcareous reef debris. Both beach and stream sand tempers were added manually to clay bodies that lacked significant aplastic constituents.

Microphanerite lithic fragments (Fig. A11) of hypabyssal intrusive igneous rock are ubiquitous in Navatu tempers ($VLi = 46–88$), with a typical grain frequency of ~15% (Table 17B). Most of the microphanerite lithic fragments are hornblende microgabbro or microdiorite, but their internal mineralogy includes clinopyroxene (most commonly as crystal cores surrounded by magmatic reaction rims of hornblende), subordinate olivine, rare quartz, and varying proportions of deuteric epidote. The hypabyssal detritus was derived from subvolcanic intrusive plugs injected into the core of the Rakiraki volcano (Fig. 33), and exposed by deep erosion of the altered interior of the volcanic edifice (Seeley and Searle, 1970). The monolithic peak of Uluinavatu, which rises above the Navatu site, is an eroded volcanic neck that marks the central feeder conduit for the Rakiraki volcano. As comparable intrusive bodies are not exposed elsewhere along the north coast of Viti Levu (Hirst, 1965; Rodda, 1976), the microphanerite grains serve as a petrographic label for the Navatu origin of exotic sherds recovered from Tuvalu far to the north (Dickinson et al., 1990).

Subordinate green-brown hornblende, sparse quartz ($QZi < 5$), rare biotite, and minor aggregate epidote mineral grains (Table 17B) in Navatu temper were probably derived from the same subvolcanic hypabyssal intrusions that contributed microphanerite lithic fragments. Epidote is also widespread, however, in altered volcanic rocks throughout the core zone of the Rakiraki volcano (Fig. 33). Temper sands in surface sherds from the offshore islet of Nananu-i-ra (Table 16B6), which is located within the annular outer zone of the Rakiraki volcano where intrusive plugs are absent (Fig. 33), lack the characteristic Navatu lithic fragments of hypabyssal derivation and also lack quartz, hornblende, or epidote mineral grains. Low contents of hornblende and epidote in placer variants of Navatu temper suggest, however, that the good

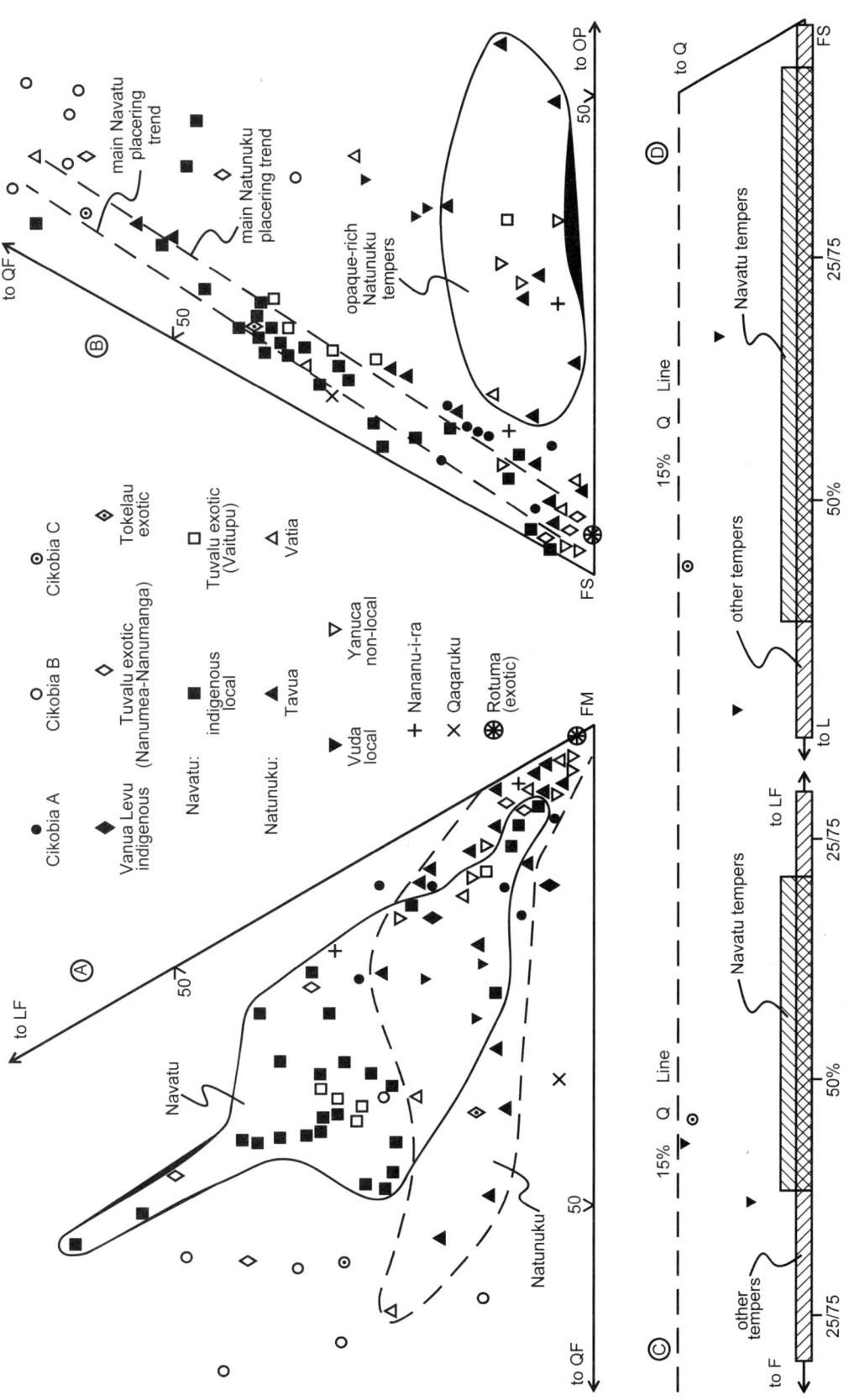

Figure 32. Compositions (grain frequency percentages) of postarc tempers in sherds from northwestern Fiji (Vanua Levu and Viti Levu including associated offshore islets). A: LF-QF-FM diagram. B: QF-FS-OP diagram. C: Q-F-LF diagram. D: Q-F-FS diagram. Strongly placered tempers (FM > 85%) omitted from plot C. See Table 5A for poles.

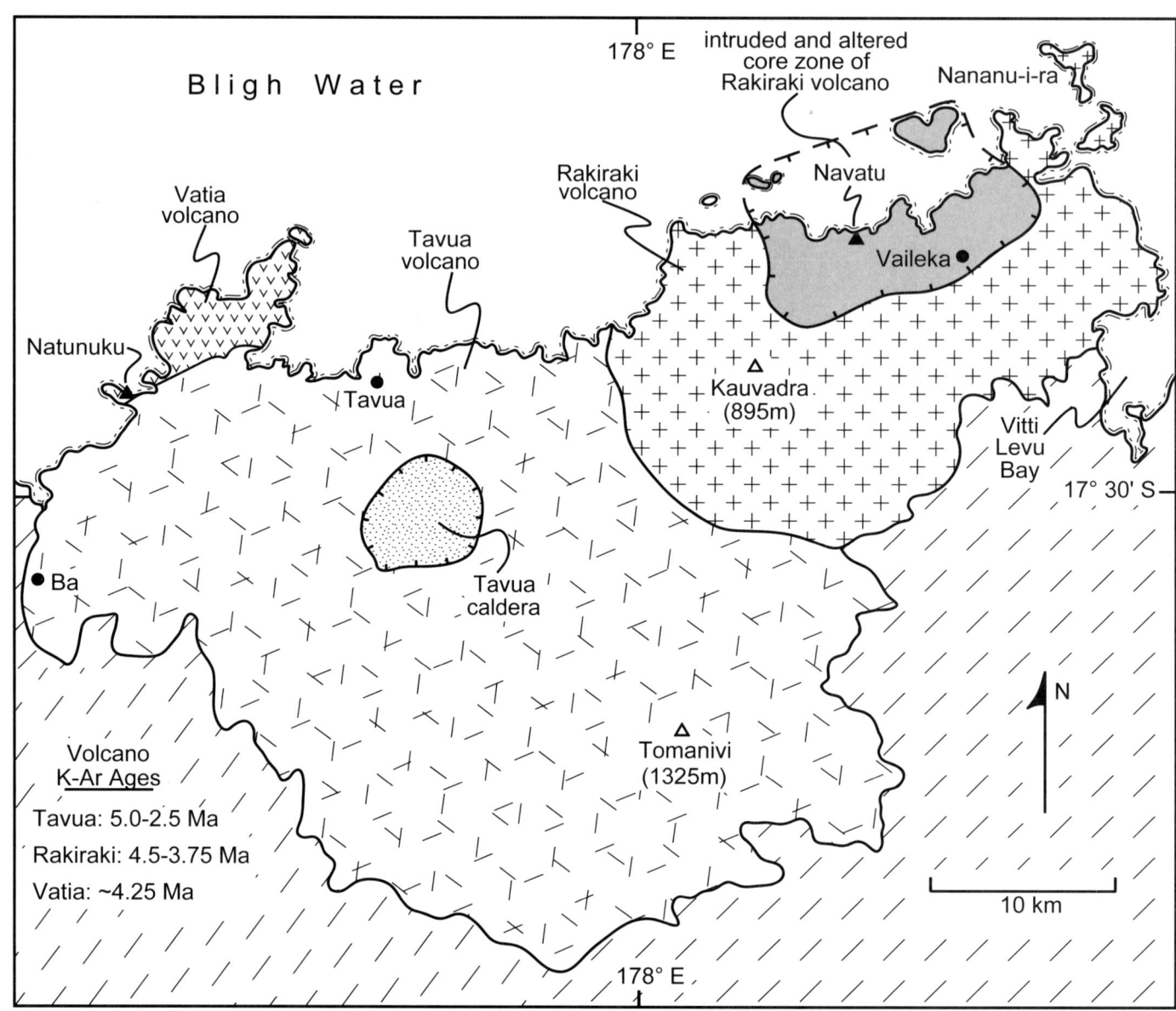

Figure 33. Geologic setting of Natunuku and Navatu archaeological sites on northeast Viti Levu in relation to Pliocene volcanic edifices depicted in contrasting patterns. Dashed diagonal rules denote prevolcano Neogene sedimentary strata (Fig. 26). Adapted after Ibbotson (1962), Hirst (1965), Seeley and Searle (1970), Rodda (1976), and Setterfield et al. (1991).

TABLE 17. KEY SUPPLEMENTAL GRAIN PARAMETERS FOR SELECTED FIJIAN POSTARC TEMPERS CONTAINING DETRITUS FROM SUBVOLCANIC IGNEOUS INTRUSIONS

Temper type	N	QZi	PFi	VLi	MPH	EPI
A - Cikobia C (exotic from Vanua Levu)	1	22	97	49	16	2
B - Navatu (indigenous on Viti Levu)	25	0–4	100	46–88	1–22	tr–8
C - Navatu (exotic on Vaitupi in Tuvalu)	4	0	100	67–85	7–9	tr
D - Vuda (indigenous on Viti Levu)	3	1–6	94	76–96	3–5	tr–2
E - Kulu Bay (indigenous on Beqa)	4	9–18	95–98	36–46	5–12	1–6
F - Kulu Bay (exotic on Ugaga)	5	11–19	92–97	35–53	9–18	2–6

Notes: N is number of sherds for which petrographic frequency counts of grain types in thin section are tabulated. See Table 6 for definitions of supplemental grain parameters (QZi, PFi, VLi) and Table 16 for other supplemental grain parameters. MPH is percentage (ranges) of hypabyssal (microgranular) igneous lithic fragments (microphanerite), and EPI is percentage (ranges) of aggregate epidote grains.

cleavage of hornblende and the polycrystalline character of the aggregate epidote grains led locally to preferential disintegration of those grain types during intense sedimentary reworking.

Natunuku Tempers

At Natunuku (Davidson et al., 1990; Davidson and Leach, 1993), lying farther west along the north coast of Viti Levu (Fig. 26) near the sulcus between the Pliocene Tavua and Vatia volcanoes (Fig. 33), Lapita sherds contain two mineralogically different temper types interpreted to reflect derivation from the Tavua and Vatia volcanic edifices, respectively (Fig. 32). Approximately half the Natunuku sherds examined in thin section contain pyroxene-rich temper sands (PYi > 95) derived from the flank of the large Tavua volcano (Table 16B12), but a quarter of the studied sherds contain hornblende-bearing temper sands (PYi = 75–90) derived instead from the coastal Vatia volcano (Table 16B13). Tavua and Vatia eruptive products are locally interdigitated in the vicinity of Natunuku (Setterfield et al., 1991).

Both Natunuku temper types contain minor olivine mineral grains in similar proportions (Table 16B12–16B13) that are comparable to the olivine content of Navatu tempers (Table 16B8). Both temper types are also variably placered, but the compositional fields for each lie along trends somewhat different from the Navatu placering trend in compositional space (Figs. 32A and 32B), and the subordinate microphanerite lithic fragments characteristic of Navatu temper are absent from Natunuku tempers. The pyroxene-rich (Tavua-derived) and hornblende-bearing (Vatia-derived) Natunuku volcanic sand tempers were probably collected from streams and coastal strands accessible from the Natunuku site on foot or by paddle canoe. A quarter of the Natunuku sherds examined in thin section contain nonlocal quartz-bearing dissected orogen tempers reflecting ceramic transfer from farther south on Viti Levu, as discussed further in a later section of this report.

Pyroxene-rich Natunuku tempers (Fig. A25) are moderately sorted and subangular to subrounded aggregates suggestive of origin as stream sand, perhaps from the Ba River, which reaches the coast near Natunuku. Uplands southeast of the site are underlain by generally basaltic lavas and breccias forming the eroded flank of the Tavua volcano (Ibbotson, 1962, 1967), which erupted a K-rich volcanic assemblage including rocks termed absarokite and shoshonite as proportions of plagioclase and clinopyroxene phenocrysts vary in relative abundance (Setterfield et al., 1991). Phenocrystic mineral grains in derivative temper sands include clinopyroxene, which is the most abundant grain type in most cases, subordinate olivine typically altered on grain margins and along internal fractures to bright red iddingsite, and variable proportions of plagioclase feldspar, which typically includes tiny inclusions of volcanic glass indicative of volcanic derivation but is only a minor constituent of placer variants. Internal textures of volcanic lithic fragments are commonly either microlitic (hyalopilitic) or vitrophyric, with plagioclase microlites or microphenocrysts set in mafic brown volcanic glass.

Hornblende-bearing Natunuku tempers (Fig. A17) are volcanic sands in which plagioclase, clinopyroxene, and olivine mineral grains are indistinguishable in appearance from counterparts in the pyroxene-rich Natunuku tempers, but in which prominent green hornblende mineral grains imply derivation from the small but nearby Vatia volcano. Volcanic rocks of the coastal Vatia volcano (Fig. 33) are termed Vatia Andesite (Rodda, 1994) because they are less mafic and more calc-alkalic in petrology than the absarokite-shoshonite assemblage of the Tavua volcano (Ibbotson, 1962; Setterfield et al., 1991). Vatia Andesite contains prominent dark green hornblende phenocrysts indistinguishable optically from the green hornblende mineral grains in hornblende-bearing Natunuku temper sands. Internal textures of volcanic lithic fragments in hornblende-bearing Natunuku tempers are most commonly microlitic (pilotaxitic), with felted plagioclase microlites and little or no interstitial volcanic glass. Although only moderately sorted and subrounded to subangular aggregates, some of the hornblende-bearing Natunuku tempers contain calcareous grains of reef debris that are diagnostic of coastal origin, perhaps from a lagoonal setting where wave energy was low.

Vuda Temper

Indigenous local temper sands in post-Lapita sherds from Vuda (Gifford, 1951) on the west coast of Viti Levu (Fig. 26) are similar compositionally to Navatu and Natunuku tempers (Fig. 32A), but differ somewhat in detail (Dickinson, 1971a). Vuda Creek and the nearby Sabeto River tap bedrock sources in a shoshonitic volcanic assemblage (Dickinson et al., 1968) more felsic in composition (Koroimavua-Sabeto volcano) than the dominantly mafic igneous rocks exposed farther east on northern Viti Levu (Dickinson, 2001a), and also acquire intrusive igneous detritus from subvolcanic monzonitic stocks intrusive into the Koroimavua volcanic assemblage in the adjacent Sabeto Range (Dickinson, 1968b). Vuda temper sands, though derived mainly from volcanic sources, contain minor quartz (Table 16B15), K-feldspar (Table 17D), and biotite (grain frequency ~1%) derived from intrusive igneous rocks. Lithic fragments include minor microphanerite from the intrusive sources (Table 17D), together with dominantly microlitic (hyalopilitic) volcanic lithic fragments, which are more abundant (VLi = 76–96). Typical Vuda temper sands are variably sorted and subangular to subrounded aggregates probably derived from placer deposits along Vuda Creek or some nearby stream.

Beqa Tempers

Indigenous tempers in sherds from multiple sites on Beqa, a large island off the south coast of Viti Levu (Fig. 26), are ferromagnesian volcanic sands that resemble tempers from the north coast of Viti Levu because the offshore Pliocene Beqa volcano (5–3 Ma) is petrologically similar to coeval volcanic edifices of northern Viti Levu (Band, 1968). Beqa tempers are typically moderately sorted and subrounded to subangular placer aggregates (Fig. 34) probably derived from local island beaches, as implied by the presence in selected sherds of sparse calcareous

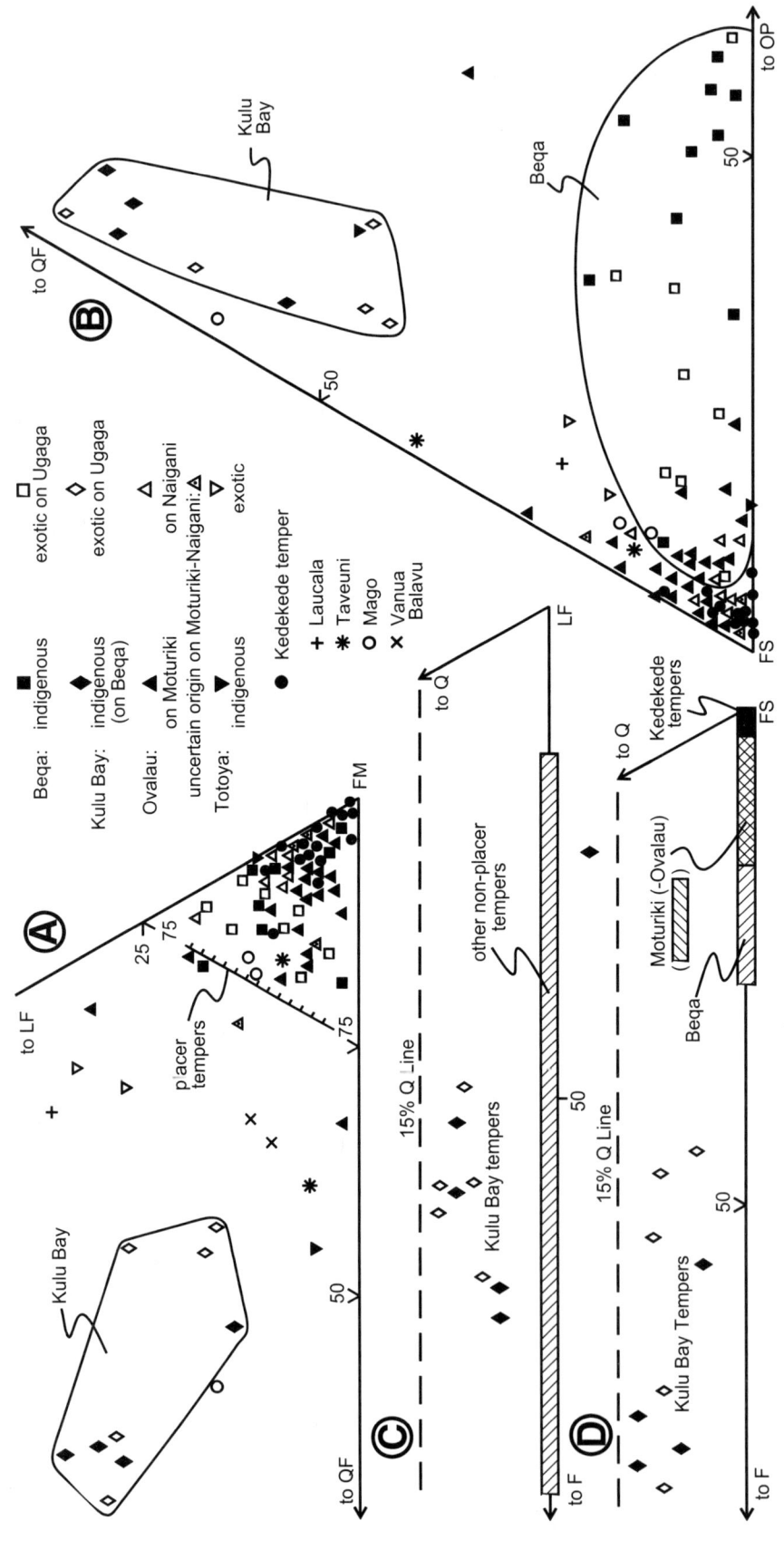

Figure 34. Compositions (grain frequency percentages) of postarc tempers in sherds from islands in southeastern Fiji (Beqa-Ugaga, Lomaiviti, Lau). A: LF-QF-FM diagram. B: QF-FS-OP diagram. C: Q-F-LF diagram. D: Q-F-FS diagram. Opaque-rich tempers (FM > 85%) omitted from plot C. Strongly placered tempers (n = 3) omitted from plots B and D. See Table 5A for poles.

reef debris or sand-sized vacuole pores formed by the dissolution of calcareous grains. Clinopyroxene mineral grains are predominant among ferromagnesian silicate constituents, although subordinate hornblende, sparse olivine, and rare biotite (trace amounts only) are also present (Table 16C7). Minor volcanic lithic fragments are dominantly vitric varieties composed of mafic brown volcanic glass.

Both Lapita and post-Lapita sherds recovered from sites at Kulu Bay on the southwest coast of Beqa (Clark, 1999) contain feldspathic temper sand that is less ferromagnesian in composition than other Beqa tempers (Figs. 34A and 34B). The Kulu Bay temper type was apparently derived from a local plug or stock of feldspar-rich intrusive shoshonite forming the only large exposure (~1.5 km^2 surrounding Kulu Bay) of an exhumed subvolcanic hypabyssal intrusion on Beqa (Band, 1968). Nonlocal placer tempers, in part hybrid, in selected Kulu Bay sherds closely resemble temper sands in sherds from elsewhere on the island, and imply ceramic transfer on some scale within Beqa.

The distinctive Kulu Bay temper type derived in large part from hypabyssal intrusive igneous rock contains more green hornblende than clinopyroxene (Table 16C9), minor quartz and K-feldspar (Table 17E–17F), and rare biotite (grain frequency ~1%) and epidote (Table 17E–17F), as well as subordinate lithic fragments that are dominantly nonvolcanic microphanerite detritus (Table 17E–17F) of hypabyssal origin (VLi = 36–46). These grain types, present also in Vuda temper, are jointly a signal of detrital contributions from intrusive igneous rock. Although the nature of the dominant plagioclase feldspar mineral grains in Kulu Bay tempers is not diagnostic of specific origin, volcanic or nonvolcanic, the accompanying hypabyssal detritus points to the plagioclase-rich Kulu Bay stock as the predominant bedrock source of the plagioclase as well. Typical volcanic lithic fragments in Kulu Bay temper were derived from volcanic rocks with groundmasses that were originally vitric, composed mainly of mafic tan to red-brown volcanic glass, but were largely converted (devitrified) to cryptocrystalline alteration products before erosion. Relict microlitic internal textures remain visible in some volcanic lithic fragments. The characteristic pervasive alteration of volcanic materials suggests metamorphic effects of the intrusive Kulu Bay stock near contacts with adjacent volcanic country rock, and some of the lithic fragments identified as microphanerite could possibly be metavolcanic rock instead.

Sherds from excavations on Ugaga (Clark, 1999), a tiny islet (<0.05 km^2) lying off Beqa only 3 km southwest of Kulu Bay, date from the immediately post-Lapita phase of Fijian prehistory. Their temper sands document ceramic transfer from Kulu Bay (Table 16C10) or other localities on Beqa (Table 16C8) including Vaga Bay on the west coast where notably hornblende-rich volcanic rocks are exposed (Band, 1968). A Vaga Bay origin for some Ugaga sherds containing abundant green-brown hornblende mineral grains best accounts for the lower pyribole index (PYi) of tempers in selected Ugaga sherds, as opposed to typical Beqa tempers (Table 16C7–16C8). Although Ugaga is a bedrock islet exposing part of the volcanic succession of the Beqa stratocone, its small size and the variety of temper types present in Ugaga sherds make it unlikely that any wares were made with local temper sand.

Cikobia Sherds

Lapita and younger sherds from Cikobia (Sand and Valentin, 1998), the northernmost Fijian island (Fig. 26), contain volcanic sand tempers (Fig. 32) interpreted to reflect ceramic transfer from nearby Vanua Levu. Cikobia is underlain primarily by uplifted Pliocene-Pleistocene limestone (Woodhall, 1985b), with exposures of subjacent volcanic rock restricted to a short segment (~2.5 km) of the northeast coast. Cikobia volcanic rocks are approximately the same age (ca. 7 Ma) as rhyodacitic volcanic rocks of the nearby Udu Peninsula on Vanua Levu but are basaltic to andesitic pillow lava and hyaloclastic breccia, with intercalated finer-grained volcaniclastic strata derived from similar eruptive sources. Quartz-bearing felsic volcanic sand tempers in some Cikobia sherds document ceramic transfer from the Udu Peninsula to Cikobia, as discussed above with descriptions of Vitiaz arc tempers within Fiji, but most Cikobia sherds examined in thin section contain more mafic volcanic sand tempers thought to reflect systematic ceramic transfer from elsewhere on Vanua Levu. Because Vanua Levu tempers are still poorly known (Parke, 2000), with only two sherds examined to date in thin section (Table 16B4), no definitive temper ties to specific locales on Vanua Levu can be suggested with present information.

The non-Udu tempers in Cikobia sherds are grouped here as Cikobia type A, type B, and type C tempers (Table 16B1–16B3), with the origin of each uncertain. If any of the three temper types is indigenous to Cikobia, the partly hybrid placer sand of Cikobia type A temper is the most logical candidate, given the geology of Cikobia. The more feldspathic Cikobia type B temper is also consistently hybrid sand probably derived from some coastal locale or locales on nearby Vanua Levu. Vanua Levu igneous assemblages are dominantly postarc in character (Fig. 26), as all except the Udu Peninsula arc assemblage are younger than 6 Ma, but their petrologic affinities are not well known. Olivine is less abundant in all three Cikobia temper types (OLi < 2) than in the well-studied Navatu and Natunuku postarc tempers from the north coast of Viti Levu (OLi > 2). Cikobia type C temper (present in only one studied sherd) contains grain types indicative of derivation in large part from hypabyssal intrusive igneous rock perhaps exposed somewhere in the interior of Vanua Levu where geologic mapping has been reconnaissance in nature (Ibbotson, 1969; Rickard, 1970; Coulson, 1971; Hindle, 1976; Woodrow, 1976).

Cikobia type A tempers are well-sorted and subrounded volcanic placer sands of beach origin in which clinopyroxene mineral grains are the predominant grain type, although orthopyroxene, olivine, and hornblende are all present in trace amounts in selected sherds. Subordinate grain types include opaque iron oxide grains, plagioclase mineral grains, and volcanic lithic fragments, dominantly vitric to microlitic. Cikobia type B temper is comparably well sorted and subrounded, but distinctly less placered, with plagioclase mineral grains the dominant grain type. Clinopyroxene is the most abundant ferromagnesian mineral grain present, with orthopyroxene present in trace amounts in selected sherds, but both olivine and hornblende are wholly absent. Other subordinate grain types include opaque iron oxide grains, polycrystalline quartzose or chalcedonic aggregate grains probably derived from amygdules or veinlets in volcanic rocks, and volcanic lithic fragments, which include felsitic as well as vitric and microlitic varieties. Generic differences in the grain types of Cikobia type A and Cikobia type B tempers argue that the placer aggregates of the former were not derived from unplacered sand similar to the latter, and imply ceramic transfer from more than one locale on Vanua Levu (or possibly indigenous Cikobia origin for the placer temper sands).

Cikobia type C temper is compositionally a more heterogeneous sand (Tables 16B3 and 17A), not derived from exclusively volcanic source rocks, and is a moderately sorted and subangular to subrounded aggregate of probable stream origin. Clinopyroxene mineral grains are also the dominant ferromagnesian grain type, however, with minor hornblende ($PYi = 95$) and orthopyroxene ($PXi = 95$) also present. Other varied grain types include monocrystalline quartz mineral grains ($QZi = 22$) of plutonic or hypabyssal quartz with trains of internal microscopic vacuoles and slightly undulatory extinction, plagioclase mineral grains (both twinned and untwinned and both fresh and altered), K-feldspar mineral grains ($PFi = 97$) including unmixed perthite, polycrystalline aggregate epidote grains, microphanerite lithic fragments of intrusive igneous rock ($VLi = 49$), and volcanic lithic fragments that are predominantly microlitic but also include minor vitric and felsitic varieties. The complete absence of sedimentary or metasedimentary lithic fragments ($SLi = 0$) precludes correlation of Cikobia type C temper with any dissected orogen tempers of Viti Levu (discussed further in later passages). Supplemental grain parameters are similar to those for Kulu Bay temper from Beqa (Table 17E–17F), except for the much greater relative abundance of hornblende in the latter (Table 16C9–16C10), and suggest derivation from a subvolcanic stock geologically similar to the local Kulu Bay stock, but located somewhere on Vanua Levu. Small intrusions of diorite and quartz-bearing tonalite are known from eastern Vanua Levu south of the Udu Peninsula (Ibbotson, 1969) and may represent the source of nonvolcanic detritus in Cikobia type C temper.

Northern Lomaiviti

Tempers in two-thirds (23 of 33) of the sherds examined in thin section from the Naitabale Lapita site (Kumar et al., 2003, 2004b) on Moturiki, an island east of Viti Levu near larger Ovalau (Fig. 26), are related volcanic sands (Table 16C5) of placer character (Fig. 34) and probable beach origin. The tempers contain both green clinopyroxene and green-brown hornblende mineral grains ($PYi = 72–90$), and are interpreted as indigenous to either Moturiki or nearby Ovalau. One of two sherds studied from the Saulevu Lapita site (Nunn, 1999) farther north on Moturiki contains a closely similar temper sand. Geologic distinction between Moturiki and Ovalau origins of sand is not feasible because the smaller island of Moturiki is an arcuate remnant of the outer fringe of the Pliocene (4–3 Ma) Ovalau volcano (Ibbotson, 1961). Unstudied Lapita sherds are also known from Ovalau (Nunn and Areki, 2004).

The mafic Ovalau eruptive assemblage is a gradational suite of pyroxenic and hornblendic lavas and breccias of basaltic to andesitic composition that could yield derivative sand comparable mineralogically to the characteristic Naitabale temper type, whereas other islands farther east in northern Lomaiviti lack hornblende-bearing rocks (Coulson, 1976). The absence of more than trace amounts of olivine mineral grains distinguishes Moturiki-Ovalau tempers from olivine-bearing postarc tempers (Navatu, Natunuku) of Viti Levu to the west (Table 16B8–16B9, 16B12–16B13), and also from Moala-Totoya-Matuku tempers (Table 16C11–16C12) of southern Lomaiviti (Fig. 26). The sparse volcanic lithic fragments in Moturiki-Ovalau temper sands are vitric and microlitic varieties composed in whole or in large part of reddish brown mafic volcanic glass.

Approximately half the sherds studied in thin section from the Matanamuani Lapita site on Naigani (Kay, 1984), a tiny islet <10 km northwest of Ovalau (Fig. 26), also contain placer tempers (Table 16C4) of Moturiki-Ovalau type, with both pyribole index and oxide index similar to those in Moturiki sherds ($PYi = 73–98$ versus $PYi = 72–90$ and $OXi = 2–10$ versus $OXi = 1–13$, respectively). The tempers could not derive from Naigani, where the exposed bedrock is largely aphyric pyroxene dacite composed dominantly of closely spaced plagioclase microlites consistently oriented in a fluidal fabric (Ibbotson, 1961; Coulson, 1976). A sample of beach sand from Naigani contains no hornblende but abundant volcanic lithic fragments (30% grain frequency) of remarkably uniform microporphyritic dacite with a microlitic groundmass composed predominantly of parallel plagioclase microlites. The Naigani beach sand closely matches the lithology of bedrock sources of terrigenous detritus on Naigani, but is markedly different from the Moturiki-Ovalau temper sand in Naigani sherds. Ceramic transfer from Moturiki or Ovalau is indicated for much of the Naigani sherd suite, but exotic sherds from elsewhere are also present on Naigani as well as at Moturiki, as described in a later section of this report discussing ceramic transfer.

Southern Lomaiviti

Most of the few sherds (four of six) studied to date from Totoya (Clark et al., 1999) contain temper sands with compositions (Fig. 34) compatible with derivation from islands in southern Lomaiviti (Fig. 26), but not closely resembling any known tempers from elsewhere in Fiji. Two surface sherds contain exotic tempers that reflect ceramic transfer from elsewhere in Fiji, and are described in a later section of this report discussing ceramic transfer.

The temper sands in two excavated Totoya sherds (Table 16C11) are well-sorted and rounded to subrounded hybrid aggregates (43% and 79% calcareous grains, respectively) inferred to represent beach sands indigenous to Totoya. The presence of minor olivine mineral grains (OLi = 7–16) and the absence of hornblende mineral gains (PYi = 100) distinguishes Totoya temper from Moturiki-Ovalau temper of northern Lomaiviti, but is in accord with the presence of olivine and the absence of hornblende in the Totoya volcanic assemblage (Coulson, 1976).

The temper sands in two surface sherds (Table 16C12) from Totoya are unplacered aggregates containing minor amounts of both olivine (OLi = 4–8) and green hornblende (PYi = 86–91) mineral grains, and are inferred to derive from nearby Moala or Matuku (Fig. 26) where hornblende occurs in dike rocks cutting the local volcanic assemblages (Coulson, 1976). One of the tempers containing sparse hornblende is hybrid beach sand (23% calcareous grains), well sorted and rounded to subrounded, but the other is a moderately sorted and subangular to subrounded aggregate of probable stream origin.

Lau Ridge

Postarc volcanic sand tempers from islands along the crest of the Lau Ridge remnant arc (Fig. 28) include indigenous tempers in sherds from the Votua Lapita site on Mago (Clark et al., 2001), presumably indigenous tempers in surface sherds from Laucala (near Taveuni) and Vanua Balavu (Nunn and Matararaba, 2000; Thomas et al., 2004), and kedekede placer temper (Table 16C15) in exotic sherds recovered from archaeological sites located both in Lau and elsewhere within Fiji (Dickinson and Shutler, 2000). Kedekede temper was named from its first discovery in sherds from the Kedekede hillfort on Lakeba (Rowland and Best, 1980), but without any implication that kedekede temper derives from Lakeba.

Sherds from Mago contain both largely calcareous sand (n = 3) and largely volcanic sand (n = 3) tempers, both unplacered and placered, with the latter containing rare calcareous grains (<3% by grain frequency). All the Mago temper sands (Table 16C14) are well-sorted aggregates interpreted as Mago beach sands, with the calcareous grains rounded to subrounded and the terrigenous grains subrounded to subangular. In the placer sands, clinopyroxene mineral grains are dominant (two-thirds of the grain aggregates by grain frequency), with sparse olivine mineral grains (OLi < 5) but no orthopyroxene or hornblende mineral grains present. Volcanic lithic fragments are exclusively microlitic or red-brown vitric varieties derived from mafic volcanic rock. The grain population is compatible with derivation from Quaternary alkalic basalt bedrock of Mago (Woodhall, 1985a). Mago placer tempers are distinguished from kedekede temper not only by their hybrid character, but by distinctly higher proportions of plagioclase feldspar mineral grains and volcanic lithic fragments (Figs. 34A and 34B). Tempers in the two surface sherds sectioned to date from nearby Vanua Balavu (Fig. 34A) are distinguished by an extreme relative abundance of opaque iron oxide grains with respect to ferromagnesian silicate mineral grains (OXi > 95), and also lack any olivine mineral grains (Table 16C13). The Vanua Balavu temper was probably derived from extensive local exposures of subalkalic Pliocene basalt forming the Korobasaga volcanic assemblage of Lau (Woodhall, 1985a).

A single sherd from Laucala, an islet north of Lau off the east end of Taveuni (Fig. 26), contains moderately sorted volcanic sand temper (Table 16C1) in which olivine mineral grains as well as clinopyroxene mineral grains are prominent (OLi = 13), and abundant volcanic lithic fragments (Fig. 34A) are dominantly (~90%) vitric varieties composed of red-brown mafic volcanic glass. The temper sand was derived from the alkalic Taveuni volcanic assemblage of Quaternary age exposed on Laucala (Woodhall, 1985b). The only Taveuni sherds available for this study are surface sherds containing volcanic sand tempers (Figs. 34A and 34B) that lack olivine mineral grains (Table 16C2). A broader qualitative survey of Taveuni tempers (Babcock, 1977) reported abundant olivine in most Taveuni sherds, implying that lack of olivine is atypical for Taveuni tempers.

DISSECTED OROGEN TEMPERS

Dissected orogen tempers (Dickinson, 1998a; Dickinson and Shutler, 2000) are found on islands where sand is derived partly from the intrusive plutonic roots of arc crust exposed by deep erosion of extinct island arcs. In early publications, the same temper types were less aptly termed "volcaniclastic orogen" tempers (Dickinson and Shutler, 1968), "volcano-plutonic orogen" tempers (Dickinson and Shutler, 1971), and "volcanic-plutonic orogen" tempers (Dickinson and Shutler, 1979).

Bedrock sources for the mineralogically complex dissected orogen temper sands (Figs. A12 and A19) include thick successions of arc volcanic rocks, associated volcaniclastic strata, granitic to dioritic and gabbroic plutons, metavolcanic and metasedimentary wall rocks of the plutons, and in some cases derivative sedimentary successions unconformably overlying the orogenic terranes. Polycrystalline and polymineralic lithic fragments include varied volcanic-metavolcanic varieties, microphanerite derived from hypabyssal and plutonic intrusions, and a sedimentary to metasedimentary spectrum of grain types including chert and metachert, argillite or shale, slate or phyllite, and hornfels from contact aureoles. Mineral grains include (1) quartz and both feldspars (plagioclase, commonly altered to albite, and K-feldspar including microcline) and (2) ferromagnesian minerals that include biotite, hornblende, clinopyroxene, and epidote in varying proportions.

In dissected orogen tempers, the proportions of detritus derived from plutonic or metamorphic and supracrustal volcanic or sedimentary sources are widely variable, and the tempers grade to andesitic arc tempers as contributions from subvolcanic source rocks decrease. In practice, the presence of plutonic or metamorphic quartz grains (QZi > 0), K-feldspar of plutonic origin (PFi < 100), combined volcanic and sedimentary lithic fragments (VLi < 100; SLi > 0), and microphanerite lithic fragments in significant quantities are taken as diagnostic of dissected orogen tempers. Alternate origins of plagioclase from plutons or their volcanic cover is problematical and difficult to gauge, although abundant diagenetic or metamorphic albite formed by alteration of original plagioclase is characteristic of dissected orogen tempers. Aggregate epidote grains from diagenetic or metamorphic sources are also typical of dissected orogen tempers, but are rare in temper sands of strictly volcanic derivation.

The salient examples of dissected orogen tempers derive from the south coast of Viti Levu in Fiji (Fig. 26) where streams draining the Eocene-Miocene Wainimala orogen of interior Viti Levu come down to the sea (Dickinson, 2001a). Limited examples also occur within the Bismarck Archipelago (Fig. 7) where assemblages of cogenetic Paleogene volcanic and plutonic rocks have been exposed to erosion (Dickinson, 2000a), on Muyua (Woodlark) in the Solomon Sea (Fig. 19) where granodiorite is locally exposed, and in the Torres Strait Islands (Fig. 4) where the Paleozoic Tasman orogenic belt of eastern Australia projects northward beneath the Arafura Sea. Dissected orogen tempers could probably be derived from several of the central Solomon islands (Fig. 7), and occur locally as exotic temper, but no indigenous ceramic assemblages are known from the appropriate islands (Reeve, 1989).

Torres Strait Temper

Three sherds from a limited collection recovered from Mask Cave (McNiven et al., 2006) on Mabuiag (David et al., 2004) in the northern Torres Strait Islands (Fig. 4) between Australia and New Guinea contain more abundant temper than is characteristic for Oceanian sherds. In parts of each sherd, clay paste is hardly more than interstitial to temper grains, and the sherds could be mistaken for samples of clayey sandstone were it not for the presence of red slip on some sherd surfaces. The tempers are well-sorted aggregates of subangular to subrounded grains of uniformly fine sand with a heterogeneous composition dominated by volcanic lithic fragments (VLi ≈ 95), mainly vitric, and monocrystalline quartz grains derived from volcanic and plutonic igneous rocks of felsic character. Plagioclase and K-feldspar mineral grains are subordinate. In each sherd, there are also sparse outsized ferruginous objects, spherical to ovoid in shape (0.5–1.25 mm diameter), which are probably pedogenic nodules collected as an integral part of the clay bodies used by the ancient potters.

Bedrock exposures in the Torres Strait Islands are igneous rocks that represent a northward continuation of the Paleozoic Tasman orogenic belt of eastern Australia (Willmott et al., 1973; Hamilton, 1979, p. 227; von Gnielinski et al., 2002). A thick (>300 m) succession of Upper Carboniferous to Lower Permian volcanic rocks is dominantly massive welded tuff intruded by cogenetic granitic bodies that also date to near the Carboniferous-Permian boundary (302–286 Ma). The Paleozoic igneous and metamorphic rocks are overlain on selected islands by undeformed Cenozoic sandstone of shelf deposits presumably derived from subjacent bedrock sources.

The dominance of quartz grains and volcanic lithic fragments in the temper sands is compatible with ultimate derivation from the volcanic-plutonic assemblage of the Torres Strait Islands, but the excellent sorting of the temper sands is striking, and seemingly precludes first-cycle derivation from bedrock outcrops. Derivation from the overlying Cenozoic shelf strata is an attractive alternative and suggests the hypothesis that the superabundant temper sand is natural temper collected, together with the clay pastes, from weathered exposures of the Cenozoic deposits, which are described as dominantly clayey sandstone. Sand fractions extracted to date from clay samples collected on Mabuiag are coarser-grained and more quartz-rich than the sherd tempers, but do not preclude derivation of indigenous temper sands from unsampled localities in the Torres Strait Islands.

Aitape Coast Tempers

Fifty sherds from the Aitape coastline (Fig. 4) of Papua New Guinea were examined (Dickinson, 2006c) to establish whether any evidence for ceramic transfer from early cultural centers along the north coast of New Guinea (Terrell and Welsh, 1990) is provided by sand tempers in prehistoric sherds from the Bismarck Archipelago to the east. To the contrary, Aitape tempers are uniformly unlike any tempers in sherds studied from the Bismarck Archipelago, and support the interpretation that pottery-making flowed inland into New Guinea from the Lapita homeland of the Bismarck Archipelago, rather than the reverse (Pétrequin and Pétrequin, 1999).

Aitape tempers from the coastal flank of the Torricelli Mountains, an intraoceanic island arc accreted to the northern flank of New Guinea in Miocene time (Fig. 9), are mineralogically complex sands in which mineral grains include plutonic quartz and both feldspars (plagioclase and K-feldspar), including both fresh plagioclase and plagioclase altered to inclusion-rich albite, and subequal proportions of pyriboles and epidote. Lithic fragments include a diverse spectrum of volcanic and metavolcanic types, microphanerite of hypabyssal and plutonic derivation, and a range of sedimentary to metasedimentary types including quartzite (polycrystalline quartz), argillite, microgranular siltstone or hornfels, tectonite, and chert-metachert. The array of grain types is appropriate for derivation of the temper sands from bedrock forming the variably metamorphosed volcanogenic core of the Torricelli Mountains and associated ranges intruded locally by comagmatic plutons and masked locally by derivative orogenic sedimentary successions (Hamilton, 1979; Pigram and Davies, 1987; Cullen and Pigott, 1989).

Aitape tempers differ generically from all volcanic sand tempers (arc, postarc, backarc) and dissected orogen tempers known from the Bismarck Archipelago in the following respects: (1) comparatively high contents of K-feldspar and epidote mineral grains, (2) both overall abundance and extreme heterogeneity of lithic fragments, and (3) the consistent presence of subordinate but prominent metavolcanic and metasedimentary lithic fragments.

Bismarck Archipelago

The Paleogene volcanogenic assemblage of the ancestral Vitiaz island arc forming the basement complex of all the principal islands of the Bismarck Archipelago (Fig. 19) is widely intruded by subvolcanic dioritic to granitic plutons of variable size that are best known on New Britain (Page and Ryburn, 1977;

Hine and Mason, 1978; Whalen and McDougall, 1980; Whalen, 1985). Although some andesitic arc tempers in exotic Mussau sherds thought to derive from the Vitiaz arc assemblage contain minor proportions of quartz probably contributed from limited exposures of subvolcanic plutons (Figs. 20B and 20C; Table 13A5–13A6), quartzose temper sands of the dissected orogen class are known to date within the Bismarck Archipelago only in sherds from New Britain (Fig. 19) or thought to derive from New Britain. Even these tempers contain a lower proportion of plutonic detritus than dissected orogen tempers from Viti Levu in Fiji, and can be regarded as gradational between the andesitic arc and dissected orogen temper classes of Oceania as a whole. The absence of K-feldspar grains and the rarity of nonigneous lithic fragments in Bismarck dissected orogen tempers (Table 18A) suggest that their nonvolcanic detritus was derived exclusively from subvolcanic dioritic to tonalitic plutons.

The principal known occurrence of dissected orogen tempers within the Bismarck Archipelago is in Lapita sherds studied from Kreslo (Specht, 1991) on the southwest coast of New Britain where streams reach the coast from interior highlands draining Paleogene volcanic and plutonic exposures of central New Britain (Fig. 19). Variably placered Kreslo temper sands (Fig. 35; Table 18A2) are well-sorted and subrounded aggregates of beach origin and are dominantly hybrid sands containing admixtures of calcareous reef debris. Quartz contents ($QZi < 25$) are highly variable, ranging downward to nil for some placer sands, and apparently reflect differing admixtures of plutonic with volcanic detritus. The ratio of pyroxene to hornblende mineral grains is also variable ($PYi = 63$–98), but generally comparable to ratios in typical Bismarck andesitic arc tempers derived from the Vitiaz arc assemblage (Table 13A1–13A3, 13A5–13A7). Orthopyroxene mineral grains are lacking ($PXi = 100$) in half the Kreslo sherds, as for all the andesitic arc tempers thought to derive from the Vitiaz arc assemblage (Table 13A1–13A12), and their presence in the other Kreslo sherds ($PXi = 78$–96) may reflect contamination of Vitiaz detritus with pyroclastic detritus from the magmatic arc of the north New Britain coast.

Elsewhere in the Bismarck Archipelago, dissected orogen tempers are known only from small islets, most of which lie close offshore from New Britain (Fig. 35). A sherd from the Arawe Islands (Table 18A1) off the coast of western New Britain contains a temper sand resembling the temper from nearby Kreslo

TABLE 18. KEY SUPPLEMENTAL GRAIN PARAMETERS FOR DISSECTED OROGEN TEMPERS FROM THE BISMARCK ARCHIPELAGO, SOLOMON ISLANDS, AND FIJI

Temper type	N	QZi	PFi	VLi	SLi	OXi	PYi	EPI
A. Bismarck Archipelago (indigenous on New Britain but exotic on other islands)								
1 - Arawe (exotic)	1	10	100	100	0	7	95	0
2 - Kreslo (New Britain)	15	14 ± 8	100	100	0	35 ± 34	86 ± 11	0
3 - Watom (exotic)	1	3	100	100	0	42*	33*	0
4 - Duke of York (exotic)	4	20 ± 10	100	100	0	39–44	67–95	tr–2
5 - Nissan (exotic)	1	21	100	48	0	23	74	3
B. Solomon Islands (Kolombangara exotic from Choiseul; Roviana exotic of unknown origins)								
1 – Kolombangara	3	23–25	100	75–78	0	8 ± 1	15 ± 7	2–4
2 - Roviana quartzose	4	~48	100	100	0	20–45	~95	0
3 - Roviana quartz-calcite	7	72 ± 4	66 ± 5	100	0	84 ± 4	–	0–2
C. Fiji (indigenous local or non-local on Viti Levu; exotic from Viti Levu on other islands)								
1 - Navatu (non-local)	2	30–44	89–99	51–80	6–14	28–35	36–92	1–2
2 - Natunuku (non-local)	7	31 ± 6	90 ± 2	66 ± 8	6 ± 2	41 ± 8	74 ± 6	tr–2
3 - Vuda (non-local)	5	49 ± 2	90 ± 2	46 ± 9	24 ± 6	18 ± 6	75 ± 10	1
4 - Yanuca (local)	14	48 ± 12	84 ± 4	44 ± 21	22 ± 7	33 ± 8	63 ± 19	1–4
5 - Yadua (local)	1	59	82	24	32	56	67*	1
6 - Sigatoka (local)	28	43 ± 4	87 ± 5	46 ± 7	12 ± 4	33 ± 9	78 ± 6	1–3
7 - Karobo (local)	5	56 ± 3	96 ± 2	60 ± 3	6 ± 2	27 ± 5	37 ± 25	2–4
8 - Nasilai (local)	20	47 ± 4	80 ± 4	34 ± 7	28 ± 6	11 ± 4	54 ± 9	1–3
9 - Beqa (exotic)	4	52 ± 3	79 ± 2	32 ± 4	30 ± 4	16 ± 4	73 ± 5	1–2
10 - Ugaga (exotic)	2	63–66	79–88	19–45	13–15	18*	100*	tr–2
11 - Moturiki (exotic)	3	35 ± 4	87 ± 3	51 ± 9	21 ± 2	31 ± 2	62 ± 10	~5
12 - Totoya (exotic)	1	48	81	47	26	14*	55*	2
13 - Lakeba (exotic)	2	26–41	62–88	38–44	19–31	12–21	67	1–2

Notes: N is number of sherds for which petrographic frequency counts of grain types made in thin section are tabulated. See Table 6 for definitions of supplemental grain parameters (or indices). EPI indicates range in grain frequency percentage of epidote grains. Hyphens indicate range of values observed for pairs of sherds. Standard deviation (±) calculated for means of multiple frequency counts (excluding extreme outlier values in each case). VLi and SLi tabulated only for tempers with LF>10%; OXi and PYi tabulated only for tempers with FM>10% or FS>10%, respectively (asterisks denote OXi and PYi values tabulated for individual sherds not meeting those criteria). OLi=0 except for Roviana quartzose temper (OLi=9-10). Local tempers from sites on south coast of Viti Levu inferred to derive from the vicinity of the indicated sites listed from west to east; non-local tempers from sites on north coast of Viti Levu listed from east to west and inferred to derive from the interior of Viti Levu; exotic tempers from sites on other Fijian islands document ceramic transfer from Viti Levu.

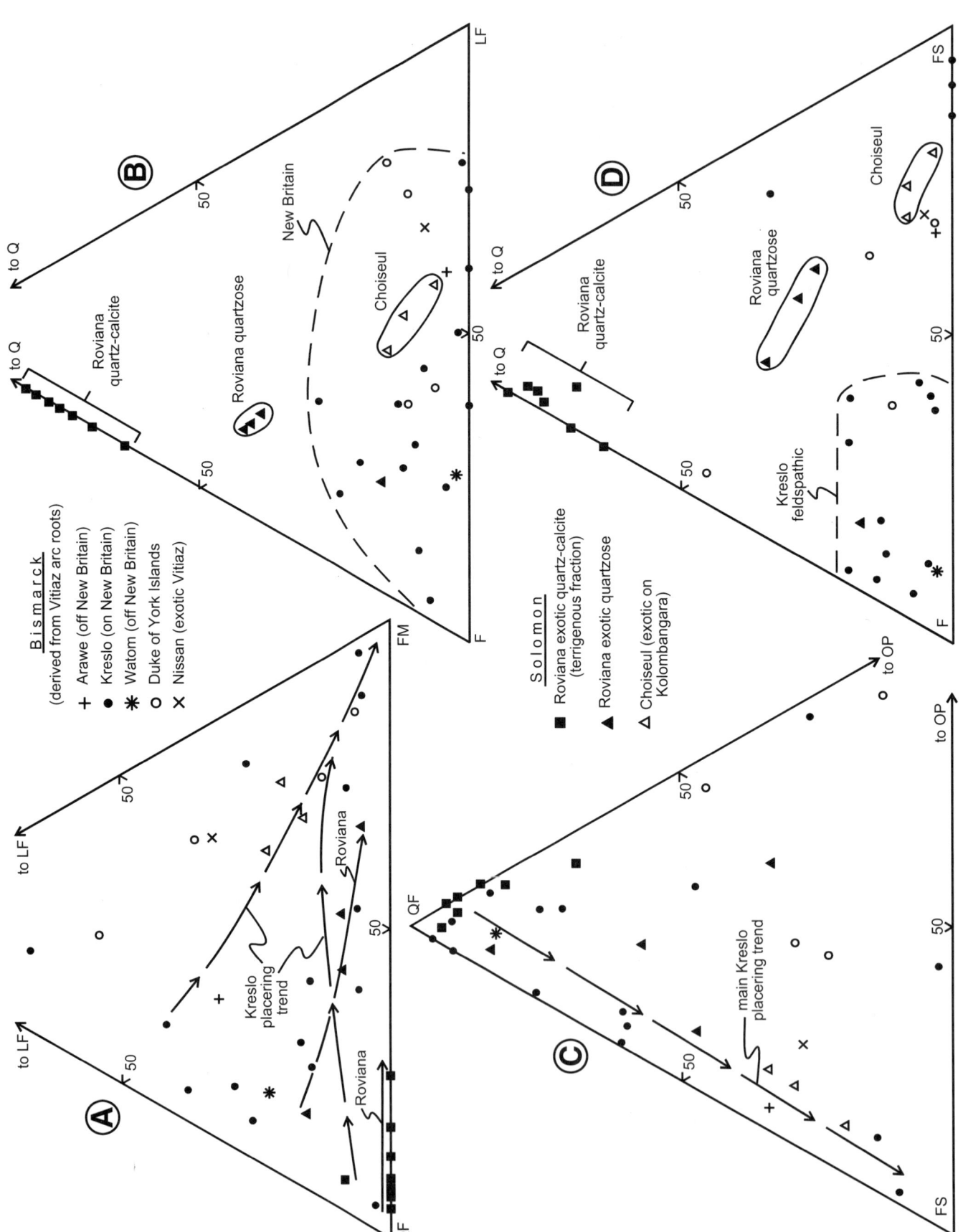

Figure 35. Compositions (grain frequency percentages) of dissected orogen tempers from the Bismarck Archipelago and Solomon Islands. A: LF-QF-FM diagram. B: Q-F-LF diagram. C: QF-FS-OP diagram. D: Q-F-FS diagram. See Table 5A for poles.

(Fig. 19). Sparse exotic sherds from Watom (Table 18A3) are interpreted as the record of wares transferred in small volume from the vicinity of Cape Pomas lying 30–60 km west of Watom on the coast of northeast New Britain (Dickinson, 2000a). Quartzose tempers, largely hybrid sands, in Lapita sherds (Lilley, 1991b) from the small Duke of York Islands (Table 18A4) midway across the narrow strait between New Britain and New Ireland (Fig. 19) were interpreted by Thomson and White (2000) as probably indigenous, derived from local exposures of Paleogene bedrock on Makada within the Duke of York Islands. Alternate derivation from nearby New Britain or New Ireland is not precluded by petrography. Other Duke of York sherds containing nonquartzose tempers are thought to derive from New Britain (Table 13A11). Quartz contents in the Duke of York sherds containing quartzose tempers are variable ($QZi = 10–33$), and the pyribole index ranges widely (<5 to >95), but pyroxene grains are exclusively clinopyroxene ($PXi = 100$) as expected for detritus from the Vitiaz arc assemblage throughout the Bismarck Archipelago. The dissected orogen temper (Table 18A5) in an exotic sherd from Nissan (Fig. 19) probably also reflects ceramic transfer from New Britain or New Ireland.

Muyua (Woodlark) Temper

Three sherds from Muyua (=Woodlark) Island on the Woodlark Rise, an eastern extension of the Trobriand platform in the Solomon Sea (Fig. 19), contain as temper moderately sorted and subangular to subrounded sand in which quartz, both feldspars (plagioclase and K-feldspar), and microphaneritic igneous lithic fragments are prominent, but no volcanic lithic fragments are present. The only abundant ferromagnesian mineral is pleochroic green hornblende. Textural relations of temper and paste suggest that the temper was manually added, probably as stream sand, to a silty clay body.

Muyua is composed dominantly of uplifted and tilted Kiriwina Limestone of Pleistocene age, but older igneous bedrock is exposed in hills and on headlands near Suloga Harbor on the southwest coast (Trail, 1967; Ollier and Pain, 1978; Joseph and Finlayson, 1991). The oldest rocks exposed are Paleogene basalt (Loluai Volcanics), overlain unconformably by Lower Miocene limestone intercalated with volcaniclastic strata. Still younger Okiduse Volcanics are pyroxene and hornblende andesites, together with minor more silicic felsite, of Middle Miocene age (15–11 Ma). Intrusive bodies emplaced ca. 11 Ma during waning phases of Okiduse magmatism include dominantly porphyritic intrusive rocks ranging from hypabyssal microdiorite and micromonzonite forming dikes and sills to coarser-grained granodiorite forming larger local stocks, notably one on Granite Point just east of Suloga Harbor (Ashley and Flood, 1981). The Muyua temper sand was apparently derived exclusively from hornblende granodiorite of the local intrusive suite, which is part of a dissected arc assemblage paired with the subduction zone of the Trobriand Trough to the north (Honza et al., 1987; Kiriwina Trench of Ollier and Pain, 1980). The Woodlark Rise was attached to the Pocklington Rise (Fig. 19), an eastern extension of the continental magmatic arc of the Papuan Peninsula, before opening of the intervening Woodlark Basin by seafloor spreading since ca. 6 Ma (Weissel et al., 1982; Taylor et al., 1995). The source rock for Muyua dissected orogen temper sand represents the eroded roots of the Woodlark-Pocklington magmatic arc, and was intruded before rifting that opened the Woodlark Basin.

Solomon Islands

Three sherds from the volcanic island of Kolombangara in the New Georgia Group (Stanton and Bell, 1969; Dunkley, 1986) contain exotic dissected orogen temper (Fig. 35; Table 18B1) in wares brought to Kolombangara from Choiseul (D.E. Yen, 1972, personal commun.), an island lying ~50 km to the north (Fig. 19). Choiseul exposes a volcanogenic pre-Miocene assemblage of igneous and metamorphic rocks, containing limited amounts of quartz, overlain by Miocene and younger volcanic and sedimentary cover (Ridgway and Coulson, 1987). The sand tempers in the Kolombangara sherds are moderately well sorted and subrounded to subangular aggregates of probable stream origin manually added to clay bodies. Heterogeneous mineralogical compositions (Table 19) reflect derivation of the sands from both partly metamorphosed volcanogenic basement and overlying volcanic cover. The presence of metamorphic detritus, coupled with the lack of detritus from ultramafic rock, suggests origin of the sherds from northern Choiseul, rather than southern Choiseul (Fig. 8 of Ridgway and Coulson, 1987).

TABLE 19. GRAIN TYPES IN DISSECTED OROGEN TEMPERS OF EXOTIC CHOISEUL SHERDS EXCAVATED ON KOLOMBANGARA IN THE NEW GEORGIA GROUP (SOLOMON ISLANDS)

Grain type	BN-5A-12	BN-5A-72	BN-5A-73	Mean (N=3)
Quartz (from intrusive and/or metavolcanic rocks)	6	6	2	5
Plagioclase feldspar (both fresh and altered)	18	20	14	18
Hornblende (~85% green, remainder brown)	39	39	52	42
Clinopyroxene (fresh and unaltered)	13	6	4	8
Epidote (of deuteric or metamorphic origin)	4	2	2	3
Opaque iron oxides (dominantly magnetite)	4	4	6	5
Volcanic lithic fragments (unmetamorphosed)	9	18	15	13
Microphanerite lithic fragments (exclusively of hypabyssal and metavolcanic igneous origin)	7	5	5	6

Note: data from traverse frequency counts of 300 grains per sherd (numbers BN-5A-12, 72, 73).

The sparse quartz in the tempers has no attributes of volcanic quartz and was probably derived from the limited amounts of quartz reported from metamorphosed volcanic rocks on Choiseul. Subordinate microphanerite lithic fragments represent detritus from mafic hypabyssal stocks (microdiorite and microgabbro) that intrude the volcanogenic basement of Choiseul, and perhaps from mafic metavolcanic rocks of coarse granoblastic texture reported from the basement assemblage. Sparse epidote grains are probably of similar derivation from deuteric or metamorphic mineral assemblages in stocks and their wall rocks. The lack of K-feldspar grains reflects the exclusively mafic character of Choiseul igneous rocks. Volcanic lithic fragments may include detritus from both the undeformed volcanic cover of Choiseul and less metamorphosed parts of the underlying basement assemblage, as may the variably altered plagioclase grains. The unaltered clinopyroxene grains were probably derived exclusively from the volcanic cover succession, although the persistence of fresh pyroxene in parts of the older pre-Miocene volcanogenic succession cannot be excluded. The more abundant hornblende grains probably derive from both volcanic and metavolcanic sources. Green hornblende, the dominant grain type in the temper sands (Table 19), is abundant (~50%) in metavolcanic amphibolites of Choiseul, and brown hornblende (also present in the temper sands) is common in andesitic to basaltic lavas and pyroclastic strata of the volcanic cover on Choiseul (Ridgway and Coulson, 1987).

Four sherds from Roviana Lagoon in the New Georgia Group, where andesitic arc tempers are prevalent (Table 13B9–13B11), contain anomalous quartz-bearing temper sands (Fig. 35; Table 18B2) probably derived from local New Georgia sources, but possibly reflecting ceramic transfer from elsewhere in the Solomon Islands. The most likely derivation of the well-sorted and partly hybrid quartz-bearing sands of beach origin is an embayment on the shoreline of southeastern New Georgia where a subvolcanic intrusive complex (Fig. 21C) of gabbro-diorite-tonalite (Dunkley, 1986) is exposed at the coast within the eroded core of a volcanic center ~40 km southeast of Roviana Lagoon. The mean composition (grain frequency percentages) of the terrigenous fraction of the most characteristic two quartz-bearing temper sands is as follows (a third is an opaque-rich placer sand and the fourth contains more hornblende than clinopyroxene): quartz, 22%; plagioclase, 25%; clinopyroxene, 24%; olivine, 4%; hornblende, 2%; opaques, 14%; volcanic lithic fragments (vitric and microlitic), 9%. The terrigenous grain population reflects mingling of volcanic and intrusive detritus, the latter including quartz grains, which was then admixed with calcareous reef debris to form hybrid beach sand temper. The relatively high content of quartz in the temper can be attributed to rapid concentration of quartz over other grain types under the regime of intense tropical weathering prevalent on New Georgia.

The exotic Choiseul sands in Kolombangara sherds and the quartzose tempers in Roviana (New Georgia) sherds are the only dissected orogen tempers known from the Solomon Islands, but similar sands could occur on Guadalcanal (Hackman, 1980; Chivas et al., 1982) and other islands farther to the southeast in the central province of the Solomon Islands (Fig. 21A). The central province is a variably dissected orogen that occupies a narrow belt, only 50–100 km wide but ~825 km long, lying between the active volcanic province to the southwest and accreted ophiolitic rocks forming the Pacific province to the northeast (Coulson, 1985; Petterson et al., 1999).

Southern Viti Levu

The best-studied dissected orogen tempers of Pacific Oceania occur in sherds from several archaeological sites on the south coast of Viti Levu in Fiji (Dickinson, 1971a) where streams draining interior highlands that expose the Eocene to Miocene Wainimala orogen reach the sea (Fig. 26). The Wainimala orogen, now oriented approximately east-west across Viti Levu, represents the dissected roots of the ancestral Vitiaz island arc, and trended northwest-southeast parallel to the Vitiaz trend along the northern Melanesian borderland before tectonic rotation of the Fiji platform (Fig. 9). Deep dissection of the Vitiaz arc assemblage on Viti Levu is attributed to island uplift (Dickinson, 1972c) related to deformation and fragmentation of the arc structure during tectonic rotation of the Fiji platform within the arc-to-arc transform belt linking the Tongan and New Hebrides arc-trench systems of opposed polarity (Hathway and Colley, 1994).

Although an Upper Eocene to Lower Oligocene component of the orogen is preserved locally (Whelan et al., 1985), the most widespread exposures of the Wainimala orogen are strongly deformed and partly metamorphosed volcanic and volcaniclastic strata of Late Oligocene to Middle Miocene (14–12 Ma) age intruded by dioritic-gabbroic to granitic plutons of Late Miocene (12–8 Ma) age (Rodda and Kroenke, 1984; Rodda and Lum, 1990; Rodda, 1994). Sources for modern detritus ultimately derived from the Wainimala orogen include (1) forearc sedimentary strata deposited along the northern (trenchward) flank of the Vitiaz arc axis (Hathway, 1994, 1995; Wharton et al., 1995) and (2) postorogenic Upper Miocene sedimentary strata that rest depositionally upon the dissected orogen along an unconformable contact dated within the interval 8–6 Ma (Hathway, 1993). Recycling of Wainimala-derived sediment through sedimentary successions overlying the volcanogenic and plutonic rocks of the orogen core probably enhanced the quartzose character of dissected orogen tempers from Viti Levu.

Heterogeneous grain populations are characteristic of all dissected orogen temper sands from southern Viti Levu (Table 20), although proportions of different grain types are variable from site to site and to a lesser extent from sherd to sherd at the same site (Fig. 36; Table 18C). Sherd assemblages containing indigenous dissected orogen tempers of local origin have been sampled from the Yanuca (Table 18C4), Sigatoka (Table 18C6), Karobo (Table 18C7), and Nasilai (Table 18C8) archaeological sites along the south coast of Viti Levu (Fig. 26), and a single similar sherd has been examined from Yadua (Table 18C5) between Sigatoka and Yanuca.

TABLE 20. SAND GRAIN TYPES IN DISSECTED OROGEN TEMPERS FROM VITI LEVU IN FIJI

Grain type	Description and origin
Non-ferromagnesian mineral grains (quartz and feldspars)	
Monocrystalline quartz	dominantly common plutonic quartz (tiny inclusions or vacuoles and undulatory extinction), but including some volcanic (or metamorphic) quartz with straight extinction
Plagioclase feldspar	monocrystalline grains (both twinned and untwinned), variably altered to secondary albite; probably derived from both volcanic to metavolcanic and plutonic sources
K (alkali)-feldspar	monocrystalline grains but including unmixed perthite; probably derived largely from plutonic sources but including some volcanic or hypabyssal sanidine with low optic axial angle
Ferromagnesian mineral grains (silicates and oxides)	
Clinopyroxene	pale green (non-pleochroic); probably derived largely from volcanic source rocks, but possibly also from dioritic-gabbroic plutons
Hornblende	pleochroic (green to pale brown); probably derived largely from dioritic to granitic plutons, but possibly also from volcanic (to metavolcanic) source rocks
Opaques	probably mainly magnetite or titanomagnetite but not specifically identified and probably also including ilmenite
Biotite (sparse)	red-brown biotite, commonly weathered or altered but typically in blocky books probably derived largely from granitic plutons
Epidote (sparse)	aggregate polycrystalline grains of epidote (or clinozoisite) probably derived from metavolcanic and deuterically altered plutonic sources
Polycrystalline lithic fragments	
Microphaneritic (microgranular)	intrusive (plutonic to hypabyssal) igneous rocks, ranging from granitic and microgranitic to more mafic varieties (microdiorite-microgabbro); probably derived from subvolcanic stocks and small batholiths
Volcanic-metavolcanic	dominantly microlitic or felsitic and vitric, but including albitic varieties strongly altered by diagenesis or metamorphism
Sedimentary-metasedimentary	chert-metachert, argillite, quartzite (or vein quartz), slate-phyllite (tectonite), contact-metamorphic hornfels, indurated siltstone

Multiple sherds (n = 27) from Bourewa (Fig. 26) on the Rove Peninsula between the Tuva River and Natadola Beach (Nunn et al., 2003, 2004; Kumar et al., 2004a), and from nearby Qoqo Island offshore, also contain indigenous dissected orogen temper sands containing the same grain types (Table 20) as other dissected orogen tempers from Viti Levu. Quartz is somewhat less abundant, however, in Bourewa tempers than in other Viti Levu dissected orogen tempers because granitic Colo plutons are sparse in the hinterland of Bourewa from which local streams drain to the coast. Bourewa tempers include nonplacer, placer, and hybrid variants. The Bourewa site is significant because it may date to 1150 ± 50 B.C. as the oldest known Lapita site in Fiji (P.D. Nunn, 2005, personal commun.). Its location is significant with its apparent antiquity in mind because it is the most westerly locale on the southern coast of Viti Levu where the barrier reef lying off western Viti Levu trends close to the island shoreline, and may have been the first place where people migrating eastward to Fiji could make landfall on Viti Levu.

From a cathodoluminescence study of quartz grains in Sigatoka temper sands, Goles (2001) suggested that the quartz was probably recycled from Miocene sedimentary strata overlying the Wainimala orogen, but derived ultimately from metamorphosed rock masses that have been rifted away or otherwise removed from the Fiji platform and no longer present on Viti Levu. Several considerations, however, make this hypothesis seem unnecessary and unlikely: (1) the widespread distribution of quartzose plutons and metamorphic rocks within the Wainimala orogen, (2) the ubiquitous presence of quartz in dissected orogen tempers from the length of the south coast of Viti Levu, and (3) the quartzose character of modern Sigatoka dune sands composed of sand reworked from the mouth of the Sigatoka River, which drains the interior of the Wainimala orogen. Ultimate derivation of quartz in Viti Levu dissected orogen tempers from the core of the Wainimala orogen is accordingly inferred here.

Selected subordinate sherds containing dissected orogen tempers at the Vuda (Table 18C3), Natunuku (Table 18C2), and Navatu (Table 18C1) sites along the north coast of Viti Levu are interpreted as representative of nonlocal wares from farther south on Viti Levu. This judgment is supported both by the nature of the geology in the immediate hinterland of the north coast sites (Fig. 26) and by empirical knowledge of the ferromagnesian nature of indigenous postarc tempers at the north coast sites (Fig. 32). Exotic sherds containing Viti Levu dissected orogen tempers have been recovered from Beqa (Table 18C9) and nearby Ugaga (Table 18C10) south of Viti Levu, from Moturiki (Table 18C11) and Totoya (Table 18C12) in Lomaiviti, and from Lakeba (Table 18C13) in Lau (Fig. 36). Protohistoric sherds from at or near Nasilai on the Rewa Delta of Viti Levu have also been found

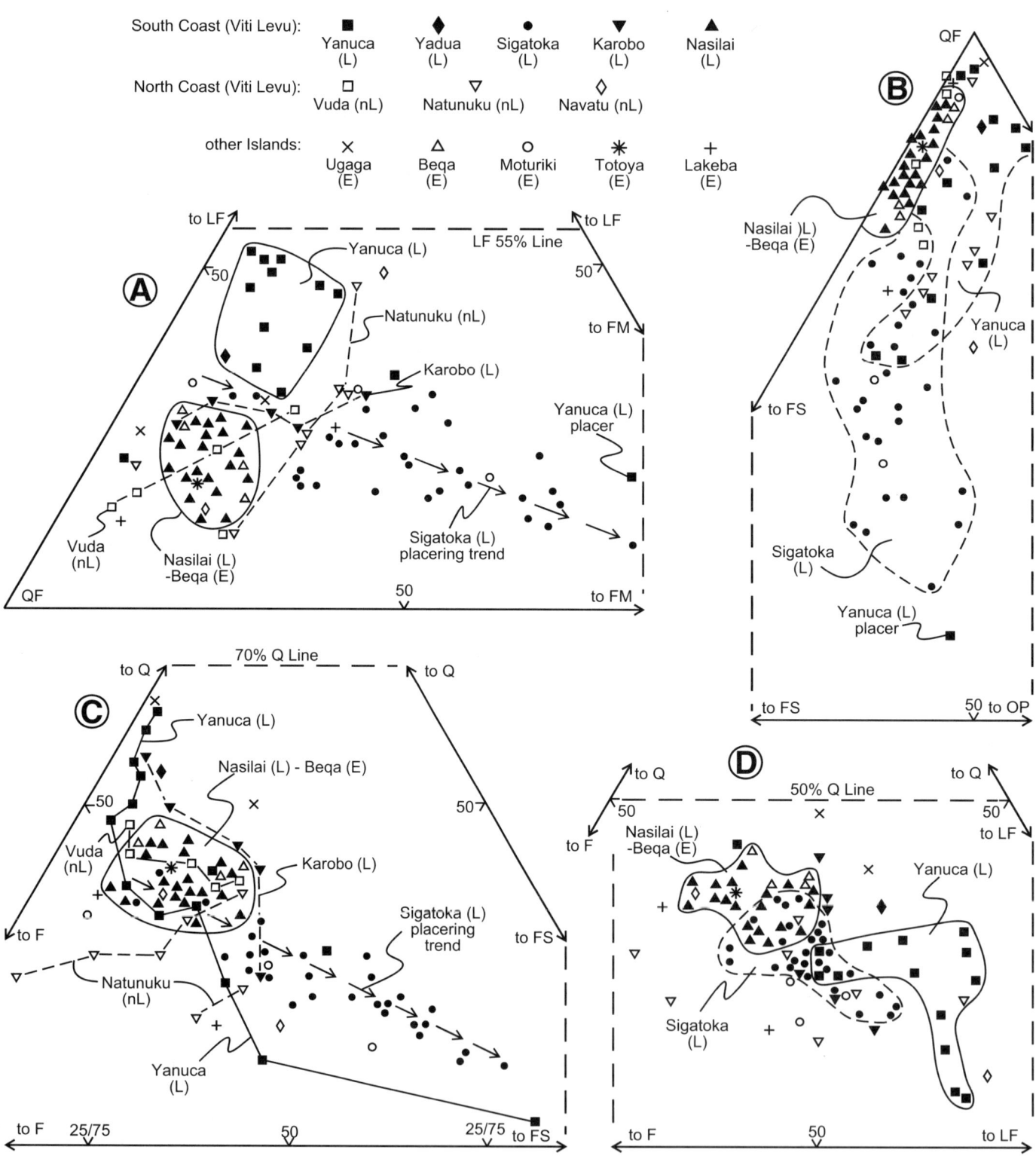

Figure 36. Compositions (grain frequency percentages) of dissected orogen tempers from Viti Levu in Fiji. A: LF-QF-FM diagram. B: QF-FS-OP diagram. C: Q-F-FS diagram. D: Q-F-LF diagram. Annotations denoting temper origins: L—local temper types at sites on south coast of Viti Levu; nL—nonlocal temper types at sites on north coast of Viti Levu; E—exotic temper types at sites on islands off Viti Levu. See Table 5A for poles.

on Tungua in central Tonga and on Nuku Hiva in the Marquesas Islands (Dickinson et al., 1998b).

Local Tempers

Dissected orogen tempers of Viti Levu are distinguishable from all other Fiji-Vanuatu-Tonga tempers by the abundance of quartz, significant amounts of both feldspars (plagioclase and K-feldspar), the near ubiquity of minor epidote, and the mingling of volcanic, hypabyssal, plutonic, and sedimentary to metasedimentary lithic fragments derived from diverse igneous, metamorphic, and sedimentary source rocks exposed within the Wainimala orogen (Dickinson, 2001a). On the other hand, distinctions among dissected orogen tempers from different locales on Viti Levu is difficult, though aided by systematic textural contrasts, minor differences in supplemental grain parameters (Table 18C), and different patterns of variation in compositional space (Fig. 36). The principal locales from which Viti Levu dissected orogen tempers are well known include a Lapita site at Yanuca (Hunt, 1980), the Lapita and post-Lapita site on the seaward edge of the Sigatoka Delta (Birks, 1973; Dickinson et al., 1998a; Burley, 2003; Burley and Dickinson, 2004), and post-Lapita sites at Karobo (Palmer, 1965) near the Navua Delta and Nasilai (Rosenthal, 1995) on the Rewa Delta (Fig. 26). At Sigatoka, there are no systematic differences between the tempers in Lapita and post-Lapita sherds.

Sigatoka tempers (Fig. A19), derived from beach or dune sands on the fringe of the Sigatoka Delta, are composed of better-sorted and more-abraded grains than Nasilai tempers (Fig. A12) derived from fluvial sands of the Rewa Delta. The array of Sigatoka temper sands also grades to more placered aggregates than the other dissected orogen temper types (Figs. 36A–36C), as expected for sands from coastal beach-dune ridges as opposed to fluvial settings. The range in composition of Sigatoka temper sands is indistinguishable from the range in composition of modern sands of the Sigatoka dune field (Dickinson, 1968a) built along the coastal edge of the Sigatoka Delta (Dickinson, 2001a).

Moderately sorted Nasilai fluvial temper sands from the largest river in Fiji are in turn better-sorted aggregates than the poorly sorted fluvial sands from smaller local drainages that served as sources for temper at Karobo and Yanuca. Yanuca tempers tend also to be more lithic in composition than the other dissected orogen tempers (Figs. 36A–36D), with lithic fragments that are typically felsitic (~70% of volcanic lithic fragments), and were apparently derived from restricted source rocks within a small drainage basin. Arrays of Yanuca, Nasilai, and Sigatoka temper sands occupy separate but adjoining fields in LF-QF-FM space (Fig. 36A), and individual tempers of marginal composition from each site are not statistically distinguishable with counting error taken into account. The elongate spread of compositions for Karobo temper sands is less compact than the other three fields. A single sherd from the otherwise unstudied Yadua site contains dissected orogen temper that plots within or near the compositional fields for temper sands in sherds from nearby Yanuca (Fig. 36). Other modal plots (Figs. 36B–36D) embody analogous ambiguities for the compositions of dissected orogen tempers from different sites on southern Viti Levu.

Supplemental grain parameters allow further discrimination among dissected orogen tempers of inferred local origin, but all differences are subtle (Table 18C4–18C8). No robust differences in quartz index (QZi) or feldspar index (PFi) are discernible, although Karobo tempers tend to be somewhat more quartzose and to contain less K-feldspar. The paucity of K-feldspar mineral grains may reflect only limited exposures of plutonic rocks in the fluvial drainage from which Karobo temper sands were derived, but a generally lower pyribole index (PYi) for the Karobo tempers might reflect disproportionately large contributions from hornblende-bearing plutonic rocks unless the volcanic rocks exposed within the drainage basin are inordinately hornblende-rich. Pyribole indices (PYi) are statistically higher for Sigatoka tempers than for Nasilai tempers, but the widely variable pyribole index (PYi) for Yanuca tempers spans between the two. Nasilai tempers also display a somewhat lower volclithic index (VLi) and a distinctly higher sedlithic index (SLi) than Sigatoka or Karobo tempers, but Yanuca tempers are widely variable by both measures and overlap with the other three local dissected orogen tempers. The VLi-SLi relations for Nasilai temper sands may reflect more widespread exposure of sedimentary cover over the Wainimala orogen in eastern Viti Levu than in western Viti Levu (Fig. 26). In detail, however, the ratio of total igneous lithic fragments (microphaneritic and volcanic) to sedimentary-metasedimentary lithic fragments is highest for Nasilai and Yanuca tempers at the eastern and western ends, respectively, of the southern Viti Levu coast, with intermediate ratios for Sigatoka and Karobo tempers from geographically intervening sites (Fig. 26). On balance, contrasting sand textures allow distinctions to be drawn among the local dissected orogen tempers with just as much confidence as any supplemental grain parameters, but neither criterion is likely to be successful for tracing the origins of individual sherds.

Nonlocal Tempers

The compositions of nonlocal dissected orogen tempers in sherds from Vuda, Natunuku, and Navatu in northern Viti Levu tend to lie outside or span across the compositional fields for local tempers at sites on southern Viti Levu (Fig. 36). This observation suggests that nonlocal sherds from the northern sites derive from undiscovered or unstudied sites in southern Viti Levu for which local temper sands remain untested. Supplemental grain parameters for the nonlocal tempers are generally ambiguous for detecting similarity to or differences from known local tempers (Table 18C1–18C3). Dissected orogen tempers indicative of origins on Viti Levu in exotic sherds from other islands (Table 18C9–18C13) are discussed further in a later section of this report dealing with ceramic transfer.

TECTONIC HIGHLAND TEMPERS

Tectonic highland tempers ("tectonic mainland" tempers of Dickinson and Shutler, 1968) are derived from uplifted and partly metamorphosed rock assemblages that formed in association with subduction zones where oceanic plates are consumed. Source terranes for terrigenous detritus include intricately disrupted mélanges composed of deformed seafloor sediments and oceanic crust, overthrust slabs of oceanic lithosphere, and underthrust segments of continental crust. Each tectonic highland temper type known to date has a somewhat different composition, as controlled by vagaries of geologic history in each specific geotectonic setting. Quartzose variants were derived dominantly from sedimentary or metasedimentary assemblages of mélanges and underthrust continental crust. Ophiolitic variants were derived dominantly from volcanic or metavolcanic and ultramafic rocks of oceanic lithosphere. Sedimentary mingling of the two kinds of detritus has been encountered to date only in the temper sands of sherds from Maluku and New Caledonia.

The varied tectonic highland tempers are characteristic of sherds from the following diverse locales: (1) Yap along the western fringe of Micronesia (Fig. 6), (2) Buru and Seram in the forearc of the Banda arc-trench system (Figs. 6A and 18), (3) the Aru Islands along the western fringe of the Sahul Shelf between Australia and New Guinea (Figs. 6A and 18), (4) the D'Entrecasteaux Islands in the Solomon Sea (Figs. 7 and 18), and (5) New Caledonia along the southwestern fringe of island Melanesia (Figs. 8 and 9). Tectonic highland tempers of ophiolitic affinity could also be expected on Malaita (and Santa Isabel) in the northeastern or Pacific province of the Solomon arc (Fig. 21A) where a segment of the oceanic Ontong Java Plateau has been accreted to the eastern flank of the Solomon chain (Hughes and Turner, 1977; Neal et al., 1997; Petterson et al., 1997), but no sherds from those islands have been studied to date.

The two end members of tectonic highland tempers are as different compositionally as any two classes of Oceanian temper, but some tectonic highland tempers are mixtures of the two. On the one hand, quartzose variants (high QZi) derived from sedimentary and metasedimentary strata (high SLi) of deformed subduction complexes and underthrust continental basement generally lack volcanic lithic fragments (Fig. A13). On the other hand, ophiolitic variants derived from overthrust or underthrust oceanic lithosphere are quartz-poor (QZi ≈ 0) and contain lithic fragments derived either from volcanic and metavolcanic rocks (high VLi; SLi ≈ 0) or from mantle rocks that are typically altered to serpentinite, with plagioclase the dominant feldspar (PFi ≈ 100).

Yap Metavolcanic Temper

Tectonic highland tempers composed of metavolcanic detritus in sherds from Yap (Gifford and Gifford, 1959; Intoh and Leach, 1985) were derived from the subvolcanic basement of Yap (Fig. 15), which is underpinned by a Tertiary subduction complex composed of variably metamorphosed oceanic crust formed by consumption of the Pacific plate downward to the west beneath the Philippine Sea plate (Hawkins and Batiza, 1977; Hamilton, 1979, p. 275–277). The close association of greenschist and amphibolite metamorphic facies in the metavolcanic rocks, without evidence for an intervening epidote-amphibolite facies, is indicative of metamorphism under the conditions of high pressure to low temperature prevalent in subduction zones (Shiraki, 1971; Shiraki et al., 1978). Low ratios of radiogenic strontium in the metavolcanic rocks are characteristic of seafloor lavas metamorphosed and uplifted within a subduction zone (Shiraki, 1971; Matsuda et al., 1977; Shiraki et al., 1978). The variable color index of Yap greenschists and amphibolites suggests, however, that metamorphosed ultramafic rocks from oceanic mantle are present in addition to metamorphosed seafloor lavas (Hawkins and Batiza, 1977). Limited fossil collections (Cole et al., 1960; Johnson et al., 1960) from locally reworked sediment overlying the metavolcanic rocks imply an Oligocene to Early Miocene age for deformation and metamorphism. Approximately half the sherds of Yap origin studied to date contain temper sands that were derived in whole or in part from Yap metavolcanic bedrock (Table 21).

Yap is the only place along >5000 km of the subduction system at the eastern edge of the Philippine Sea plate (Fig. 5) where components of the subduction complex have been uplifted above sea level. Uplift is probably due to subduction of buoyant rift shoulders and young oceanic crust along the spreading center of the Sorol Trough, which intersects the subduction zone opposite Yap (Fig. 6A). Comparatively buoyant seafloor along the Caroline Ridge paralleling the Sorol Trough is <3000 m deep, whereas seafloor to the north and south lying east of the Mariana and Palau segments of the subduction system is >4000 m deep.

Temper Sources

The metamorphic grade within the Yap subduction complex increases eastward from largely greenschist on Yap Island to largely amphibolite on the eastern or Gagil Peninsula of Gagil-Tomil Island (Shiraki, 1971), reflecting greater structural depth in the direction toward the Yap Trench where successive components of the subduction complex were underthrust. The distribution of structurally interleaved lenses of serpentinite, derived from harzburgitic dunite (as identified in thin section), indicate that the overall structure of the Yap subduction complex is dominated by three fault-bounded structural panels, separated by

TABLE 21. DISTRIBUTION OF SHERDS WITH YAP METAVOLCANIC TEMPER VARIANTS

Temper type	Yap	Ngulu	Ulithi	Fais	Lamotrek
Yap natural metavolcanic temper (*n*=17)	10	2	3	2	-
Yap manually added metavolcanic temper (*n*=6)	2	-	-	4	-
Yap hybrid metavolcanic-calcareous temper (*n*=22)	8	3	-	9	2
Yap grog temper with metavolcanic paste (*n*=4)	-	-	1	3	-
[total Yap tempers of tectonic highland class, *n*=49]	[20]	[5]	[4]	[18]	[2]

imbricate thrusts (Divis, 1975) that dip westward away from the Yap Trench (Fig. 15).

The structurally highest Weloy-Fanif thrust panel above the Rumung thrust fault is exposed over most of Yap and Rumung islands where it is composed of phyllonitic greenschist (Yap Formation of Johnson et al., 1960). Mesoscopic pinch-and-swell fabrics of foliation anastomosing around less deformed domains of phacoidal shape are characteristic, and discrete shear zones marked by narrow belts of intense structural dislocation or by bands of scaly clay juxtapose domains of differently oriented schistosity. These features show that the Yap Formation is broken formation developed by complexly superposed shear within the Yap subduction zone. The structurally underlying Tomil-Maap and Gagil-Riquen thrust panels, lying respectively above the Madalai and Gachpar thrusts (Fig. 15), expose more pervasively dislocated broken formation and polymictic breccia of both tectonic and sedimentary origin (Map Formation of Johnson et al., 1960) on Maap and Gagil-Tomil Islands. Meter-scale blocks in tectonic breccia (mélange) include diorite-gabbro, peridotite, pyroxenite, serpentinite, hornblendite, amphibolite, and rare plagiogranite derived from oceanic lithosphere. Associated with the tectonically brecciated broken formation and mélange are isolated domains of bedded sedimentary breccia representing slope-basin deposits that accumulated on a substratum of deformed rock still undergoing deformation on the trench slope. The dominant lithologic component of the Tomil-Maap and Gagil-Riquen thrust panels is tectonic breccia akin to broken formation in that the prevalent clasts are similar greenschist-amphibolite, but akin to mélange in that clasts are separated by a sheared matrix of scaly clay. Pervasively sheared scaly clay of tectonic origin is more common as breccia matrix than finely comminuted sedimentary detritus, and indicates that most of the Map Formation is tectonically brecciated, rather than being structurally dislocated sedimentary breccia.

In roadcuts on northern Gagil-Tomil Island, on the grade south of the bridge to Maap Island, a packet of ~5 m of turbidites only slightly disrupted by fracture surfaces is exposed within the Tomil-Maap thrust panel. Stratigraphic intervals of finely laminated mudstone, containing interbeds of fine sandstone in graded beds ~1 cm thick, form successions 50–75 m thick separated by coarse graded sandstone beds 25–50 cm thick. The coarser sandstone beds have scoured and load-casted bases but gradational tops, and contain rip-up mudstone clasts of penecontemporaneous sediment. The association of sedimentary structures is reminiscent of interchannel or basin-plain turbidite sedimentation, and the turbidite packet may represent trench fill that was incorporated into and uplifted with the subduction complex. The fossils from the Yap subduction complex were collected not far to the northwest on Maap Island, and probably came from a similar sedimentary assemblage. Rare quartz grains of uncertain derivation are present in the coarser sandstones, but the bulk of the sandy detritus is petrographically indistinguishable from Yap metavolcanic bedrock.

Temper Variants

Yap metavolcanic tempers include four principal variants (Table 21). Most characteristic is natural temper of poorly sorted and generally unabraded sand composed dominantly of foliated greenschist or amphibolite lithic fragments, together with mineral grains of plagioclase or albite and actinolitic amphibole (pale green coloration) derived from metavolcanic bedrock. The natural temper is embedded within clay paste apparently derived from residual soils or retransported soil materials. This temper variant is especially characteristic of laminated ware made during the past few hundred years (Fig. 3), but occurs in older sherds as well. Subordinate numbers of sherds contain well-sorted and rounded sand of beach origin containing a similar mix of grain types derived from metavolcanic source rock but manually added to clay bodies. A third and common variant of metavolcanic temper is hybrid beach sand composed of terrigenous and calcareous grains admixed in variable proportions, but with calcareous grains commonly dominant. Predominantly calcareous temper sands, which are especially prevalent in the oldest cultural horizons (Intoh and Dickinson, 2002), are linked conclusively to an origin on Yap by subordinate metavolcanic lithic fragments among the temper grains or by metavolcanic detritus embedded as silt within the clay paste, or both. Limited numbers of Yap sherds containing grog temper also contain metavolcanic detritus embedded in the clay pastes as evidence for affinity with the natural metavolcanic temper of Yap.

Modern hybrid beach sands that closely resemble the hybrid tempers of Yap sherds were collected from five coastal sites where beach placering has concentrated terrigenous grains relative to the prevalent reef debris that dominates Yap beach faces (Fig. 15). Fresh terrigenous detritus is derived from prominent bedrock headlands and carried by longshore currents to nearby beaches, which lie mostly on the windward shores of Maap and Gagil-Tomil Islands where surf action is strong, but are also present locally on the leeward shore of Yap Island (Fig. 15). The sands from all five sites are petrographically so similar as to preclude the tracing of Yap beach sand tempers to specific origins within Yap. Relative proportions of terrigenous and calcareous grains in Yap beach sands depend partly upon the care with which placer concentrations of terrigenous placer sand are scraped from thin veneers of black sand on beach faces or berms, and their relative proportions may also vary seasonally and with variable storminess.

Outer Banda Arc Tempers

Seram, the southwestern half of Buru to the west, and the islet of Gorom to the southeast in the southern Molucca Islands (Fig. 18) are composed dominantly of uplifted and partly metamorphosed mélange formed by intricate dislocation and internal thrusting of Mesozoic-Cenozoic sedimentary assemblages deformed in the forearc subduction complex of the Banda arc-trench system (Hamilton, 1979, p. 133–137). The subduction zone lies to the north in the Seram Trough where the subduction complex has overridden a western projection of the New Guinea continental block. Thrust slivers of ophiolitic oceanic basement derived from seafloor consumed at the subduction zone, metamorphic continental basement derived from underthrust New Guinea crust, and overlying sedimentary successions that were deposited on both types of substratum are incorporated into the mélange as structural slices of varying dimensions and lithology (Audley-Charles et al., 1979; Charlton, 2004). The northeastern half of Buru is underlain by a domain of Precambrian metamorphic rocks, either incorporated into or thrust beneath the mélange, which include phyllite, chlorite and mica schists (partly garnetiferous), amphibolite, and granitic gneiss.

The tempers in nine indigenous sherds from Gorom include a spectrum of quartzose, calcareous, and hybrid sands with terrigenous fractions composed dominantly of quartz with minor feldspar and chert grains. The calcareous and hybrid tempers are clearly beach sands, but three wholly noncalcareous temper aggregates are not as well sorted or as abraded, and may represent local stream sands. Similar but more heterogeneous tectonic highland tempers occur in a sherd from Seram, and in five sherds from nearby Ambon that probably derive from Seram. The quartz, feldspar, and chert grains are associated with mica flakes (both biotite and muscovite) and lithic fragments of argillite and tectonite (metasedimentary slate and phyllite), as well as clinopyroxene and pale magnesian amphibole mineral grains and lithic fragments of serpentinite derived from metamorphosed ophiolitic rocks. The sources for these quartzose temper sands were the varied structural components of the mélange belt forming the outer fringe of the Banda arc-trench system.

Typical sherds from Buru contain subangular aggregates of uniquely tectonite-rich coarse temper sand of probable fluvial origin embedded in silty and micaceous clay paste. On sherd surfaces, many of the temper grains display a glossy sheen typical of phyllite or schist. The large contrast in grain size between the coarsest silt particles in the clay paste and the finest temper grains indicate that the temper was manually added, rather than being natural temper embedded in clay bodies as collected. One Buru sherd contains a much better sorted hybrid beach sand and less silty clay paste, but the terrigenous fraction of the hybrid sand is indistinguishable compositionally from the coarser detritus in the fluvial temper sands. A composite count of 240 total grains in five Buru sherds containing tectonite-rich coarse temper sand yielded the following frequency percentages of grain types reflective of sand derivation exclusively from exposures of Precambrian basement on Buru: lithic fragments of foliated micaceous quartzite (tectonite), 31%; lithic fragments of quartzite (aggregate grains of polycrystalline quartz), 29%; lithic fragments of foliated quartz-mica schist and phyllite (tectonite), 26%; biotite mica flakes, 6%; quartz mineral grains, 5%; garnet mineral grains (equant and isotropic with high relief but pale coloration), 2%.

Aru Islands Temper

Indigenous hybrid temper sand in sherds from the Aru Islands near the western edge of the Sahul Shelf in the Arafura Sea between Australia and New Guinea (Fig. 18) is composed of terrigenous quartz grains mixed in varying proportions with calcareous reef debris (Dickinson, 2005). The Sahul Shelf is an integral part of the Australia–New Guinea continental block and was subaerially exposed during Pleistocene lowstands in sea level. The more calcareous of the temper sands in prehistoric sherds closely resemble the hybrid beach sand used today by modern Aru potters. Staining for feldspars reveals only trace amounts of either plagioclase or K-feldspar in the tempers. The terrigenous fraction of Aru temper was apparently derived from quartzose sediment in subdued exposures of subhorizontal Neogene strata (Verstappen, 1959) that form the Aru Islands. The Aru landmass is essentially one large island broken into segments by narrow channels with orientations controlled by fracture systems that probably developed during uplift of the edge of the Sahul Shelf on the crest of the flexural forebulge in front of the Banda arc-trench system, which is approaching the flank of the continental block from the west. The sedimentary succession rests on continental basement rocks that underpin the Sahul Shelf (Hamilton, 1979), and the ultimate provenance for the quartz grains in local Neogene strata was probably the interior of the Australian craton. Because of its probable continental derivation, indigenous Aru temper is atypical of the tectonic highland temper class, but uplift of the Aru Islands on the forebulge of the Banda subduction zone (Fig. 18) was responsible for subaerial exposure of the temper sources.

D'Entrecasteaux Temper

The principal D'Entrecasteaux Islands in the Solomon Sea (Fig. 19) are composed dominantly of domical core complexes of gneissic continental basement and associated granitic intrusions (Ollier and Pain, 1980; Davies and Warren, 1992; Hill et al., 1992; Baldwin et al., 1993). The core complexes formed by the rise of basement during Pliocene to Quaternary tectonic denudation by detachment faulting that removed supracrustal cover, including an ophiolitic slab of overthrust oceanic lithosphere (Martínez et al., 2001), during superextension caused by propagation of the Woodlark Basin oceanic rift system into the New Guinea continental block (Benes et al., 1994; Mutter et al., 1996). Deformed ophiolitic rocks of the D'Entrecasteaux Islands are an eastern continuation of the ophiolitic slab thrust over the Papuan Peninsula to the west (Fig. 9). The crustal extension recorded by the core complexes represents an early phase of continental breakup by intra-arc rifting, with a record of preceding Miocene to Pliocene arc magmatism preserved in the detached tectonic cover of the D'Entrecasteaux core complexes (Smith and Compston, 1982; Smith and Milsom, 1984). Synextensional peralkaline rhyolites characteristic of intracontinental rift magmatism occur along Moresby Strait between Goodenough and

Fergusson Islands and along Dawson Strait between Fergusson and Normanby Islands (Smith, 1976; Smith et al., 1977; Smith and Johnson, 1981).

Two sherds from Goodenough Island contain poorly sorted and angular gneissic detritus as a quartzo-feldspathic temper that is compositionally unlike any other temper sand known from Pacific Oceania (Fig. A14). Whether the temper is natural or manually added is indeterminate from textural relations between temper and paste. Quartz and both feldspars (plagioclase and K-feldspar) are present as separate mineral grains, together with granitic and gneissoid (foliated) lithic fragments and a variety of unusual mineral grains including abundant pale monoclinic amphibole (tremolite) and prominent coarse flakes of muscovite in addition to brown biotite. Subordinate mineral grains include pale jadeitic pyroxene with anomalous (nonspectral) interference tints, pale orthorhombic amphibole (anthophyllite), blue-green sodic amphibole (glaucophane), and isotropic garnet. The quartz-feldspar, granitic-gneissoid, and garnet grains were undoubtedly derived from massive core zones and flanking foliated shear zones of the local Goodenough core complex, and the pale magnesian amphiboles presumably from metamorphosed ultramafic rock of amphibolite sheaths that bound the core complex and form the structural floor of structurally detached core-complex cover. Jadeitic pyroxene and sodic amphibole indicative of high-pressure metamorphism were derived from eclogitic rocks that are incorporated tectonically into the flanks of the core zone (Davies and Warren, 1992; Hill and Baldwin, 1993).

Two of six sherds from the Trobriand Islands in the Solomon Sea to the northeast (Fig. 19) contain tempers indistinguishable from those in the Goodenough sherds and reflect ceramic transfer from the D'Entrecasteaux Islands. Islands on the Trobriand platform expose only Pleistocene Kiriwina Limestone (Davies and Smith, 1971; Francis et al., 1987), which could not serve as a source for the exotic temper. Two other Trobriand sherds contain similar but finer-grained quartzo-feldspathic tempers inferred also to derive from the D'Entrecasteaux Islands. As discussed in a later section dealing with ceramic transfer, still other Trobriand sherds contain volcanic sand tempers thought to derive from the Papuan Peninsula to the south of the D'Entrecasteaux Islands (Fig. 19).

New Caledonia Tempers

Tectonic highland tempers from New Caledonia (Figs. 8 and 9) include quartzose and ophiolitic variants, and tempers of intermediate composition representing mixtures of those two end members of the local temper spectrum (Galipaud, 1990; Dickinson, 2002b). Reconnaissance work has shown that New Caledonia sherds with quartzose tempers provide generally satisfactory results from thermoluminescence dating, whereas sherds with ophiolitic tempers yield problematical results (Huntley et al., 1983). Hybrid temper sands with admixtures of calcareous reef debris occur in many New Caledonia sherds and contain terrigenous fractions of variously quartzose, ophiolitic, and mixed character. Sherds with exclusively or dominantly calcareous temper sands occur as well in New Caledonia collections, but have been largely ignored for petrographic analysis.

Quartzose tempers, which are composed dominantly of monocrystalline quartz grains and quartzose sedimentary and metasedimentary lithic fragments (Fig. A13), contain largely recycled detritus derived from several tectonic elements: (1) the Triassic-Jurassic forearc basin and subduction complex that formed along the active continental margin of Gondwana before rifting separated New Caledonia from that continental landmass to the west (Cluzel et al., 1994; Meffre et al., 1996; Cluzel and Meffre, 2002); (2) Cretaceous cover strata that were deposited upon the Gondwanan assemblages, and are composed largely of sediment reworked from them, during drift of the New Caledonia microcontinent away from Gondwana as the intervening Tasman Sea opened by seafloor spreading (Gaina et al., 1998); and (3) the Paleogene subduction complex and foreland basin, containing sediment again largely reworked from older local strata, associated with early phases of the subduction of New Caledonia beneath an ophiolitic slab overthrust from the northeast (Cluzel et al., 2001). The ophiolitic tempers, composed dominantly of detritus from peridotite (or derivative serpentinite) and associated basalt, were derived from the overthrust slab of oceanic lithosphere emplaced structurally above the New Caledonia microcontinent after the latter had reached an intraoceanic position and was drawn into a subduction zone of reversed polarity lying to the northeast (Collot et al., 1987; Aitchison et al., 1995; Cluzel et al., 1997; Eissen et al., 1998).

Grain Types

Table 22 indicates the varied grain types observed in New Caledonia sherds, and Table 23 indicates the known distribution of New Caledonia temper types in sherds from archaeological sites on the mainland (Grand Terre) and on nearby islands within ~100 km of the mainland. There are no apparent areal variations in temper type of statistical significance, nor are there any consistent differences in the tempers of decorated Lapita sherds, undecorated sherds from Lapita horizons, or non-Lapita sherds. Chemical analysis of 42 sherds from the classic Lapita (=Koné) site of Grand Terre by X-ray fluorescence (Chiu, 2003b) indicates that chemical and petrographic groupings of sherds with quartzose and ophiolitic tempers are coordinate. Clay pastes in New Caledonia sherds are consistently silty, but most or all temper sands appear to have been manually added. A third of the tempers examined in thin section are hybrid beach sands (Table 23), but many of the others, particularly nonhybrid lithic and ophiolitic variants, are more poorly sorted and less abraded aggregates that are probably stream sands. The heterogeneity of New Caledonia tempers in both texture and composition makes their sedimentological classification uncertain.

In selected quartzose tempers, sparse detritus derived from blueschists within the partly metamorphosed Paleogene subduction complex imparts a diagnostic facet to some New Caledonia tempers (Table 23). Grain types of blueschist derivation include

TABLE 22. GRAIN TYPES IN TECTONIC HIGHLAND TEMPERS FROM NEW CALEDONIA

Grain category	Quartzose variants	Ophiolitic variants
Light minerals (monocrystalline)	quartz (abundant), feldspars (dominantly though not exclusively plagioclase)	quartz (rare), plagioclase
Aggregate grains (monominerallic)	epidote, vein quartz	serpentinite
Heavy minerals (monocrystalline)	biotite	clinopyroxene, olivine, orthopyroxene, opaque iron oxides (mainly magnetite), chrome-spinel (translucent ruby red and isotropic)
Lithic fragments (polycrystalline)	chert-metachert, argillite, slate-phyllite, metaquartzite	basalt (lathwork), mafic glass (brown to red), gabbro-diorite (plagioclase and pyriboles), peridotite (olivine and pyroxenes) granitic (quartz and feldspars)

TABLE 23. DISTRIBUTION OF TEMPER TYPES IN 65 SHERDS FROM NEW CALEDONIA (NC) AND NEARBY ISLANDS: ÎLE DES PINS (IdP) AND LOYALTY ISLANDS (OUVEA AND MARÉ)

Site: Island:	Lapita NC	other NC	Kapume IdP	Vatcha IdP	Mouly Ouvea	Wabao Maré
Quartzose (Q>L) with blueschist detritus	1 (+3)	-	-	1 (+1)	-	-
Quartzose (Q and L subequal)	4 (+2)	3	-	-	-	-
Quartzose (Q<L)	5	2	-	-	4	-
Mixed quartzose-ophiolitic	4 (+7)	4	-	3 (+2)	1	2
Ophiolitic	(+5)	3	2	1 (+1)	1	3
Total sherds examined	31	12	2	9	6	5

Notes: Q, monocrystalline quartz grains; L, polycrystalline lithic fragments; terrigenous fractions of hybrid tempers (n=22) in parentheses.

mineral grains of actinolite (pale green monoclinic amphibole), isotropic garnet, glaucophane (soda-amphibole pleochroic in blue to yellow tints), lawsonite (straight extinction with moderate birefringence), pumpellyite (resembling green epidote but with lower birefringence), and muscovite (or phengite) white mica, as well as lithic fragments of blueschist (tectonite containing the unusual minerals glaucophane, lawsonite, pumpellyite, and stilpnomelane in varying proportions), and foliated amphibolite. Translucent red chrome-spinel derived from peridotite is common in most ophiolitic variants of New Caledonia temper (Table 22) but is not present in other Oceanian temper suites, and provides an additional diagnostic guide to temper origins on New Caledonia.

Temper Sources

Petrographic work has dealt with only fragmentary components of the rich ceramic heritage in New Caledonia (Fig. 3), with a focus on selected Lapita wares, and the known distribution of temper types does not yet allow effective discrimination between different places of ceramic origin on New Caledonia. Both quartzose and ophiolitic source rocks are widely distributed in New Caledonia (Dickinson, 2002b), with both types of bedrock exposed within the drainage basins of approximately half the ~50 short streams and rivers that transport sediment from interior highlands to the coast along the length of the island (~375 km). Dozens of ceramic-bearing archaeological sites are distributed around the coast of New Caledonia (Gifford and Shutler, 1956; Galipaud, 1996; Sand, 1999a), and occur near the mouths of streams tapping virtually every type of bedrock exposed on the island in varying combinations from site to site.

Present knowledge of New Caledonia tempers is inadequate to support detailed interpretations of temper origin. For example, essentially all known New Caledonia temper types occur in sherds from the classic site of Lapita (Chiu, 2003a), whence the name for Lapita pottery derives (Sand et al., 1998a). Whether this diversity implies that prehistoric pottery reached Lapita from multiple places of origin within New Caledonia or that multiple types of temper were accessible to local Lapita potters is not yet known. The areal geology of New Caledonia provides no conclusive guidance to the geographic origins of temper types on the island in the absence of detailed empirical information on temper distribution (Dickinson, 2002b). All New Caledonia tempers are distinctive from all other temper types within the southwest Pacific arena, because no other islands within the region of ceramic cultures are microcontinental fragments overthrust by ophiolite, but tracing New Caledonia tempers to specific localities within New Caledonia is not yet feasible.

Loyalty Islands

Five sherds examined from the Kurin site (Sand, 2002) on Maré contain calcareous tempers that may be indigenous to the Loyalty Islands, but sherds from Maré and Ouvea containing quartzose and ophiolitic tempers (Table 23) are inferred to derive from New Caledonia because the Loyalty Islands are paleoatolls uplifted on the flexural forebulge of the New Hebrides Trench (Dickinson, 2001b). The islands are composed almost exclusively of limestone, although weathered outcrops of underlying or intercalated lava and tuff are exposed locally (Paris, 1981, p. 162). There is independent archaeological evidence for regular ceramic transfer of Nera, Oundjo, Plum, and even older wares (Fig. 3) from New Caledonia to the Loyalty Islands (Sand, 1998c).

Île des Pins

Île des Pins is similarly formed mainly by emergent limestone terraces that form a continuous annular ring around restricted exposures of deeply weathered peridotite in the island interior (Paris, 1981). Sherds from Île des Pins (Sand, 1999b) containing ophiolitic temper (Table 23) may derive from the southeast end of New Caledonia (Sand et al., 2001), nearest Île des Pins, where the bedrock is exclusively overthrust peridotite (Guillon, 1975). A majority of the sherds studied from Île des Pins, however, contain quartzose or mixed quartzose-ophiolitic tempers (Table 23) presumably derived from farther north on New Caledonia. Examples of both quartzose and quartzose-ophiolitic temper types in sherds from Île des Pins contain detritus from blueschists exposed mainly near the far northwest end of New Caledonia (Paris, 1981).

Summary Appraisal

Temper relationships on Île des Pins and in the Loyalty Islands suggest widespread prehistoric ceramic transfer from multiple localities on New Caledonia to nearby islands. The likelihood of both local and nonlocal tempers in sherds from sites on the mainland of Grand Terre suggests significant ceramic transfer between multiple sites on Grand Terre as well. The task of sorting New Caledonia tempers into valid geographic groups seems more formidable than any other problem posed by Oceanian temper analysis. The large size of New Caledonia, coupled with its similar geology over long distances, violates the fundamental premise adopted here for Oceanian temper analysis—namely, that islands represent limited sources of sand derived from bedrock of restricted composition. For the large landmass of New Caledonia, this assumption is invalid, and the challenge of New Caledonia tempers underscores the inherent limitations of my methodology for temper analysis used so successfully on smaller islands of Pacific Oceania.

COMPARATIVE TEMPER COMPOSITIONS

Regional comparisons of temper compositions between and within temper classes reveal (1) systematic variations that reflect contrasting geotectonic settings or different stages in the local evolution of island arcs and orogenic belts and (2) more serendipitous differences that reflect vagaries of local bedrock exposures on individual islands. Relative proportions of heavy ferromagnesian mineral grains, light nonferromagnesian mineral grains, and polycrystalline-polymineralic lithic fragments are dependent upon local sedimentological processes rather than regional or subregional geology. Relative abundances of those broad categories of grain type are generally not diagnostic of temper variations controlled by differences in bedrock sources for the temper sands. On the other hand, relative abundances of different kinds of heavy minerals, light minerals, and lithic fragments reflect contrasts in the geologic sources of Oceanian temper types and classes.

Heavy Minerals

The hornblende-pyroxene-olivine (hbl-pyx-olv) ternary plot (Fig. 37) displays generic differences among quartz-free or quartz-poor volcanic sand tempers of the oceanic basalt, andesitic arc, and postarc-backarc temper classes in which heavy ferromagnesian silicate mineral grains are prominent constituents. By contrast, the distribution of opaque ferromagnesian grains in Oceanic ceramic tempers is typically dependent upon local placering effects restricted to specific sand sources, and does not provide a reliable criterion for regional temper variations.

Temper types containing abundant olivine mineral grains (OLi > 25) are restricted to the oceanic basalt and backarc temper classes (Table 24A), the latter where basaltic eruptions occur in oceanic marginal seas. Lesser proportions of olivine (OLi < 25) are present in selected andesitic arc and postarc temper types (Table 24B–24C), in some of which hornblende mineral grains are also present. Temper types that contain both hornblende and olivine in significant proportions, in addition to pyroxene, include postarc tempers from Fiji and selected andesitic arc tempers from the Solomon Islands (Roviana Lagoon) and Bismarck Archipelago (Fig. 37). Olivine-bearing arc and postarc tempers occur locally on selected islands in nearly all Oceanian island arcs. Although diagnostic of tempers from those specific islands, they afford no regional guide to temper origins. Clinopyroxene is the dominant ferromagnesian mineral in all the olivine-bearing arc and postarc temper types, although orthopyroxene is also present in Boduna-Garua sherds derived from nearby New Britain (Table 13A14). The ratio of olivine to pyroxene in all olivine-bearing arc and postarc tempers is less than for oceanic basalt tempers in which olivine in some temper types exceeds clinopyroxene in abundance (Fig. 37).

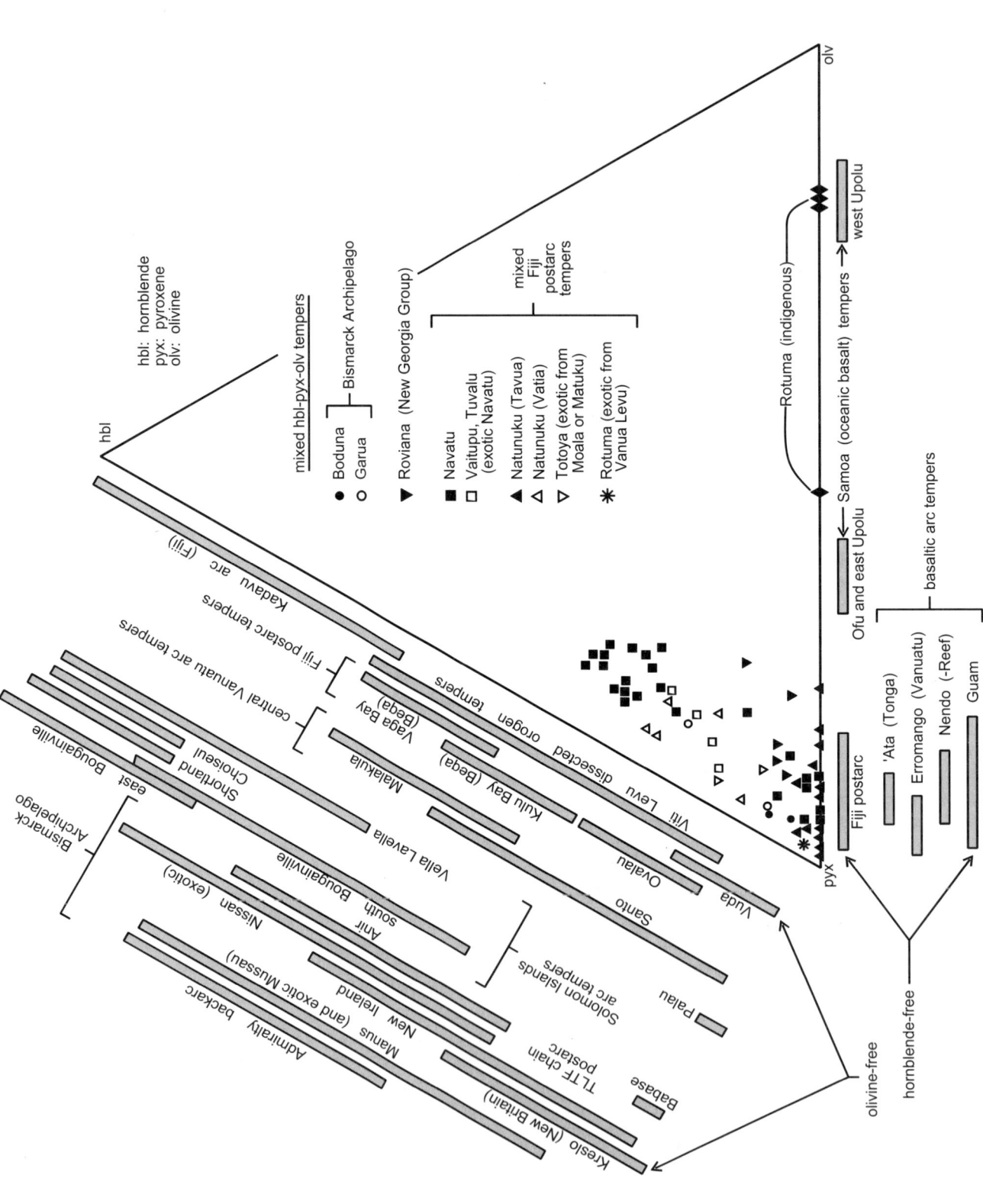

Figure 37. Hornblende (hbl)–pyroxene (pyx)–olivine (olv) triangular compositional diagram for hornblende-bearing and olivine-bearing Oceanian tempers (tempers plotted left of hbl-pyx leg lack olivine; tempers plotted below pyx-olv leg lack hornblende). Unplotted tempers lack appreciable proportions of either hornblende or olivine (inadequately sampled tempers also unplotted). Ovalau (Fiji postarc) tempers in sherds from nearby Moturiki and Naigani. Hornblende-free olivine-bearing Fiji postarc tempers occur in sherds from Cikobia, Lau (kedekede temper), Laucala, Mago, Navua Delta, Qaqaruku rockshelter, Totoya, and Vanua Levu. See Table 5B for poles.

TABLE 24. FERROMAGNESIAN SILICATE MINERAL GRAINS IN MAFIC (QUARTZ-FREE) OCEANIAN VOLCANIC SAND TEMPERS OF DIFFERENT TEMPER CLASSES

Grain types(s)	Oceanic basalt or backarc (BA)	Andesitic arc	Postarc
A - pyx (cpx) + olv	BA-Tikopia (*Van*) BA-Rotuma (*nMb*) Upolu (*Sam*) Ofu (*Sam*)	none	none
B - pyx (cpx) + olv (OLi<25)	Marquesas Is.	Guam (*Mar*) Nendö (*Van*) Erromango (*Van*) Navua Delta (*Fij*) 'Ata (*Ton*)	Natunuku Tavua (*Fij*) Totoya (*Fij*) Laucala (*Lau*) Mago (*Lau*) Kanacea (*Lau*)
C - pyx + olv + hbl	none	Boduna-Garua (*Bis*) New Georgia (*Sol*)	Natunuku Vatia (*Fij*) Navatu (*Fij*)
D - pyx (cpx only), PXi > 95	none	Babeldaob (*Pal*) Banda (*Mal*) Huon Peninsula (*Pap*) Vanikolo-Banks (*Van*) Efate-Shepherd (*Van*) Yasawa (*Fij*) Lakeba (*Lau*)	Feni Is. (*Bis*) Tanga Is. (*Bis*) Cikobia (*Fij*)
E - pyx (cpx + opx), PXi < 95	none	Saipan (*Mar*) Arawe Is. (*Bis*) Watom (*Bis*) Tonga (except 'Ata)	Vanua Levu (*Fij*)
F - pyx (cpx) + hbl (grn)	none	Manus (*Bis*) New Ireland (*Bis*) east Bougainville (*Sol*) Shortland Is. (*Sol*) Vella Lavella (*Sol*) Santo-Malakula (*Van*)	Vuda (*Fij*) Beqa, Kulu Bay (*Fij*)
G - pyx (cpx) + hbl (brn)	none	south Bougainville (*Sol*)	Anir, Feni Is. (*Bis*) Beqa, Vaga Bay (*Fij*) Ovalau (*Fij*)
H - hbl (brn) + oxy + bio	none	Kadavu (*Fij*)	none

Note: Grain types: bio, biotite mica; cpx, clinopyroxene; hbl, hornblende (grn, green; brn, brown or green-brown); olv, olivine; opx, orthopyroxene; oxy, oxyhornblende; pyx, pyroxene (hbl untabulated where PYi>95; opx untabulated where PXi>95); bio and oxy untabulated where <1% by grain frequency count; QZi<1 for all tabulated temper types; see Table 6 for definitions of OLi, PYi, and PXi. Island groups: *Bis*, Bismarck Archipelago; *Fij*, Fiji (exclusive of Lau); *Lau*, Lau Archipelago (eastern Fiji); *Mal*, Maluku (Molucca Islands of eastern Indonesia); *Mar*, Mariana Islands; *nMb*, northern Melanesian borderland; *Pal*, Palau; *Pap*, Papua New Guinea mainland; *Sam*, Samoa (including American Samoa); *Sol*, Solomon island arc (including Bougainville in Papua New Guinea politically); *Ton*, Tonga; *Van*, Vanuatu (New Hebrides) island arc and backarc (including political outliers of Solomon Islands).

Some andesitic arc and postarc temper types lack either hornblende or olivine in more than trace amounts (Table 24D–24E) and would plot at the pyroxene (pyx) pole of the hbl-pyx-olv diagram (not shown on Fig. 37). At least some temper types containing exclusively pyroxenic populations of ferromagnesian silicate mineral grains occur at selected locales along all island arcs with the exception of the Solomon arc. Pyroxenic tempers are especially characteristic of the active Neogene Banda and Tonga arcs, as well as the Lau remnant arc of the Tongan arc-trench system, and are also typical of volcanic sands derived from relict Paleogene arc assemblages exposed along the modern Mariana and Palau island arcs. Pyroxene is exclusively clinopyroxene in many of the pyroxenic temper types (Table 24D), but orthopyroxene accompanies clinopyroxene in others (Table 24E), most notably on Saipan in the Mariana Islands, at Watom off New Britain in the Bismarck Archipelago, and throughout Tonga.

Both clinopyroxene and hornblende mineral grains occur jointly in selected andesitic arc tempers from the Bismarck Archipelago, Bougainville and the Shortland Islands of the Solomon arc, and the central Santo-Malakula segment of the New Hebrides island arc in Vanuatu, and in selected postarc temper types from Fiji and from the TLTF chain northeast of

New Ireland in the Bismarck Archipelago (Table 24F–24G). Combined hornblende-clinopyroxene populations of ferromagnesian mineral grains are also characteristic as subordinate grain types in dissected orogen tempers from the Bismarck Archipelago, the Solomon arc (Choiseul), and Viti Levu in Fiji (Fig. 37). Proportions of hornblende and pyroxene are highly variable in Oceanian tempers, with hornblende more abundant than clinopyroxene in about half the temper types in which it occurs (Fig. 37). In general, the presence or absence of hornblende is more diagnostic of temper origins than its absolute abundance in a given temper, although temper types from the Solomon arc tend to be more hornblende-rich than tempers from elsewhere.

In most Oceanian temper suites, including dissected orogen tempers, hornblende is pleochroic in green tints (Table 24F), but brown to green-brown hornblende occurs as well, and is the dominant variety of hornblende in selected temper types (Table 24G). In detail, the pleochroic schemes of hornblende in Oceanian tempers are varied, with the yellowish tints displayed by some green hornblendes in appropriate crystallographic orientation gradational to the brownish tints of green-brown hornblende, which is in turn gradational in its pleochroic displays to exclusively brown hornblende. Many tempers contain a variety of hornblende types displaying slightly different pleochroic schemes. Deeply colored oxyhornblende is abundant relative to brown hornblende only in temper sands derived from the short segment of a reversed-polarity Pliocene arc assemblage exposed on Kadavu and neighboring islets in southern Fiji (Table 24H). Hornblende coloration in the phenocrystic mineral grains of volcanic rocks is probably a function of the oxidation state of magma before eruption, rather than fundamental differences in magma petrochemistry.

Biotite (black mica) flakes are prominent as temper grains in (1) nonquartzose andesitic arc tempers from Kadavu (Fig. 26), (2) quartzose andesitic arc tempers from the Lease Islands of Maluku (Molucca Islands) in eastern Indonesia (Fig. 18) where biotite is the only ferromagnesian grain type, (3) quartzose felsic backarc tempers of the southern Admiralty Group in the Bismarck Archipelago (Fig. 19) where biotite accompanies both clinopyroxene and hornblende, (4) selected postarc temper sands from the alkalic Tanga Islands of the TLTF chain northeast of New Ireland (Fig. 19), (5) some temper types from Buka at the northern end of the Solomon arc (Fig. 21B), (6) exotic sherds from the calcareous Trobriand Islands (Fig. 19) containing andesitic arc temper that probably reflects ceramic transfer across the Solomon Sea from the Papuan Peninsula to the south, and (7) exotic sherds from the Mochong site on Rota in the Mariana Islands (Fig. 13) containing dacitic Saipan temper. Minor biotite is also present in selected dissected orogen and tectonic highland tempers.

Muscovite (white mica) flakes are prominent as temper grains only in exotic sherds from the Murray Islands apparently derived from Papua New Guinea, in tectonic highland tempers derived from gneissic rocks exposed in the D'Entrecasteaux Islands of the Solomon Sea (Fig. 19), and in quartzose variants of New Caledonia tectonic highland tempers.

Light Minerals

The quartz–plagioclase–K-feldspar (QPK) ternary plot (Fig. 38) displays generic differences among dissected orogen tempers and quartzose variants of andesitic arc tempers in which varied light nonferromagnesian mineral grains are present. Quartzose variants of tectonic highland tempers would plot near the Q pole of the QPK diagram, whereas ophiolitic variants of tectonic highland tempers would plot near the P pole, together with all quartz-free or quartz-poor volcanic sand tempers.

K-feldspar-free quartzose volcanic sand tempers are typically plagioclase-rich ($QZi < 35$), although Saipan dacitic tempers grade to more quartzose variants (Fig. 38). Dissected orogen tempers from southern Viti Levu and postarc Kulu Bay (Beqa) and Vuda (Viti Levu) tempers from Fiji occupy a field in QPK space analogous to the field defined by other orogenic sand suites of the circum-Pacific region as a whole (Dickinson, 1982c). Rhyodacitic volcanic sands from the Lease Islands at the curved northern end of the Banda arc in eastern Indonesia (Fig. 18) plot within the QPK field of dissected orogen tempers from Viti Levu (Fig. 38), but other volcanic sand temper types and most dissected orogen temper types from outside Fiji lack any K-feldspar mineral grains.

Lithic Fragments

The distribution of different generic types of lithic fragments is a prime criterion for recognition of different temper classes, and comes into play as well for identification of selected temper types within each geotectonic class. Useful distinctions among various kinds of volcanic lithic fragments have been discussed in descriptions of specific temper types and are not treated further here. The Mph-Lvm-Lsm ternary plot (Fig. 39) embodying proportions of nonvolcanic as well as volcanic lithic fragments codifies salient contrasts among the key geotectonic temper classes. Strictly volcanic sands derived exclusively from exposures of volcanic bedrock plot at the Lvm pole, whether representative of the oceanic basalt, andesitic arc, or postarc-backarc temper classes. Quartzose variants of tectonic highland tempers, known principally from New Caledonia, plot at or near the Lsm pole. Ophiolitic variants of the tectonic highland temper class plot at or near the Mph-Lvm leg of the diagram, moving from near the Mph pole toward the Lvm pole as the ratio of basaltic volcanic-metavolcanic to ultramafic peridotite-serpentinite lithic fragments increases. Mixed quartzose-ophiolitic tectonic highland tempers from New Caledonia would plot variously near the Mph-Lsm leg of the diagram.

Andesitic arc, postarc, and dissected orogen tempers from 'Eua (Tonga), Fiji, the Solomon arc, and the Bismarck Archipelago lacking any nonigneous lithic fragments plot along the Mph-Lvm leg of the diagram (Fig. 39). Dissected orogen tempers from southern Viti Levu containing a more varied suite of lithic fragments occupy partly discrete and partly overlapping compositional fields within the Mph-Lvm-Lsm plot, which serves as an

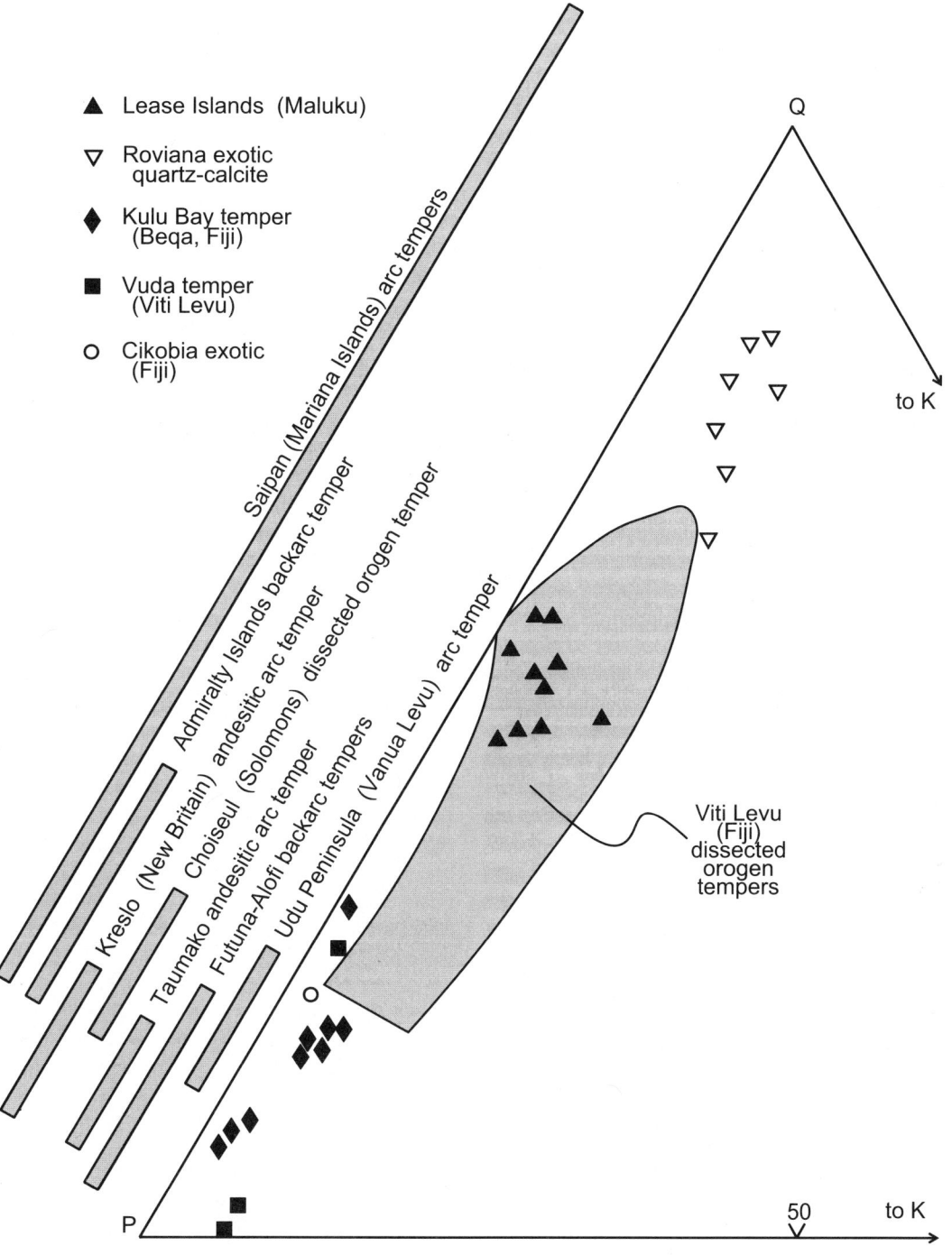

Figure 38. QPK (quartz–plagioclase–K-feldspar) triangular compositional diagram for quartz-bearing Oceanian temper suites (tempers plotted left of Q-P leg lack K-feldspar; unplotted tempers lack quartz or K-feldspar). See Table 5B for poles.

additional means to discriminate among the related temper types (Fig. 39). Petrographic distinctions among dissected orogen temper types based upon populations of lithic fragments are difficult to reproduce, however, both because different kinds of lithic fragments are inherently difficult to identify consistently and because identifications are potentially prone to operator variance.

PATTERNS OF CERAMIC TRANSFER

An initial aim of Oceanian temper analysis was to trace patterns of human migration across the seascape of Pacific islands from the successive transfer of ceramic vessels from island group to island group (Dickinson and Shutler, 1968, 1971, 1979). This

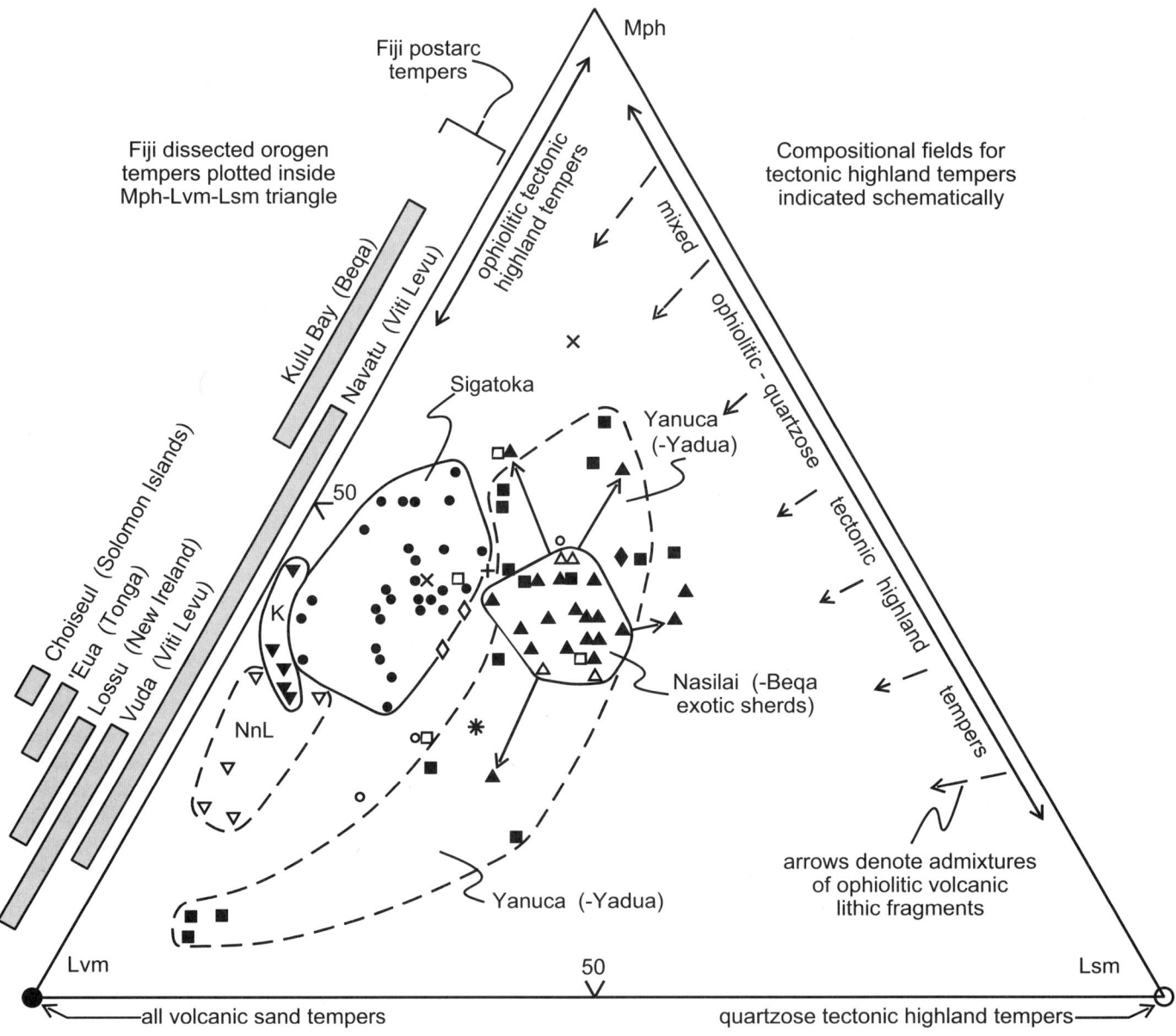

Figure 39. Triangular compositional diagram (Mph-Lvm-Lsm) showing relative proportions of generic types of lithic fragments in Oceanian temper sands (see Fig. 36 for symbols of Fiji dissected orogen tempers from southern Viti Levu; K—Karobo; NnL—Natunuku nonlocal): Mph—microphaneritic igneous (plutonic-hypabyssal) lithic fragments; Lvm—volcanic-metavolcanic lithic fragments; Lsm—sedimentary-metasedimentary lithic fragments (ranges of tempers plotted left of Mph-Lvm leg lack Lsm lithic fragments). See Table 5B for poles.

goal has been largely frustrated by the lack of evidence for systematic ceramic transfer over long interisland distances (Green, 1996; Green and Kirch, 1997; Dickinson and Shutler, 2000). Exploring voyagers and early settlers apparently found it easier to carry potters on long sea passages, and make needed ceramic vessels at landfall, rather than transport bulky and fragile earthenware in open canoes traversing rough seas (Dickinson et al., 1994). In most locales, ceramic suites were made predominantly at or near sites of recovery using local temper sands and presumably local clays. Although ceramic raw materials could have been collected from any places readily accessible, on foot or by paddle canoe, to sites of fabrication and firing, the difficulty of transporting heavy raw materials requiring sturdy containers to carry argues against systematic transport of either temper or clay over long distances unless special circumstances or technical considerations justified the effort involved.

Nevertheless, an extensive record of minor or incidental ceramic transfer over moderate to long interisland distances, and selected instances of ceramic transfer at large scale over shorter interisland distances, has been built up over the years through temper analysis. Petrographic evidence for widespread and locally voluminous ceramic transfer documents instances of close prehistoric cultural ties difficult to sense by other means.

Approximately a third of the sherds examined in thin section (Appendix 1) are fragments of pottery exotic to islands of recovery. This percentage is not a valid measure of the percentage of exotic sherds collected from Pacific archaeological sites because sherds of unusual appearance at each site have been studied preferentially in a deliberate attempt to detect exotic sherds among more abundant indigenous sherds.

Ceramic transfer in prehistoric Pacific Oceania involved movement of wares from high volcanic islands, where terrigenous sands and clay deposits abound, to low-lying islets of atolls where clays are rare and sands are exclusively calcareous; from major islands, serving as cultural centers, to satellite islands of the same island group; and more rarely between bedrock islands of different island groups. Table 25 is a summary of all inferred instances (more than 100) of prehistoric ceramic transfer within Pacific Oceania divided into groupings (A–F) of geographic and cultural significance.

Anomalous Sherds

The following cases of historic or protohistoric movement of pottery into Pacific island groups, either from other islands or from outside the Pacific realm, lie beyond the scope of this report and are not discussed further: (1) protohistoric transport of wares made on the Rewa Delta of Viti Levu in Fiji to Tungua and Lifuka of the Ha'apai Group in central Tonga and to the Ha'atuatua site on Nuku Hiva in the Marquesas Islands (Dickinson and Shutler, 1974; Dickinson et al., 1998b); (2) Spanish pottery from the A.D. 1595 Mandaña expedition recovered from Nendö in the Santa Cruz Group and from Makira in the Solomon Islands (Dickinson and Green, 1974); (3) individual Spanish or other non-Oceanian sherds recovered from Taumako in the Duff Islands, Vanikolo in the Santa Cruz Group, and Pakea in the Banks Islands where they were possibly introduced by the A.D. 1606 Quiros expedition (unpublished data); (4) Japanese Jomon sherds dating to ca. 3000 B.C. that were adventitiously mingled by uncertain means during the past few decades with indigenous sherd collections from Mele Plain on Efate in Vanuatu (Sinoto et al., 1999; Dickinson et al., 1999a); (5) two sherds containing non-Oceanian tempers that probably arrived adventitiously in modern times on Ontong Java atoll (unpublished data); (6) a sherd of probable modern Japanese origin recovered from Kapingamarangi atoll in the eastern Caroline Islands (Dickinson, 1981); (7) a historic sherd of Rume type recovered on Ulithi atoll but most probably derived from Japan or the Philippine Islands (unpublished data); (8) a sherd of uncertain age and affinity recovered from tiny Tobi Island between New Guinea and Palau, but containing biotite-bearing quartzo-feldspathic temper sand unlike any known Oceanian tempers (unpublished data); (9) a sherd of modern kiln-fired ware collected together with recent cultural materials from a marae platform on Atiu in the southern Cook Islands (Dickinson et al., 1998b, p. 130); and (10) an anomalous sherd of uncertain derivation recovered from Anuta in the Vanuatu backarc region (Dickinson, 1973d, p. 110–111), but containing quartzo-feldspathic sand of non-Oceanian aspect.

Micronesian Region

Temper analysis documents widespread ceramic transfer within the Mariana Islands, and dispersal of pottery from Yap and Palau to other Micronesian islands, but there is no evidence for ceramic transfer from ceramic centers farther east in the Caroline Islands.

Intra-Mariana Transfer

As discussed in descriptions of Mariana temper types, prehistoric wares were transported in still indeterminate volume from Saipan (Table 25B1, 25D1–25D2, 25E1) to multiple Mariana islands (Alamagan, Pagan, Tinian, Aguijan, Rota, Guam), and from Guam to Rota (Table 25D2). The cultural reasons for widespread intra-Mariana ceramic transfer from Saipan but not from Guam remain unknown. There is no petrographic evidence for ceramic transfer from the Mariana Islands to other island groups, or vice versa (Dickinson et al., 2001).

From Yap and Palau

Sherds from Babeldaob or a nearby bedrock islet in Palau (Table 25A2, 25D4–25D5) are known from multiple limestone islands in Palau (Ulong, Peleliu, Angaur), nearby Kayangel atoll to the north, and Fana in the Sonsorol Islands to the south (Fitzpatrick et al., 2003), as well as from Ngulu and Fais closer to Yap than to Palau (Table 12). Initial interpretation (Dickinson, 1984) that grog-tempered sherds from Lamotrek in the central Caroline Islands also derive from Palau was faulty, however, because it was made at a time when the use of grog temper on Yap was not appreciated. Reappraisal of the Lamotrek sherds shows that the clay pastes and grog fragments are markedly silty, with a texture characteristic of grog-tempered sherds from Yap rather than Palau (Intoh and Dickinson, 2002). Moreover, the grog-tempered sherds from Lamotrek (Table 11) are associated with other sherds containing hybrid sand temper of Yap origin (Table 21).

Prehistoric Yap wares (Table 25A1, 25D3) were carried to Ulithi, Ngulu, and Fais in the western Caroline Islands (Table 11), as well as to Lamotrek in the central Caroline Islands (Fujimura and Alkire, 1984), documenting a prehistoric pattern of ceramic transfer that mimics the *sawei* exchange network of trade and tribute known from modern ethnographic studies (Takayama, 1981; Intoh, 1996). The presence of Yap sherds in cultural horizons dating back to the early part of the first millennium A.D. on both Ngulu and Fais (Intoh, 1981; Intoh and Dickinson, 2002) indicates the long time depth of intra-Caroline ceramic transfer. Importation of Yap wares to calcareous islets of far-flung atolls

TABLE 25. PATTERNS OF PREHISTORIC CERAMIC TRANSFER IN PACIFIC OCEANIA

Origin island or island group	Destination island(s)	Interisland distance (approximate in km)
A. Limited ceramic transfer (small volume) from high bedrock islands to distant calcareous islands lacking any record of indigenous pottery containing calcareous temper		
1. Yap (western Caroline Islands)	Lamotrek (central Caroline Islands)	900
2. Palau (Caroline Islands)	Ngulu-Fais (Caroline Islands)	350, 700
3. D'Entrecasteaux Islands (*PNG*)	Trobriand Islands (*PNG*)	100
4. Papuan Peninsula, New Guinea (*PNG*)	Trobriand Islands (*PNG*)	175
5. Buka, Solomon Islands (*PNG*)	Ontong Java atoll	550
6. New Georgia (Solomon Islands)	Bellona (Solomon Islands)	400
7. Navatu site, Viti Levu (Fiji)	Temei site, Vaitupu (Tuvalu)	1100
8. Viti Levu or Vanua Levu (Fiji)	Nanumanga-Nanumea (Tuvalu)	1200, 1300
9. Fiji or Lau Archipelago (*FIJ*)	Atafu (Tokelau)	1200
B. Limited ceramic transfer (small volume) from high bedrock islands to isolated or distant bedrock islands lacking any record of indigenous pottery containing local temper		
1. Saipan (Mariana Islands)	Alamagan-Pagan (Mariana Islands)	75, 90
2. New Guinea Highlands (*PNG*)	Dauar-Maer (Murray Islands)	250
3. Udu Peninsula, Vanua Levu (Fiji)	Ugaga (Fiji)	300
4. Viti Levu south coast (Fiji)	Ugaga (Fiji)	20
5. Tonga (locale uncertain)	Mauke (Cook Islands)	1750
C. Limited ceramic transfer (small volume) from high bedrock islands to isolated or distant bedrock islands with abundant indigenous pottery containing local tempers		
1. Lou (Admiralty Group)	New Ireland (Lasigi)	500
2. Manus (Admiralty Group)	Watom (Bismarck Archipelago)	600
3. New Britain (*PNG*)	Watom (Bismarck Archipelago)	30
4. New Caledonia (Grand Terre)	Santo-Malo (Vanuatu)	600
5. New Caledonia (Grand Terre)	Erromango (Vanuatu)	450
6. Vanua Levu (Fiji)	Rotuma (*FIJ*)	450
7. Udu Peninsula, Vanua Levu (Fiji)	Upolu (Samoa)	900
D. Systematic ceramic transfer (large volume) from large bedrock islands to nearby calcareous islands or to smaller bedrock islands lacking any record of indigenous pottery containing local terrigenous tempers		
1. Saipan (Mariana Islands)	Tinian-Aguijan (Mariana Islands)	5, 30
2. Guam-Saipan (Mariana Islands)	Rota (Mariana Islands)	60, 105
3. Yap (western Caroline Islands)	Ngulu-Ulithi-Fais (Caroline atolls)	150, 200, 275
4. Babeldaob (Palau)	Ulong-Peleliu-Kayangel-Angaur	25, 40, 55, 60
5. Babeldaob (Palau)	Fana (Sonsorol Islands)	350
6. Huon Peninsula, New Guinea (*PNG*)	Siassi Islands (Vitiaz Strait)	35
7. New Britain (*PNG*)	Boduna-Garua (offshore)	1, 1
8. New Britain (*PNG*)	Arawe Islands (offshore)	~1
9. New Britain and New Ireland (*PNG*)	Nissan atoll (*PNG*)	150, 225
10. Buka, Solomon Islands (*PNG*)	Nissan atoll (*PNG*)	75
11. Buka, Solomon Islands (*PNG*)	Islet of Teop (off Bougainville)	50
12. Bougainville, Solomon Islands (*PNG*)	Islet of Teop (off Bougainville)	<10
13. Nendö, Santa Cruz Group (*SOL*)	Reef Islands (Santa Cruz Group)	50
14. Santo (Vanuatu)	Torres Islands (Vanuatu)	150
15. Santo (Vanuatu)	Malo (Vanuatu)	3
16. Malakula (Vanuatu)	small offshore islets (n=6)	<1–3
17. Efate (Vanuatu)	small offshore islets (n=2)	1, 4
18. New Caledonia (Grand Terre)	Île de Pins (New Caledonia)	50
19. New Caledonia (Grand Terre)	Ouvea-Maré (Loyalty Islands)	100, 125
20. Vanua Levu (Fiji)	Cikobia (Fiji)	40+
21. Beqa (Fiji)	Ugaga (Fiji)	2.5
22. Ovalau or Moturiki (Fiji)	Naigani (Fiji)	10–15

(*continued*)

was probably to allow collection and simmering of coconut sap to make coconut syrup, and for cooking otherwise toxic varieties of taro (Intoh, 1992b). The stratigraphic distribution of Yap ceramic styles and tempers on Fais and Ulithi is the same as on Yap itself (Craib, 1981; Intoh, 1996; Descantes, 1998; Intoh and Dickinson, 2002) and reflects essentially continuous cultural contacts with Yap through much of Caroline prehistory. Chemical analyses of 225 Yapese sherds from Yap, Ulithi, Fais, and Satawal (near Lamotrek), and of 25 clay samples from Yap, imply that all originated from Yap, although exact localities of manufacture cannot be specified with present information (Descantes et al., 2004).

TABLE 25. PATTERNS OF PREHISTORIC CERAMIC TRANSFER IN PACIFIC OCEANIA (continued)		
Origin island or island group	Destination island(s)	Interisland distance (approximate in km)
E. Incidental ceramic transfer (indeterminate volume) between bedrock islands within culturally coherent island groups (indigenous pottery present at both origins and destinations in most but not all cases)		
1. Saipan (Mariana Islands)	Guam (Mariana Islands)	180
2. Seram (Maluku Islands)	Ambon (Lease Islands)	<10
3. Ambon or Haruku (Lease Islands)	Seram (Molucca Islands)	<10
4. Manus (Admiralty Group)	Lou-M'Buke-Baluan	25, 25, 50
5. Manus (Admiralty Group)	Mussau (Bismarck Archipelago)	250
6. Lou (southern Admiralty Group)	Mussau (Bismarck Archipelago)	275
7. TLTF Chain (Bismarck Archipelago)	Mussau (Bismarck Archipelago)	275
8. New Britain (*PNG*)	Duke of York Islands	18
9. Anir (Feni Islands), TLTF Chain	Tanga Islands, TLTF Chain	75
10. Bougainville, Solomon Islands (*PNG*)	Alu (Shortland Islands)	9
11. Choiseul, Solomon Islands	Kolombangara, Solomon Islands	55
12. Vella Lavella, Solomon Islands	New Georgia, Solomon Islands	85
13. Makira, Solomon Islands	Santa Ana, Solomon Islands	<20
14. Vanikolo or Banks Islands	Tikopia, Vanuatu (*SOL*)	215
15. Santo (Vanuatu)	Vanikolo-Tikopia (*SOL*)	325, 350
16. Santo (Vanuatu)	Malakula-Banks-Efate (Vanuatu)	35, 130, 250
17. Viti Levu north coast (Fiji)	Yanuca site, Viti Levu (Fiji)	60
18. Viti Levu south coast (Fiji)	Viti Levu north coast sites (Fiji)	90
19. Rewa Delta, Viti Levu (Fiji)	Beqa-Totoya (Fiji)	50, 175
20. Viti Levu south coast (Fiji)	Moturiki-Naigani-Lakeba (Fiji)	100, 100, 275
21. Waya (Yasawa Islands, Fiji)	southwest Viti Levu (Fiji)	75
22. Kadavu (Fiji)	Karobo-Nasilai, Viti Levu (Fiji)	75, 100
23. Kadavu (Fiji)	Moturiki, Lomaiviti (Fiji)	125
24. Kadavu (Fiji)	Lakeba, Lau Archipelago (*FIJ*)	300
25. Moala or Matuku (Fiji)	Totoya, Lomaiviti (Fiji)	35
26. Kanacea "kedekede" temper (Lau)	Vanua Balavu-Lakeba-Kabara	15, 100, 170
27. Kanacea "kedekede" temper (Lau)	Totoya-Moturiki-Naigani	200, 230, 235
28. Lakeba (Lau Archipelago)	Aiwa-Nayau (Lau Archipelago)	12, 28
29. 'Eua (Tonga)	Tongatapu (Tonga)	35
F. Incidental ceramic transfer (indeterminate volume) between bedrock islands of different island groups		
1. Banda (Banda Islands)	Seram-Gorom (Molucca Islands)	145, 200
2. Banda (Banda Islands)	Aru Islands	525
3. unknown (quartz-calcite temper)	New Georgia Group	?
4. Vanuatu (locale unknown)	Nukuleka site, Tongatapu (Tonga)	2300
Note: FIJ, PNG, and *SOL* denote islands tabulated by geologic-geographic affinity but politically part of Fiji, Papua New Guinea, and Solomon Islands, respectively.		

Molucca-Arafura Region

Ceramic transfer on an uncertain scale is documented as follows by temper analysis of sherds from small islands in the Banda Sea of eastern Indonesia and the Arafura Sea covering the shallow Sahul Shelf between Australia and New Guinea.

From Banda

Exotic sherds from Seram and Gorom in the southern Molucca Islands and from the Aru Islands (Dickinson, 2005) farther to the southeast (Fig. 18) contain volcanic sand tempers of andesitic arc type that are statistically indistinguishable from island to island in both texture and composition (Table 26), and are set in clay pastes rich in fresh and unweathered shards (Fig. A30) of felsic volcanic glass (vitroclastic texture). The exotic tempers are both texturally and compositionally indistinguishable from indigenous temper on Banda (Table 26), and the ash-rich pastes are also visually indistinguishable from comparable clay pastes in Banda sherds. The combined match of both temper and paste argues that the exotic Seram-Gorom-Aru sherds reflect ceramic transfer from Banda eastward to multiple destinations (Table 25F1–25F2).

The occurrence of unaltered glass shards within clay bodies derived from products of rock weathering is unusual and demands special conditions to form ash-rich clay. The intimate and uniform admixture of fine ash with clay could not be achieved by forcefully stirring ash and clay together without damaging the fragile glass shards, and manually sprinkling ash into clay is precluded as a mode of origin because there are no visible compositional gradients of ash content in the ash-rich clay pastes. The most likely source of the ash-rich clay is an alluvial deposit into which clayey sediment and fine ash jointly settled quietly from slack water. Suitable ash blankets from which the ash in the pastes could be derived are unknown and unexpected

TABLE 26. FREQUENCY PERCENTAGES OF GRAIN TYPES IN BANDA TEMPER AND EXOTIC SHERDS FROM SERAM, GOROM, AND THE ARU ISLANDS

Grain type	Banda (N=4)	Seram (HS-5) (n=222)	Gorom (N=7), (mostly from Giru Gaja, N=6)	Aru (N=10) (mostly from Wangil, N=9)
quartz	tr	tr	tr	tr
plagioclase	35 ± 5 (29-43)	37 ± 3	34 ± 6 (25-40)	37 ± 4 (31-43)
clinopyroxene	4 ± 1 (2-6)	1 ± 1	4 ± 1 (3-5)	6 ± 2 (4-9)
opaque iron oxide	7 ± 2 (5-9)	15 ± 2	10 ± 4 (6-15)	10 ± 5 (4-22)
vitric lithic fragments	52 ± 6 (42-59)	42 ± 3	49 ± 3 (44-54)	44 ± 3 (40-52)
microlitic lithic fragments	2 ± 1 (1-3)	5 ± 2	2 ± 1 (1-3)	3 ± 2 (1-5)

Notes: Figures are means and SD (±) with total observed ranges (for multiple sherds) in parentheses; n is number of grains counted for individual sherd from Seram; N is number of sherds from Banda (180-235 grains counted per sherd), Gorom (235-475 grains counted per sherd), and Aru (80-235 grains counted per sherd); for single sherd (Seram), SD is standard deviation of counting error of a single grain count, as given by the expression $[p(100-p)/n]^{1/2}$ where p is percentage of a given grain type in a single sherd (Van der Plas and Tobi, 1965); for multiple sherds (Banda, Gorom, Aru), SD is standard deviation of variance of the mean for multiple grain frequency counts.

for Seram, Gorom, and Aru because the active volcanic islands of the Banda arc stand >100 km away in the Banda Sea.

Grains of temper sand embedded in the ash-rich clay pastes contrast markedly in size with any paste constituents, and were evidently added manually to the clay bodies by ancient potters. The most abundant temper grains are microvesicular vitric lithic fragments of pumiceous volcanic glass. Subordinate mineral grains of plagioclase (dominant) and clinopyroxene (minor) are inferred to derive from sand-sized phenocrysts in the pumiceous volcanic source rocks. Evident abrasion of subrounded lithic fragments, coupled with lesser abrasion of the harder mineral grains that tend to be subangular, suggests that the source of the temper was reworked volcanic ash of coarser grain size than the glass shards embedded in the clay pastes. Both the paste shards and the temper sand were probably derived from different textural components of the same ash blanket mantling the landscape near the active Banda volcano. Varying proportions of opaque iron oxide grains of high specific gravity in the temper sands probably reflect differing degrees of sedimentological placering during transport and deposition of reworked volcanic ash (OXi range is 36–94 for exotic Seram-Gorom-Aru tempers and 57–71 for indigenous Banda tempers). Small numbers of microlitic volcanic lithic fragments in the tempers may represent limited detrital contributions from volcanic rocks older than the ash eruptions. The vitric lithic fragments are far more abundant, however, amounting by grain frequency to an average of 95% ± 3% (89%–98%) of the total lithic fragments. Traces of quartz in nearly all the thin sections suggest partly dacitic rather than entirely andesitic ash.

Before a close match of both temper and paste with Banda sherds was demonstrated for the exotic Seram-Gorom-Aru sherds, comparative examination of selected Halmahera sherds was undertaken to seek a potential correlation. Halmahera is a feasible geologic source for the exotic wares both because volcanic centers on Halmahera display complex morphology, with calderas reflecting massive pyroclastic eruptions capable of triggering massive ash deposition, and because the Halmahera volcanic suite is known to range petrologically from basalt and basaltic andesite to dacite (Morris et al., 1983; Hall et al., 1988a, 1988b). A sherd with ash-rich paste from Pulau Kumo on Halmahera (Irwin et al., 1999) contains a temper sand closely similar in mineralogical composition to volcanic temper sands in exotic Seram-Gorom-Aru and indigenous Banda sherds. Derivation of the exotic Seram-Gorom-Aru wares from Halmahera was initially inferred from temper analysis. The Pulau Kumo temper is much finer-grained, however, and texturally unlike the tempers with which comparison was sought. The closer match of the exotic tempers with Banda temper is more satisfactory than an origin on Halmahera, which is more distant from Seram, Gorom, and Aru (Fig. 18). The temporarily attractive but erroneous Halmahera connection is an apt reminder that conclusive temper matches require both textural and compositional aspects of comparative tempers to be visually identical and statistically indistinguishable, respectively. Andesitic arc tempers throughout Oceania bear broad generic resemblances that do not imply ceramic transfer.

Within Maluku

As discussed in descriptions of Moluccan temper types, selected exotic sherds from Seram and Ambon (only ~10 km apart) were apparently derived from the other island (Table 25E2–25E3), or in the case of presumed Ambon origin from nearby Haruku or Saparua along the trend of the Lease Islands. The scale of this close interisland ceramic transfer cannot be estimated with available data. Seram and Ambon are close enough together to raise the possibility of temper transport, but the apparent two-way travel suggests incidental ceramic transfer instead.

Murray Islands Sherds

Sparse sherds (n = 4) recovered from the Murray Islands in the Coral Sea east of Torres Strait represent exotic wares probably derived from the New Guinea coast to the north where detritus from the New Guinea Highlands in the interior of Papua New Guinea washes down toward the sea (Dickinson, 2004b). The Murray Islands (Willmott et al., 1973) are formed by a cluster of erosionally degraded basaltic tuff cones and associated alkalic basalt lavas that could yield only volcanic sands of the oceanic basalt temper class. By contrast, coarse sand tempers in two of the Murray Islands sherds are mineralogically complex aggregates (Table 27) derived in part from heterogeneous, though largely mafic, volcanic sources but in part also from quartzose metasedimentary rocks.

The volcanic lithic fragments (VLi = 18–53), all or most of the ferromagnesian silicate mineral grains, many of the feldspar grains, and some of the opaque grains are probably detritus from volcanic sources forming half or more of the temper aggregates. The sedimentary and metasedimentary lithic fragments (SLi = 47–82), most or all of the quartz grains, the muscovite flakes of white mica, and probably some of the opaque grains were derived, however, from nonvolcanic sources forming at least a quarter to nearly half the temper aggregates. Rare composite quartz-feldspar lithic fragments associated with the metasedimentary detritus reflect derivation from associated granitic rocks, from which some of the quartz mineral grains may have been derived as well. Finer temper sands in the other two Murray Islands sherds consist dominantly of quartz and feldspar mineral grains, probably recycled from sedimentary strata, together with sparse volcanic lithic fragments but no metasedimentary detritus. Both types of exotic temper are moderately sorted, subangular to subrounded aggregates of probable stream origin, but the two are mineralogically different. Pleochroic brown hornblende is predominant among the ferromagnesian silicate mineral grains in the finer-grained tempers, whereas clinopyroxene is dominant (PYi > 90) in the coarser-grained tempers (Table 27).

The Torres Strait Islands lying to the west off the tip of the Cape York Peninsula are not a feasible origin for the exotic Murray Islands sherds because indigenous tempers of the dissected orogen temper class in the Torres Strait Islands lack either the mafic volcanic detritus or the metasedimentary detritus observed in Murray Islands sherds. Because prehistoric pottery is unknown from the mainland of Australia, attention is drawn to Papua New Guinea as the potential source of the exotic Murray Islands sherds.

Appropriate sources for the heterogeneous volcanic detritus in Murray Islands sherds occur widely in the Highland volcanic province of Neogene age in the interior of Papua New Guinea (Mackenzie, 1976). Remnants of the Highland volcanic assemblage are distributed over an immense area, 300 km east-west by 200 km north-south. Most of the volcanic edifices rest unconformably on the Papuan foldbelt of deformed middle Mesozoic to middle Cenozoic sedimentary strata from which metasedimentary or granitic detritus could not be derived except by sedimentary recycling. Lithologic descriptions of sedimentary strata exposed within the foldbelt suggest

TABLE 27. FREQUENCY PERCENTAGES OF GRAIN TYPES IN COARSE SAND TEMPERS OF EXOTIC MURRAY ISLANDS SHERDS

Grain type	ORM-2 (n=375)	MER-2 (n=415)
Quartz (monocrystalline; probably derived from granitic rock)	3	10
Feldspar (monocrystalline; predominantly or exclusively plagioclase)	13	23
Pyroxene (monocrystalline; predominantly or exclusively clinopyroxene)	36	16
Hornblende (monocrystalline; red to brown pleochroic tints)	3	-
Olivine (monocrystalline; partly altered to reddish iddingsite)	1	tr
Muscovite (monocrystalline: white mica of metamorphic derivation)	-	3
Opaque iron oxide grains (probably dominantly magnetite)	10	10
Lathwork volcanic lithic fragments (polycrystalline)	4	1
Microlitic volcanic lithic fragments (polycrystalline)	7	4
Felsitic volcanic lithic fragments (polycrystalline)	1	1
Vitric volcanic lithic fragments (partly altered cryptocrystalline)	3	3
(total volcanic lithic fragments summed from above)	(18)	(7)
Chert-metachert (microcrystalline chalcedonic quartz lithic fragments)	1	1
Argillite (optically murky clay-rich clastic sedimentary lithic fragments)	3	3
Hornfels (microgranular metasedimentary lithic fragments)	2	3
Metaquartzite (foliated polycrystalline quartzose lithic fragments)	tr	6
Tectonite (foliated quartz-mica slate-phyllite lithic fragments)	10	18
(total sedimentary-metasedimentary lithic fragments summed from above)	(16)	(31)

Notes: Sherd ORM-2 from excavation on Dauar Island; sherd MER-2 from surface on Maer Island; *n* is number of grains counted in each sherd thin section.

no attractive source strata for recycled metasedimentary and granitic detritus of coarse grain size.

The metasedimentary detritus in the temper sands of Murray Islands sherds draws special attention to the Kubor basement block (Bain et al., 1975; Van Wyck and Williams, 2002) in the headwaters of the Purari (-Wahgi) River where pre-Mesozoic metamorphic and granitic rocks of the Tasman orogen are exposed in the Kubor Anticline beneath Mesozoic and Cenozoic sedimentary cover for a distance of ~75 km along the trend of the Kubor Range (Fig. 7). Descriptions of metamorphic rocks (argillite, slate, phyllite, schist, hornfels) exposed in the Kubor Range closely match the metasedimentary grain types observed in Murray Islands sherds (Table 27), and a large local stock of Kubor Granodiorite could have provided the subordinate granitic detritus in the sherd tempers. As tributary streams within the Purari-Wahgi drainage tap multiple volcanic edifices of the Highland volcanic province, mingling of volcanic and metasedimentary detritus in the coarser-grained temper sands of Murray Islands sherds could be achieved readily within the Purari-Wahgi fluvial system. The apparent dominance of recycled sedimentary detritus in the finer-grained temper sands of Murray Islands sherds may reflect derivation of sand from a separate fluvial system draining the foldbelt along the southwest flank of the Highland volcanic province.

There is currently no archaeological evidence to support derivation of the Murray Islands sherds from the New Guinea Highlands, but no alternate source of combined volcanic and quartzose metasedimentary detritus comes to mind for Papua New Guinea. Although the suggested origin of the Murray Islands tempers is speculative, geologic ties to the Kubor block of the New Guinea Highlands are uniquely attractive (Table 25B2). As no pottery predating ca. A.D. 1000 is known from the New Guinea Highlands, the tempers in Murray Islands sherds may represent Highland-derived sands from the southern New Guinea coast where pottery was introduced ca. 0 B.C./A.D. in close accord with the age of archaeological sites that have yielded prehistoric sherds in the Murray Islands (Carter, 2004).

Bismarck-Solomon Region

Multiple paths of ceramic transfer (Fig. 19) are documented or suggested by temper analysis for the Bismarck-Solomon region lying east of New Guinea at the extremity of Near Oceania. No systematic tendencies for ceramic transfer are evident, except for a general tendency for movement of wares from central to peripheral parts of the overall region. No evidence for ceramic transfer eastward from central or interior New Guinea into the realm of island Melanesia has been detected by temper analysis.

Across Vitiaz Strait

Prehistoric type X sherds (Specht et al., 2006), a distinctive post-Lapita ware with hard shiny surfaces (both exterior and interior), occurs in the region of Vitiaz Strait between the Huon Peninsula of mainland New Guinea and New Britain to the east (Lilley, 1988a). Type X pottery predates modern systems of ceramic exchange in the Vitiaz Strait region known from ethnographic studies to have been operative since A.D. 1650–1700, and was probably made at some time during the interval A.D. 400–1200 (Lilley, 1988b). Type X sherds occur as far east as the Arawe Islands off the southwest coast of New Britain (Fig. 19), and Poi Island off the north coast of New Britain west of the Willaumez Peninsula (Lilley, 1991a), but are most abundant at coastal sites on the Huon Peninsula and in the coralline Siassi Islands lying within Vitiaz Strait south of the Schouten arc (Fig. 19).

Type X sherds examined in thin section contain dominantly grog tempers in which the dominant terrigenous grains are subspherical to ovoid ferruginous particles probably derived from pedogenic nodules and providing no evidence of specific geographic origin. In three type X sherds containing composite tempers, however, the terrigenous grains are volcanic sand of the andesitic arc class (Fig. 20) lacking quartz, K-feldspar, or olivine, containing only minor opaque grains, and with clinopyroxene the dominant grain type (Table 13A16). The volcanic lithic fragments are a heterogeneous but undiagnostic population with internal microlitic and vitric textures in subequal proportions (50% ± 10% each). By inference, the volcanic sand was derived from exposures of Middle Eocene to Lower Miocene volcanic and volcaniclastic strata in the interior of the Huon Peninsula (Jaques, 1976; Abbott et al., 1994a). Volcanogenic bedrock is overlain along the coastal flank of interior uplands by Middle Miocene and younger limestone including coastal terraces of emergent Quaternary reef complexes (Jaques and Robinson, 1977; Abbott et al., 1994a). The limestone terraces block most transport of volcanic sand to the coast, and the use of grog as temper in type X sherds may reflect the paucity of terrigenous sand deposits along the coast of the Huon Peninsula.

The volcanogenic Huon Peninsula bedrock is a westward continuation of the ancestral Vitiaz arc assemblage exposed on New Britain in the Bismarck Archipelago (Figs. 7 and 9), and petrographic distinction between derivative detritus from mainland and island sources is not feasible. Postulated derivation of type X sherds from the Huon Peninsula (Table 25D6) is in accord, however, with studies of clay pastes suggesting type X manufacture somewhere along the east coast of the Huon Peninsula (Lilley, 1988a; Specht et al., 2006).

Intra-Bismarck Transfer

As outlined in discussions of Bismarck andesitic arc and dissected orogen tempers, wares derived from Manus (Table 25E4) are widespread on smaller islands lying farther south within the Admiralty Group (Fig. 19), rare exotic sherds from Watom document limited ceramic transfer from nearby New Britain (Table 25C3) and more distant Manus (Table 25C2), and exotic Lapita wares probably from New Britain or New Ireland or both have been recovered from the Duke of York

Islands (Table 25E8) lying east of Watom in the strait between New Britain and New Ireland, and from Nissan atoll to the east of New Ireland (Table 25D9). A single anomalous sherd from Lasigi on New Ireland containing hybrid quartzose temper sand was probably derived from the Admiralty Group 500 km to the northwest (Table 25C1), although whether ware was brought directly to New Ireland or arrived as a result of down-the-line exchange (Green and Kirch, 1997) through intervening and still-unknown archaeological sites on New Hanover and New Ireland is indeterminate with present information. Most other sherds recovered from the Bismarck Archipelago represent indigenous wares from widespread local pottery manufacture if ceramic transfer from New Britain (Table 25D7–25D8) to small offshore islets (Apalo-Adwe in the Arawe Islands; Boduna-Garua off the Willaumez Peninsula) is discounted (Thomson and White, 2000; Dickinson, 2000a).

Mussau Suite

Sherd collections from the Mussau (or Saint Matthias) Group at the northern fringe of the Bismarck Archipelago (Fig. 19) provide a notable exception to the rule that most Bismarck wares are indigenous (Green and Kirch, 1997; Dickinson, 2006a). Archaeological sites on the coral islets of Eloaua and Emananus lying 3–6 km south of the main island of Mussau (Kirch, 1987, 1988b, 2001a) include the oldest known Lapita sites, dating to ca. 3500 B.C. (Kirch, 1988a, 2001b; Kirch et al., 1991) and among the most prolific in terms of ceramic yield. Scanning electron microscopy (SEM) of clay pastes from Mussau sherds imply that most represent wares brought to Mussau from elsewhere because only 4% of 172 analyzed clay pastes closely resemble the compositions of clay tiles prepared from local Mussau clays (Kirch et al., 1991).

Many Mussau sherds contain calcareous temper that is undiagnostic of origin, but petrographic study of 53 sherds containing hybrid and terrigenous temper sands confirms that all are exotic to Mussau (Dickinson, 2006a). The varied temper types in Mussau sherds are predominantly well-sorted aggregates of rounded to subrounded sand implying manual addition of beach sands to clay bodies. The prevalent use of beach sands as temper accords well with the coastal settings of nearly all known Lapita sites in the Bismarck Archipelago (Gosden et al., 1989).

The islets of Eloaua and Emananus are composed entirely of Quaternary limestone and derivative sediment and are flanked by beaches of exclusively calcareous sand. The subtidal waters of Malle Channel separating the offshore islets from the Mussau mainland effectively block transport of sand from strandlines of Mussau to reefs and beaches surrounding the offshore islets. On Mussau, an island core of igneous rock is flanked by elevated coastal terraces underlain by Neogene limestone (Kirch and Catterall, 2001). The nature of Mussau bedrock is poorly known, but a dozen control samples collected from bedrock exposures to guide interpretation of manuports (portable tools) in Mussau artifact collections are dominantly plutonic diorite and gabbro (Dickinson, 2006b). The absence of plutonic detritus in the temper sands of Mussau sherds studied in thin section argues against derivation of the Eloaua-Emananus sherd assemblage from nearby Mussau.

As noted in discussions of Bismarck tempers, the prevalence of Admiralty backarc tempers (Table 16A2) and Vitiaz arc tempers (Table 13A3–13A6) in Mussau sherds containing terrigenous or hybrid tempers implies ceramic transfer (Table 25E5–25E6) principally from Manus and other islands of the Admiralty Group (Dickinson, 2006a). This inference is supported by data showing that clay pastes in selected Mussau sherds are geochemically indistinguishable from clay bodies used by modern potters on M'Buke Island in the Admiralty Group south of Manus and on tiny Ahus islet off the north coast of Manus (Kirch et al., 1991). On strictly petrographic grounds, however, the Vitiaz arc tempers in Mussau sherds might derive from New Hanover (Lavongai) or New Ireland to the south, rather than or as well as from Manus to the west (Fig. 19). Bismarck postarc tempers containing the clinopyroxene aegirine-augite include selected Eloaua sherds (Table 16A3) containing distinctive volcanic sand temper derived from the TLTF island chain (Table 25E7) east of New Ireland (Fig. 19). The TLTF sherds at Mussau provide petrographic evidence for ceramic transfer to Mussau from the southeast as well as from the west, possibly via Lavongai as a way station (Fig. 19).

Tanga Suite

The majority (80%–85%) of Lapita sherds from the Tanga Islands (Fig. 19) studied in thin section contain indigenous feldspathic stream sands or ferromagnesian beach sands in which clinopyroxene (aegirine-augite) is the only ferromagnesian silicate mineral present. Sand samples from beaches on Tanga islets have similar feldspathic or placer compositions, and lack any grains of hornblende. The presence of significant hornblende mineral grains in subordinate numbers (~15%) of Tanga sherds examined in thin section provides evidence for ceramic transfer from elsewhere.

The hornblende-bearing tempers in Tanga sherds closely resemble, both visually and compositionally, tempers in selected sherds from the Balbalankin and Malekolon Lapita sites on Anir in the Feni Islands, lying 75 km to the southeast (Fig. 19). Similarities include the dominance of honey-brown volcanic glass grains among the volcanic lithic fragments in selected pairs of Tanga and Anir sherds in which 85%–95% of the volcanic lithic fragments are the characteristic glassy grains. The hornblende mineral grains in both exotic Tanga and indigenous Anir tempers are hastingsitic varieties with a hint of bluish green tones in their pleochroic schemes. The pyribole indices for indigenous Anir tempers and hornblende-bearing exotic tempers in Tanga sherds are closely comparable (Table 16A6–16A8). Although quantitative temper matches are not exact (Fig. 31), the close qualitative relationships of the two subsets of sherd tempers is strong evidence that the Tanga sherds containing exotic hornblende-bearing tempers reflect ceramic transfer, during Lapita times, from Anir in the Feni Islands to the Tanga Islands (Table 25E9).

Across the Solomon Sea

Six exotic sherds from the Trobriand Islands collected decades ago by the late C.O. Key (Australian National University) are surface sherds from uncertain archaeological contexts that document ceramic transfer northward across the Solomon Sea at some unknown stage of prehistory. As noted in the discussion of tectonic highland tempers from the D'Entrecasteaux Islands, four of the exotic Trobriand sherds had their origins in that island group lying ~100 km to the south (Table 25A3). Their occurrence in the Trobriand Islands documents some level of ceramic transfer northward across the Solomon Sea.

The temper in two other Trobriand sherds is poorly sorted and subangular to subrounded volcanic sand (Fig. A18) that was probably natural temper embedded in a clayey volcaniclastic deposit. The temper aggregate contains no contributions from granitic or gneissic rocks of the D'Entrecasteaux core complexes, and must derive from elsewhere. Frequency percentages of grain types reflect a simple mineralogy for the source rocks: plagioclase, 57%; green hornblende, 26%; biotite, 5%; opaque iron oxides, 1%; volcanic lithic fragments (dominantly vitric and vitrophyric), 11%. The origin of the volcanic sand temper is indeterminate with present information, but the abundance of biotite provisionally suggests a continental arc assemblage, and derivation from the Papuan Peninsula just to the south of the D'Entrecasteaux Islands seems most likely from a geographic standpoint (Table 25A4). Unfortunately, no petrographic data are presently available for abundant ceramic collections from the south side of the Papuan Peninsula (Irwin, 1991; Bickler, 1997; Bulmer, 1999).

Outward from Buka

Tempers in sherds from a range of ceramic traditions extending from Lapita to recent times on Buka (Fig. 3F), an island just north of Bougainville (Fig. 21B), were first studied petrographically by Key (1969). Multiple local terrigenous temper types contain characteristic microphaneritic lithic fragments of dolerite, the hypabyssal intrusive equivalent of basalt or mafic andesite. Identical microphanerite grains are present in unfired clay bodies obtained from modern Buka potters. Analogous temper grains are rare, where present at all, in other known tempers of either the Bismarck Archipelago or the Solomon Islands. Ceramic transfer from Buka to both nearby and distant islands can be detected without ambiguity for sherds containing terrigenous temper sands with the diagnostic microphanerite lithic fragments as a major constituent. An estimated 80% of indigenous Buka sherds, however, contain exclusively or dominantly calcareous tempers that are undiagnostic of origin (Table 4.10 of Wickler, 2001a).

Buka tempers are typically feldspathic, with plagioclase feldspar the most abundant grain type in most variants. Subordinate but ubiquitous grain types, in order of overall abundance, include the microphanerite lithic fragments, clinopyroxene mineral grains, opaque iron oxide grains, and volcanic lithic fragments, most commonly with microlitic internal textures but typically deeply weathered. Rare temper grains, absent from some variants, include weathered red-brown biotite, occurring both as flakes and as compact books of mica, and dominantly green hornblende.

Much of Buka is an undulating plateau underlain by Pleistocene Sohano Limestone, but the rugged Parkinson Range and Talof islet to the south are formed by exposures of the volcaniclastic Buka Formation of Paleogene to Middle Miocene age (Blake and Miezitis, 1967; Hilyard and Rogerson, 1989; Rogerson et al., 1989). The source for the volcaniclastic detritus was an assemblage of Eocene to Oligocene arc volcanic rocks now largely buried beneath Neogene volcanic edifices on nearby Bougainville. Modern Buka potters reside mainly along the east coast, where only Sohano Limestone is exposed, but obtain temper by scraping inland outcrops of lithic tuff or volcanic sandstone in the Buka Formation with mollusk shells, adding the temper to clays lacking sand-silt impurities (Specht, 1972). The clay bodies are collected from a swamp occupying an infilled tidal channel (Key, 1969). At least one of the hilltops in the Hahan Hills flanking the Parkinson Range is described as pockmarked with temper pits, and local lithologic variability within the Buka Formation is evidently sufficient to allow multiple temper variants with the characteristic microphanerite lithic fragments to be collected at one locale or another on Buka. The poor sorting and subangular nature of Buka temper sands suggest natural temper from textural considerations alone, but are also compatible with derivation from the volcaniclastic Buka Formation. The abundance of hypabyssal intrusive detritus, as microphanerite lithic fragments of dolerite, can be ascribed to deep dissection of Eocene-Oligocene volcanic edifices when the Buka Formation was deposited, well before later eruption of widespread Neogene volcanic cover on Bougainville.

Characteristic Buka tempers are present in exotic post–A.D. 1000 sherds of Malasang, Mararing, and Recent ceramic styles (Fig. 3) from the islet of Teop (Dickinson, 1975c) off northeast Bougainville (Table 25D11) and from Nissan atoll (Kaplan, 1976) to the north of Buka (Table 25D10), and also occur in five sherds of uncertain age collected from the surface on Ontong Java atoll (Table 25A5) lying 500 km east of Buka (Fig. 4).

Intra-Solomon Transfer

As noted in discussions of andesitic arc and dissected orogen temper types, documented ceramic transfer within the Solomon island chain involved varying distances of <100 km from Bougainville, Choiseul, and parts of the New Georgia Group to neighboring islands (Table 25D12, 25E10–25E12). Ceramic transfer of at least 20 km is implied also for sherds from Santa Ana (Davenport, 1972; Dickinson, 1978), a dominantly calcareous island at the southeastern end of the Solomon chain (Fig. 21A). Two of the five sherds examined from Santa Ana contain exclusively calcareous tempers, but two other tempers are hybrid beach sands with mean proportions of terrigenous grain types indicative of volcanic derivation (frequency percentages): plagioclase, 28%; clinopyroxene, 52%; opaques, 2%; volcanic lithic fragments, 18%. The temper in a fifth sherd (from Feru rockshelter) is a mineralogically more complex sand in which abundant volcanic detritus is admixed with minor intrusive detritus (frequency percentages of grain types): composite quartz grains and chalcedonic quartz grains, 7%; fresh and altered plagioclase, 20%; clinopyroxene, 7%; opaques, 5%; volcanic lithic fragments, 60%; microphanerite lithic fragments, 1%. None of the terrigenous tempers in the Santa Ana sherds are diagnostic, but origin on Makira (Fig. 21A) in the central province of the Solomon chain is inferred from overall geographic and geologic relations (Table 25E13).

New Georgia to Bellona

A sherd from an excavated archaeological site dating to ca. 0 B.C./A.D. (Polach, 1972) on Bellona (Poulsen, 1972), contains pyroxene-rich placer temper indistinguishable compositionally from placer tempers in sherds from Roviana Lagoon in the New Georgia Group (Fig. 22). Bellona is an isolated island composed entirely of emergent reef limestone uplifted on the flexural forebulge of the San Cristobal Trench (Dickinson, 2001b). Ceramic transfer (Table 25A6) of ~400 km is indicated by the temper correlation (Fig. 4).

Exotic Roviana Sherds

From Roviana Lagoon in the New Georgia Group of the Solomon Islands (Fig. 21C), 5%–10% of the sherds in an extensive collection of late-Lapita and post-Lapita wares (Felgate, 2001) contain an unusual hybrid quartz-calcite temper sand (Fig. A29) of unknown derivation (Felgate and Dickinson, 2001). The well-sorted and rounded to subrounded aggregates are beach sands collected from some tropical shoreline of uncertain location. Most Roviana sherds contain andesitic arc tempers indigenous to the New Georgia Group, but the quartz-calcite tempers exotic to New Georgia (Table 25F3) occur in sherds of a typology atypical for the Roviana sherd suite as a whole.

Although the quartz-calcite tempers contain variable admixtures (10%–80%) of calcareous reef debris, all are closely related generically, with terrigenous fractions that vary within the following compositional limits, expressed as frequency percentages of terrigenous grains determined from traverse counts of seven thin sections stained for both feldspars: quartz grains, 58%–76%; plagioclase feldspar grains, 10%–22%; K-feldspar grains, 5%–9%; opaque iron oxide grains, 2%–16%; aggregate epidote grains, 0%–6%. The consistent presence of both plagioclase and K-feldspar ($PFi = 56-71$) suggests ultimate derivation of the sand from granitic rock, and the minor epidote in the temper sands could derive from deuteric or hydrothermal alteration within a plutonic complex. Rare grains of hornblende and clinopyroxene (approximately one mineral grain of one or the other per thin section) offer no diagnostic information on geologic sources. The abundance of quartz relative to feldspar ($QZi = 64-82$), in comparison to the observed abundances of those minerals in granitic rocks, leaves open the question whether the terrigenous detritus is first-cycle debris from tropical weathering of granitic bedrock or was generated by repetitive sedimentary reworking of granitic detritus over multiple cycles of weathering and erosion. The terrigenous fraction of the quartz-calcite temper is more quartzose than any other Oceanian temper sands derived in whole or in part from plutonic bedrock (Fig. 38).

Attractive geologic sources for the exotic Roviana temper are not apparent. As discussed elsewhere in this report, plutonic complexes of the Solomon Islands, the Bismarck Archipelago, and Vanuatu are generally mafic in overall character, and not expected to yield either abundant quartz or appreciable K-feldspar to derivative detritus. From the exclusively plutonic provenance of the terrigenous detritus in the temper sands, an island exposing only granitic rock surrounded by reef tracts would provide the most feasible origin for the hybrid temper sand, but no island of that description comes to mind within Pacific Oceania. Islets off the Queensland coast of Australia display attractive geologic relations, but no ceramic cultures are known from that region. Some special island locale not yet described geologically may be the answer to the puzzle.

New Caledonia–Vanuatu Region

Temper analysis reveals evidence for widespread ceramic transfer within New Caledonia, systematic ceramic transfer from New Caledonia to nearby islands, and ceramic transfer among islands within Vanuatu, but only limited evidence for ceramic transfer from New Caledonia to Vanuatu, and none for ceramic transfer from Vanuatu to New Caledonia.

From New Caledonia

The discussion of New Caledonia tectonic highland tempers highlighted widespread ceramic transfer on a large scale from New Caledonia to the nearby Loyalty Islands and Île des Pins within ~100 km of New Caledonia (Table 25D18–25D19). Four sherds of New Caledonia origin from Vanuatu document ceramic transfer in at least limited quantities for much greater distances of 400–600 km to the north and northeast (Dickinson, 2001a, 2002b). One exotic sherd from Malo (Table 25C4) containing spinel-bearing New Caledonia temper (Fig. A28) of mixed quartzose-ophiolitic character is definitely of Lapita age (Dickinson, 1971b), and an undecorated exotic sherd from Erromango (Table 25C5) containing typical quartzose New Caledonia temper (Dickinson, 2001a) may have been reworked from a Lapita horizon. Two surface sherds of indeterminate age from Santo (Table 25C4) contain especially quartz-rich New Caledonia temper (Dickinson, 2001a).

Intra-Vanuatu Transfer

Ceramic transfer on variable scales of both distance and volume has been discussed for andesitic arc tempers of the New Hebrides island arc (Fig. 23) during various phases of Vanuatu prehistory. Nevertheless, all the larger principal islands (Nendö, Santo, Malakula, Efate, Erromango) that have yielded voluminous sherd collections supported indigenous ceramic industries that account for the overwhelming bulk of ceramic assemblages in each case.

The most massive ceramic transfer of both Lapita and post-Lapita wares occurred from the bedrock island of Nendö to the low-lying coral islets of the Reef Islands (Table 25D13) within the Santa Cruz Group (politically part of the eastern Solomon outliers). Systematic ceramic transfer from the large islands of Efate, Malakula, and Santo to small offshore islets nearby was also widespread (Table 25D15–25D17), and of particular significance for Lapita sites on Malo near Santo. There was more limited ceramic transfer of post-Lapita Mangaasi wares from Vanikolo or the Banks Islands (or both) eastward to Tikopia (Table 25E14), and from Santo northward to the Torres Islands (Table 25D14). There is further limited temper evidence for ceramic transfer of post-Lapita wares in uncertain volume from Santo to nearby Malakula, southward to Efate, and northward to the Banks Islands, Tikopia, and Vanikolo (Table 25E15–25E16). Taken together, multiple lines of evidence for ceramic transfer along the New Hebrides island arc, and into the backarc region, imply that at least incidental transport of pottery over distances as great as 250–350 km was common during Vanuatu prehistory.

Fiji-Lau Region

The temper evidence for prehistoric ceramic transfer among multiple islands within the integrated cultural sphere of Fiji, including the Lau Archipelago, is more extensive than for any other island cluster of comparable size within Pacific Oceania (Dickinson and Shutler, 2000). Wares were transported among different temper provinces of the large central island of Viti Levu, from Viti Levu and Vanua Levu to satellite islands, from selected satellite islands to Viti Levu, and among several of the satellite islands lying at varying distances (50–250 km) from Viti Levu. The record of compound ceramic transfer that has thus far come to light suggests that further evidence for ceramic transfer within Fiji will be detected by future temper analysis as archaeological research within Fiji is expanded.

Within Viti Levu

Selected sherds from north coastal sites (Vuda, Natunuku, Navatu) contain dissected orogen tempers reflective of ceramic transfer from locales farther south on Viti Levu where the Wainimala orogen is exposed (Fig. 26). The nonlocal tempers do not closely match tempers from any well-studied south coastal sites (Fig. 36) and are inferred to derive from elsewhere in southern Viti Levu (Table 25E18). Conversely, selected nonlocal sherds (Dickinson, 1980b) from the south coastal site of Yanuca (Table 25E17), where a distinctive variety of dissected orogen temper is predominant in local sherds (Fig. 36), contain volcanic sand of the postarc temper class (Fig. 32). Derivation from the northern part of Viti Levu is indicated (Dickinson, 1971a), but the nonlocal Yanuca temper is not a close match for postarc temper sands from known north coastal sites because it wholly lacks hornblende or olivine (Table 16B14). Accurately specifying patterns of prehistoric ceramic transfer within Viti Levu must await analysis of Viti Levu ceramic assemblages from as yet undiscovered archaeological sites, which are probably abundant on such a large island.

From Viti Levu

Dissected orogen tempers reflecting ceramic transfer from the south coast of Viti Levu are present in selected exotic sherds from Beqa and Ugaga (Table 25B4) lying only 12–20 km offshore, from Lomaiviti (Naigani, Moturiki, Totoya), and from Lakeba in Lau (Figs. 26 and 28). The exotic Beqa and Totoya sherds contain temper sands texturally and compositionally indistinguishable from temper sands in Nasilai sherds (Fig. 36; Table 18C8–18C9, 18C12), and are inferred to derive from the Rewa Delta on Viti Levu (Table 25E19). The exotic Viti Levu sherds from Ugaga, Naigani, Moturiki, and Lakeba do not as closely match any known temper sands from southern Viti Levu (Fig. 36; Table 18C), and may derive from other parts of the Rewa Delta or even elsewhere along the south coast of Viti Levu (Table 25E20). Selected sherds from Waya in the Yasawa Islands west of Viti Levu have been interpreted independently, on the basis of combined petrographic and geochemical data (Aronson, 1999; Bentley, 2000; Cochrane and Neff, 2006), to reflect ceramic transfer from southwestern Viti Levu (Table 25E21).

From Vanua Levu

Evidence for ceramic transfer across a span of ≥40 km from the Udu Peninsula and elsewhere on Vanua Levu to nearby Cikobia (Table 25D20) was discussed with Fijian postarc temper suites. Two anomalous sherds from Ugaga south of Viti Levu (Fig. 26) contain felsic volcanic temper sand inferred provisionally to derive also from the Udu Peninsula of Vanua Levu (Table 25B3). The temper in one of the two sherds is a moderately sorted and subangular to subrounded aggregate interpreted as stream sand, but better-sorted volcanic temper sand in the other sherd is hybrid beach sand mixed with subequal proportions of calcareous reef debris. The composition of the terrigenous sand implies derivation exclusively from dacitic volcanic rock (frequency percentages of grain types): quartz, 56%; plagioclase, 38%; felsitic volcanic lithic fragments, 3%; opaque iron oxide grains, 2%; clinopyroxene, 1%. The feldspar is untwinned and optically negative sodic oligoclase that can be confused with K-feldspar without staining. Although the temper sand is more quartzose than Udu Peninsula tempers in sherds from Cikobia (Fig. 27; Table 13D1), with $QZi > 50$ instead of $QZi < 25$, the Udu Peninsula of Vanua Levu exposes the only voluminous dacitic volcanic assemblage in Fiji.

From Kadavu

Hornblende-rich and biotite-bearing temper sands reflecting ceramic transfer from Kadavu (Fig. 26) to other islands occur in selected subordinate sherds (Table 25E22–25E24), including some in Lapita horizons, from Lakeba in Lau, Moturiki in Lomaiviti, and both the Rewa Delta (Nasilai) and the Navua Delta on Viti Levu (Fig. 27; Table 13D5). The widespread occurrence of exotic Kadavu wares suggests that archaeological surveys of Kadavu should be undertaken to search for as yet unknown ancient ceramic sites.

Beqa to Ugaga

Both Lapita and post-Lapita ceramic transfer from Beqa to the small islet of Ugaga, lying only 2.5 km to the southwest (Fig. 26), is well documented by the presence of several distinctive Beqa temper types in Ugaga sherds (Table 25D21), as detailed in discussions of Fijian postarc temper suites. Sherds derived from Viti Levu and Vanua Levu at Ugaga are much less abundant than the Beqa-derived sherds, which reflect systematic ceramic transfer to a tiny islet where pottery was probably not made locally. The presence in Ugaga sherds of tempers from multiple locales on Beqa, as well as from more distant Viti Levu and Vanua Levu, suggests that finished wares, rather than temper sands, were brought to Ugaga.

Within Lomaiviti-Lau

As noted in discussions of Fiji postarc temper suites, temper analysis has documented systematic Lapita ceramic transfer on a large scale from Moturiki or Ovalau to the nearby small islet of Naigani (Table 25D22), lying only 10–15 km to the northwest in northern Lomaiviti (Fig. 26). Ceramic transfer from Lakeba in Lau to the smaller and calcareous nearby islands of Aiwa and Nayau (Table 25E28) was also apparently on a massive though still indeterminate scale. The scale of ceramic transfer from Moala or Matuku to Totoya (35–40 km distant) in southern Lomaiviti is unclear from present information (Table 25E25). Unlike tiny Naigani farther north, or Aiwa and Nayau to the east in Lau, Totoya apparently supported local ceramic manufacture, and ceramic transfer from elsewhere may have been limited.

Kedekede Temper

Distinctive *kedekede* placer temper (Table 16C15) of beach origin but uncertain derivation is widespread in sherds from Lau (Fig. 28) and islands along the eastern fringe of the Fiji platform in Lomaiviti (Fig. 26). Well-sorted but nonhybrid kedekede temper is composed predominantly of clinopyroxene mineral grains, with minor olivine (OLi < 10) but no hornblende or orthopyroxene present. Examples are known in selected sherds from Kabara (Smart, 1965), Lakeba, Naigani, Moturiki, Totoya, and Vanua Balavu. Derivation of kedekede detritus from the subalkalic Pliocene postarc basaltic assemblage of the Lau Ridge seems assured, but the exact island (or islands) of origin remains uncertain. Following Best (1984), derivation of kedekede temper from Kanacea in northern Lau (Table 25E26–25E27) is inferred provisionally because Kanacea is the only island in Lau or Lomaiviti where indigenous beach placers of appropriate pyroxene-rich character are known. Best (2002, p. 35) has further noted that prehistoric caches of black sand temper discovered on both Lakeba and Naigani suggest that kedekede temper may have been transported in bulk to multiple islands as a raw material for ceramic manufacture. In all of Pacific Oceania, kedekede temper is the most attractive candidate for transfer of temper rather than finished wares, perhaps because ancient potters on small islands were unable to locate adequate local sources of terrigenous temper sands. Among the closely spaced small islands of Lomaiviti and the Lau Archipelago, seaborne transport of a variety of cultural materials may also have been more common than elsewhere in Pacific Oceania.

Outward from Fiji

Conclusive or suggestive temper evidence for ceramic transfer northward from Fiji to far-flung island groups implies that prehistoric Fijian wares were dispersed in limited quantities for distances of 500–1000 km to the north of Fiji.

Tuvalu

Sparse sherds (n = 4) recovered from the deepest cultural horizon (dating to ca. A.D. 1000) at the Temei site on Vaitupu in Tuvalu (Fig. 4) contain temper sands compositionally and texturally indistinguishable from Navatu temper of northeast Viti Levu (Table 25A7). Microprobe analyses of clinopyroxenes in Navatu and Vaitupu sherds has shown that their internal compositions are also indistinguishable (Dickinson et al., 1990). Most diagnostic petrographically of the temper correlation is the prominence in both sherd suites of microphanerite lithic fragments that are identical in optical appearance and occur in statistically indistinguishable proportions (Table 28). Well-sorted and subrounded hybrid volcanic sand tempers of beach origin in two other Temei sherds from Vaitupu, and in sparse sherds from excavations on other atolls in northernmost Tuvalu (five sherds from Nanumanga; one sherd from Nanumea), do not contain Navatu temper (Table 28). Terrigenous fractions of the temper sands have frequency compositions compatible, however, with derivation from elsewhere along the north coast of Viti Levu, or quite possibly from Vanua Levu, which lies somewhat closer to Tuvalu (Table 25A8). The occurrence of sparse Fijian sherds on several calcareous atolls of Tuvalu documents limited but widespread ceramic transfer during post-Lapita phases of Fijian prehistory.

Tokelau

A single sherd containing volcanic sand temper has been recovered from Atafu atoll in Tokelau (Best, 1988), lying east of Tuvalu to the north of Samoa (Fig. 4). Frequency percentages of grain types (clinopyroxene mineral grains, 45%; plagioclase feldspar mineral grains, 35%; volcanic lithic fragments, 15%; opaque iron oxide grains, 5%) preclude derivation of the sherd from nearby Samoa where all indigenous tempers contain significant proportions of olivine. Derivation from Lau or elsewhere in Fiji is suggested by the temper composition, but derivation from Tonga is precluded by the absence of orthopyroxene. The proportions of different kinds of volcanic lithic fragments in the temper sand are compatible with Fijian derivation (lathwork, 7%; microlitic, 33%; vitric, 47%; vitrophyric, 13%). The sherd

TABLE 28. FREQUENCY PERCENTAGES OF GRAIN TYPES AND SUPPLEMENTAL GRAIN PARAMETERS
FOR SAND TEMPERS IN NAVATU AND TUVALU SHERDS

Grain types, parameters	Selected (n=4) Navatu tempers (Viti Levu, Fiji)	Navatu tempers (n=4), Vaitupu sherds	Placer temper, Nanumanga (n=3) and (Vaitupu (n=1)	Hybrid temper, Nanumea (n=1) and Vaitupu (n=1)	Hybrid temper, Nanumanga sherds (n=2)
Mineral grains					
Quartz	0	0	2 ± 1	0	0
Plagioclase	27 ± 6	25 ± 2	1	7-8	22-35
Clinopyroxene	35 ± 7	29 ± 2	84 ± 2	29-34	15-18
Hornblende	7 ± 1	7 ± 1	0	tr	0
Olivine	5 ± 1	3 ± 1	2 ± 1	tr	0
Opaque FeOx	5 ± 1	6 ± 1	3 ± 2	27-29	8-10
Lithic fragments					
Microphanerite	7 ± 1	8 ± 1	0	0	0
Lathwork	3 ± 1	5 ± 1	-	1-2	-
Microlitic	5 ± 2	9 ± 1	-	3-5	-
Vitrophyric	2 ± 1	2 ± 1	-	3-5	-
Vitric	5 ± 1	6 ± 2	-	2-5	-
(Total)	(22 ± 5)	(30 ± 2)	(8 ± 2)	(11-20)	42-50
Grain parameters					
Calcareous (%)	0	0	[2 ± 1]	[16-19]	[~60]
VLi	66 ± 7	72 ± 4	100	100	100
OXi	11 ± 2	14 ± 2	3 ± 2	48	36-53
OLi	10 ± 3	8 ± 2	2 ± 1	~0	0
PYi	83 ± 3	82 ± 4	100	~100	100

Notes: Vaitupu, Nanumanga, and Nanumea are atolls in Tuvalu. Frequency percentages tabulated are means and standard deviations (±) or ranges (hyphen) recalculated free of calcareous grains for terrigenous fractions of hybrid tempers (bulk percentages of calcareous grains in brackets). See Table 6 for definitions of supplemental grain parameters (VLi, OXi, OLi, PYi). Categories of volcanic lithic fragments not tabulated for placer or hybrid tempers for which count statistics of limited numbers of lithic fragments are unreliable. Summary data for Navatu tempers in selected sherds from Temei (Vaitupu, Tuvalu) and Navatu (Viti Levu, Fiji) extracted from more complete frequency compositions tabulated by Dickinson et al. (1990).

apparently reflects post-Lapita ceramic transfer from Fiji (or Lau) to Tokelau as well as to Tuvalu (Table 25A9).

Samoa

A single sherd containing anomalous quartz-bearing volcanic sand temper that could not be derived from basaltic rocks of Samoa occurs among indigenous sherds from the Mulifanua Lapita site on Upolu (Petchey, 1995; Dickinson and Shutler, 2000). The temper sand is a poorly sorted aggregate of probable stream sand containing the following grain types (frequency percentages): quartz, 6%; plagioclase, 6%; sanidine K-feldspar, 6%; clinopyroxene, 54%; orthopyroxene, 8%; opaque iron oxides, 10%; volcanic lithic fragments, 10%. The quartz index (QZi = 33) and feldspar index (PFi = 50) of the temper sand are diagnostic of rhyodacitic source rock. Because the only extensively exposed rhyodacitic eruptive assemblage in Vanuatu, Fiji, Lau, Tonga, or Samoa occurs on the Udu Peninsula of Vanua Levu (Fig. 26), where both clinopyroxene and orthopyroxene occur as phenocrysts, limited Lapita ceramic transfer from Fiji to Samoa is provisionally inferred to explain the anomalous Mulifanua sherd (Table 25C7).

Rotuma

A single exotic sherd (Table 16B5) recovered from Rotuma, lying to the northwest of Fiji along the northern Melanesian borderland (Fig. 8), contains olivine-poor volcanic sand temper unlike indigenous Rotuma volcanic sand tempers of the oceanic basalt class (Ladefoged et al., 1998). The temper sand is a well-sorted and subrounded hybrid aggregate of beach origin, containing the following grain types (frequency percentages) exclusive of ~2% calcareous grains: clinopyroxene, 77%; orthopyroxene, 13%; hornblende, 2%; olivine, 2%; opaque iron oxides, 4%; volcanic lithic fragments, 2%. The pyroxene index (PXi = 86) would be permissive of derivation from Tonga (Table 13E1–13E5) were it not for the presence of hornblende, which is wholly lacking in Tongan tempers. Alternatively, the composition of the temper is permissive of ceramic transfer (Table 25C6) from somewhere on Vanua Levu, which lies along the northwestern fringe of Fiji closest to Rotuma but where knowledge of indigenous tempers is still severely limited. The few sherds studied to date from Vanua Levu (Table 16B4) and sherds on nearby Cikobia inferred to derive from Vanua Levu (Table 16B1–16B3) contain orthopyroxene mineral grains, although in proportions somewhat less (PXi = 91–99) than in the exotic Rotuma sherd.

Tonga Relations

Ceramic transfer within Tonga was rare, but several instances of ceramic transfer into Tonga or outward from Tonga are known from temper analysis.

Fiji to Tonga

Well-known ethnographic ties between Tonga and Fiji, and especially between Tonga and the Lau Archipelago of eastern Fiji (Fig. 8), are reflected by the occurrence of protohistoric Fijian sherds of Vuda and Ra typology (Fig. 3) as surface artifacts on Niuatoputapu (Kirch, 1988c) in northern Tonga (Fig. 29A), and on Tungua and Nomuka of the Ha'apai Group (Fig. 29B) in central Tonga (Kaeppler, 1973a, 1973b; Dye, 1987). The presence of sherds as artifacts from late in Tongan prehistory is otherwise anomalous because the making of pottery had been abandoned in Tonga approximately a millennium earlier (Fig. 3). The exotic Fijian sherds share various incised, paddle-impressed, and applied-relief design elements well known for ceramic assemblages from late in Fijian prehistory (Dye and Dickinson, 1996).

The exotic Fijian sherds contain four different kinds of temper sands (Dye and Dickinson, 1996), suggesting that protohistoric ties between Tonga and Fiji were extensive and systematic, involving more than fortuitous contact with some particular locale in Fiji. Most distinctive are quartz-bearing dissected orogen tempers derived from the Rewa Delta on Viti Levu (Dickinson et al., 1998b). Also present are feldspathic to lithic temper sands typical of Lakeba tempers derived from exposures of the andesitic volcanic rocks of the Lau remnant arc lying directly across the Lau Basin from Tonga. Some of the exotic sherds contain placer sand temper similar to the kedekede temper type distributed from sources on Kanacea(?) in Lau to multiple islands in Lau and Lomaiviti. Finally, hornblende-bearing tempers in selected exotic sherds were erroneously attributed by Dye and Dickinson (1996) to an origin within Lau because similar tempers occur in selected sherds from Lakeba in Lau. The lack of hornblende in all indigenous Lau tempers suggests instead that the hornblende-bearing tempers record ceramic transfer from Kadavu to Tonga as well as to Lakeba and other islands in Fiji.

Intra-Tonga Transfer

The only present evidence for ceramic transfer within Tonga is provided by a single sherd derived from nearby 'Eua in a collection of sherds from the Pea site (Fig. 29B) on Tongatapu (Key, 1987). The 'Eua origin of the sherd is unmistakable from the abundance of angular lithic fragments of uralitic gabbro (microphanerite) in its temper sand (Tables 13E6 and 25E29).

Tonga to Cook Islands

A single sherd recovered from the Anaio site on Mauke in the Cook Islands (Walter and Dickinson, 1989) contains a pyroxenic temper sand (Table 25B5) indistinguishable from typical Tongan placer tempers (Fig. 30). The pyroxene index (PXi = 96) of the sherd temper (Table 13E8) is unusually high for Tongan tempers and may imply an origin from the Vava'u Group (Table 13E2). The significance of this sherd find on Mauke remains uncertain because it derives from a cultural horizon dating to A.D. 1340–1430, approximately a millennium after pottery-making had ended in Tonga (Fig. 3) (Dickinson et al., 1998b).

Island Melanesia to Tonga

The single most intriguing instance of Oceanian ceramic transfer documented by temper analysis is the occurrence of several anomalous sherds from ware made in the western Lapita province of modern Melanesia at Nukuleka, the oldest Lapita site of the eastern Lapita province on Tongatapu in West Polynesia (Burley and Dickinson, 2001). The exotic sherds occur only in the lowest cultural horizon at Nukuleka, and only in sherds with a tan clay paste that contrasts with the reddish pastes typical of indigenous Tongan sherds. The exotic sherds further bear design motifs characteristic of the western Lapita province rather than the eastern Lapita province (Fig. 2). The presence of similar design motifs on associated indigenous sherds containing Tongan temper sand implies that migrants moving from the western Lapita province to the eastern Lapita province carried some ceramic vessels with them, and then made additional ceramic vessels of the same typology shortly after arrival in Tonga before the more simplified design motifs of the eastern Lapita province had evolved.

The tempers in the exotic Nukuleka sherds are related volcanic sands lacking in the orthopyroxene mineral grains that are ubiquitous in Tongan volcanic sand tempers, but containing minor hornblende mineral grains that are lacking in Tongan volcanic sand tempers. During initial petrographic study of the exotic Nukuleka sherds (Dickinson, 1987), the significance of the pyroxene index (PXi = 100) of the sand tempers was not appreciated, and the tempers were erroneously appraised as indigenous to Tonga because their total pyroxene contents and overall compositions are closely similar to those of indigenous Tongan tempers (Figs. 30A and 30B). Reanalysis of the anomalous sherds in the light of subsequent experience indicates that the temper sand in one is identical in visual appearance and texture to, and statistically indistinguishable in composition (Table 29) from, the temper sand in an exotic sherd from Nendö in the Santa Cruz Group lying 2300 km to the west-northwest (Fig. 2). Particularly striking is the uncommon association of quartz and clinopyroxene mineral grains as the most abundant constituents of the exotic Nukuleka and Nendö temper sands, for this association is unknown from any other Oceanian temper suites studied to date. Although the exact origin of the Nendö sherd remains uncertain, derivation from a source among the island arcs of Melanesia seems assured (see previous discussion of the Nendö sherd suite). The tempers in other exotic tan-paste sherds from Nukuleka are not as quartzose (Table 29), but are interpreted as sands closely related to the more quartzose temper, and derived from the same general locale in modern Melanesia (Table 25F4). The indicated ceramic transfer from Vanuatu to Tonga is the longest distance documented by Oceanian temper analysis.

Transfer Distances

Figure 40 is a summary diagram of the distances of interisland ceramic transfer indicated by Table 25. A systematic quasi-logarithmic decline in known instances of ceramic transfer with

TABLE 29. GRAIN TYPES IN EXOTIC NENDÖ AND NUKULEKA SHERDS

Grain type	Nendö sherd SZ8-68/12	Nukuleka sherd 99-10	Average of 3 other Nukuleka sherds
Quartz mineral grains	19 ± 2	19 ± 2	2 ± 1
Plagioclase feldspar mineral grains	14 ± 2	12 ± 2	15 ± 6
Clinopyroxene mineral grains	52 ± 3	58 ± 3	71 ± 6
Hornblende mineral grains	2 ± 1	1	1
Opaque iron oxide grains	6 ± 1	4 ± 1	8 ± 2
Volcanic lithic fragments	7 ± 1	6 ± 1	5 ± 1

Notes: Data after Burley and Dickinson (2001). Standard deviation (±) denotes counting error for individual sherds, as given by the expression $[p(100-p)/n]^{1/2}$ where p is observed percentage of grain type and n is number of grains counted (van der Plas and Tobi, 1965), or variance of different counts for multiple sherds (right column).

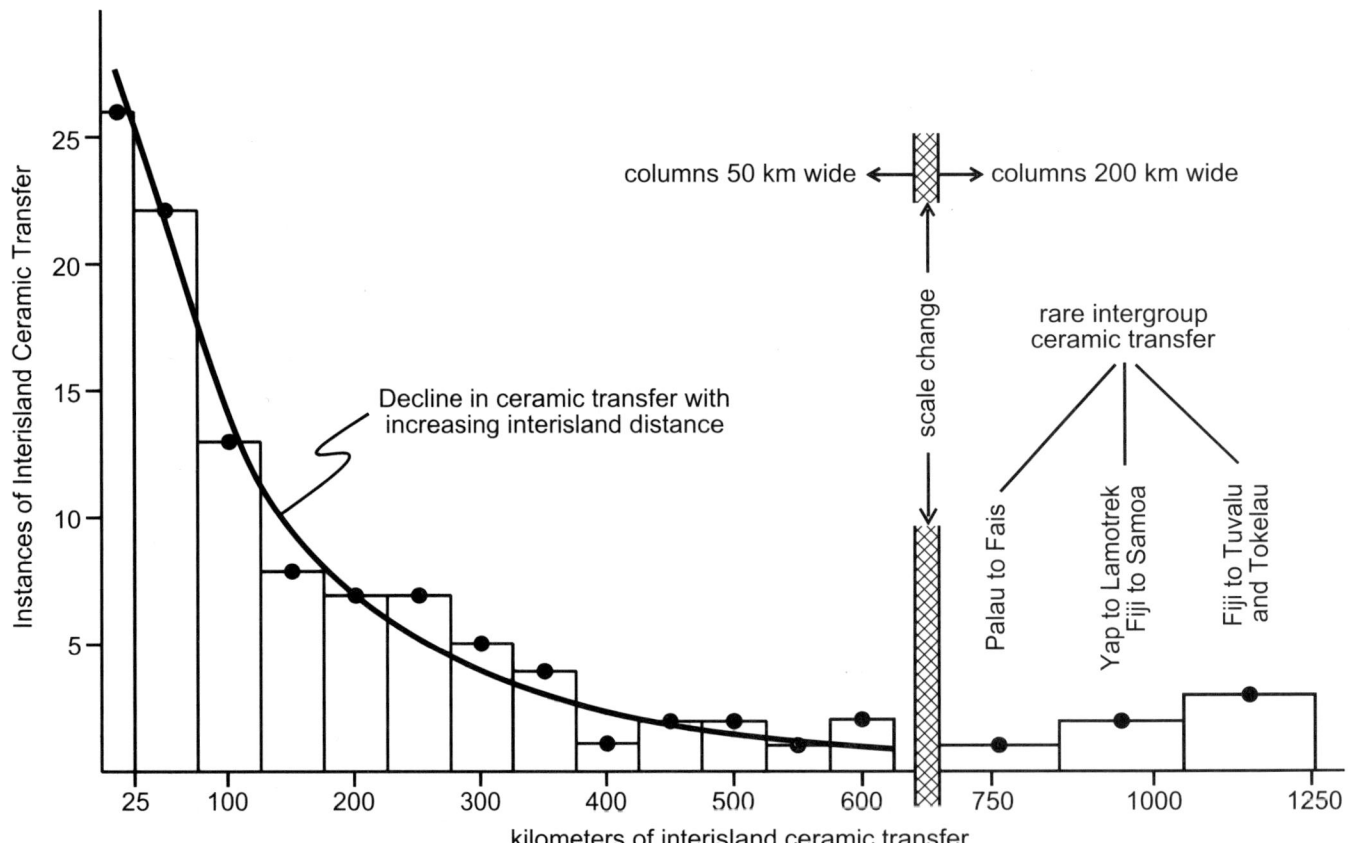

Figure 40. Interisland distances of 105 cases of prehistoric ceramic transfer documented by temper analysis (note scale change at 650 km). Data from Table 25 (distances grouped into increments of 50 km from 25 to 625 km, with distances <25 km plotted together, and distances >650 km grouped into increments of 200 km). Ceramic transfer over distances >1250 km from Vanuatu to Tonga (Table 25F4) and Tonga to Cook Islands (Table 25B5) not plotted for reasons of scale.

increasing interisland distance up to ~600 km is apparent and is perhaps expected from the difficulty of transporting ceramic vessels on sailing canoes. More isolated cases of ceramic transfer over distances >600 km do not follow the same quasi-logarithmic pattern for ceramic dispersal. Ceramic transfer over long interisland distances was apparently more fortuitous and subject to random effects, which may include fortuitous discovery of exotic sherds in collections from archaeological sites.

SUMMARY AND CONCLUSIONS

Tracing of temper sands to bedrock sources for detritus is feasible for small islands where bedrock assemblages are restricted in character, but becomes difficult on larger islands where bedrock exposures are more varied and sand from multiple source rocks is mixed within large drainage basins. The methodological approach developed during my study could be applied to advantage in any region of small islands, but would require modification to be effective on continental landmasses where sources of individual sands are spread over large areas, and sand derived from disparate source rocks is dispersed over long distances by extensive river systems.

Except where cultural traditions such as disuse of calcareous temper or use of grog temper evolved within selected island groups, the compositions of terrigenous temper sands used in Oceanian pottery did not vary significantly over time during prehistory. Despite changes in specific places for sand accumulation as island paleoenvironments were modified over time, compositional signals defined by the nature of local bedrock detritus remained constant on any given island.

Understanding the nature of geologic sources for terrigenous temper sands in many island groups requires tectonic reconstructions to recover the paleotectonic settings of the various islands because exposed bedrock may not relate closely to their present tectonic settings. Few island groups have remained in the same geotectonic position throughout Cenozoic time.

Distinctions between beach and stream sand tempers are commonly equivocal except for hybrid or placer beach sands, as are distinctions between natural and manually added stream sand tempers, but tracing a given temper sand to bedrock sources is not dependent upon detailed sedimentological interpretations or full understanding of local tempering practices.

Grain types in Oceanian temper suites can be classified according to standard precepts of sedimentary petrology, but standard triangular compositional diagrams are unsatisfactory for quantitative depiction of temper compositions because ferromagnesian grains are so prominent in many Oceanian placer tempers. Moreover, Oceanian temper compositions are most conveniently determined by frequency counts rather than point counts. Consequently, the Gazzi-Dickinson convention for apportioning sand-sized crystals within lithic fragments to other grain categories during summation of point counts cannot be applied.

Oceanian temper classes correlate in detail with the tectonic or paleotectonic settings of specific Pacific islands. Distinction between andesitic arc tempers and some postarc tempers, or between oceanic basalt tempers and some backarc tempers, relies in part upon knowledge of local geotectonic settings, rather than upon temper composition alone. Within temper classes, identification of bedrock sources for indigenous temper sands on individual islands is possible wherever adequate information is available from geologic mapping. Recognition of exotic tempers can thus be based upon the incompatibility of specific temper compositions with derivation from local bedrock, as well as where temper sand of a different temper class is observed in exotic sherds.

In general, different temper classes and types cannot be distinguished on the basis of gross compositional variations marked by different proportions of light minerals, heavy minerals, and lithic fragments as might be gauged by megascopic scrutiny, but varying proportions of grain types within each of those grain groups provide reliable petrographic criteria for temper recognition. Contrasting types of lithic fragments, which can only be classified with confidence by microscopic study, provide the most diagnostic basis for distinguishing among volcanic sand (oceanic basalt, andesitic arc, postarc-backarc), dissected orogen, and tectonic highland tempers. The specific nature of heavy mineral species, only identifiable with precision microscopically, is the most generally useful criterion for distinguishing among nonquartzose volcanic sand tempers. Despite its ubiquity in most continental sand accumulations, quartz is present only in dissected orogen, tectonic highland, and silicic volcanic sand tempers of Pacific Oceania.

Temper analysis has documented >100 instances of ceramic transfer between specific islands or island groups and provides insight into cultural relations not apparent from other investigations. Prehistoric ceramic transfer in Pacific Oceania involved the movement of wares from high bedrock islands to low coral islands, between different bedrock islands forming various island groups within which strong cultural ties were maintained throughout prehistory, and between distant island groups for which close cultural ties might otherwise not be suspected. The full significance of the ceramic transfer indicated by temper analysis can be understood only after further archaeological analysis, but the geographic distribution of exotic sherds derived from places of origin specified by the character of their temper sands provides a valuable physical database independent of cultural interpretations.

ACKNOWLEDGMENTS

I am indebted to all providers of sherds, or sherd thin sections, listed in Appendix 1 (86 individuals and four museums). Without their collaboration, none of my petrographic work would have been possible, but any errors in petrographic descriptions and interpretations of the archaeological contexts of sherds are my own.

An indispensable supporter of my studies from their outset has been Richard Shutler Jr., now of Simon Fraser University. It was his contacts with colleagues that initially enabled me to acquire sherds from key Pacific archaeological sites, and to establish liaison with investigators of Pacific prehistory based in many different countries. My wife Jacqueline Dickinson has aided every stage of inquiry since our courtship began in 1967. My former wife Peggy Dickinson, through her pioneering ethnographic research on modern pottery-making in Fiji (Dickinson and Sykes, 1965), was the prime instigator of my Pacific geoarchaeological work at its inception in 1965.

I have benefited for long periods extending in some cases for several decades from the advice and counsel of the following archaeological collaborators, but in singling them out for a special salute I do not mean to denigrate the contributions of many other archaeologists whose help also furthered my project: Wallace R. Ambrose, J. Stephen Athens, Lawrence and Helen Birks (now deceased), David V. Burley, Roger C. Green, Michiko Intoh, Patrick V. Kirch, Darlene R. Moore, Yoshihiko H. Sinoto, Jim Specht, Matthew Spriggs, Jun Takayama, John Terrell, and

J. Peter White. I also owe special thanks to the following, students at the time, for close liaison during their thesis research at various universities: Karen Aronson, Stuart Bedford, Simon Best, Melissa Carter, Scarlett Chiu, Geoffrey Clark, Christophe Descantes, Moira Doherty, Matthew Felgate, Scott Fitzpatrick, Stephanie Garling, Elizabeth Hinds (nee Shaw), Terry Hunt, Geoffrey Irwin, Sharyn O'Day, Sarah Phear, and Epi Suafo'a.

Correspondence over the years with Simon Best has been particularly helpful in resolving petrographic questions and issues of temper origins in Fiji, Samoa, and Tokelau, and discussions of geoarchaeology in Fiji with Patrick D. Nunn clarified several points with respect to Fijian archaeological sites. Special thanks are also due Terry Hunt and Simon Best for culling selected nonlocal sherds from the Yanuca collection of Lawrence and Helen Birks. Correspondence and petrographic collaboration with the late Colin A. Key and with Thomas S. Dye have expanded my knowledge of Tongan tempers. Correspondence with Roger C. Green clarified several points about the ceramic history of Samoa and the Marquesas Islands, and discussions with David Addison clarified the probable origin of Samoan crushed-rock tempers for me.

Thanks are due the following museum curators and managers for access to museum collections of archaeologists noted in parentheses: Paul Beelitz, Richard A. Gould, Belinda Kay, and David Hurst Thomas of the American Museum of Natural History in New York (R.C. Suggs); Shannon E. Kemp of the Bernice Pauahi Bishop Museum in Honolulu (Y.H. Sinoto); Christine Burke and Kate Vusoniwailala of the Fiji Museum in Suva (M.E. Rosenthal); Frank A. Norick of the Lowie (now Hearst) Museum of Anthropology at the University of California, Berkeley (E.W. Gifford). The chance to examine sherd thin sections in the collection of the late C.A. Key at the Australian National University, and the extended loan of some of his thin sections by W.R. Ambrose, are also gratefully acknowledged.

Most thin sections for my studies were prepared expertly from difficult materials by Ruperto Laniz (now deceased) of Stanford University and Los Altos, and Ray Lund of Quality Thin Sections in Tucson. Photomicrographs for Appendix 2 were taken with equipment graciously made available by George E. Gehrels in the Department of Geosciences at the University of Arizona. All figures were prepared by Jim Abbott of SciGraphics in Tucson.

An anonymous review improved my presentation, and an especially detailed and painstaking review by Kathleen Marsaglia helped me clarify a number of points.

REFERENCES CITED

Abbott, L.D., Silver, E.A., and Galewsky, J., 1994a, Structural evolution of a modern arc-continent collision in Papua New Guinea: Tectonics, v. 13, p. 1007–1034, doi: 10.1029/94TC01623.

Abbott, L.D., Silver, E.A., Thompson, P.R., Filewicz, M.V., Schneider, C., and Abdoerrias, 1994b, Stratigraphic constraints on the development and timing of arc-continent collision in northern Papua New Guinea: Journal of Sedimentary Research, v. B64, p. 169–183.

Aitchison, J.C., Clarke, G.L., Meffre, S., and Cluzel, D., 1995, Eocene arc-continent collision in New Caledonia and implications for regional southwest Pacific tectonic evolution: Geology, v. 23, p. 161–164, doi: 10.1130/0091-7613(1995)023<0161:EACCIN>2.3.CO;2.

Allen, J., 1996, The pre-Austronesian settlement of island Melanesia: Implications for Lapita archaeology, in Goodenough, W.H., ed., Prehistoric settlement of the Pacific: American Philosophical Society Transactions, v. 86, p. 11–27.

Allen, J., and Gosden, C., editors, 1991, Report of the Lapita homeland project: Canberra, Australian National University Occasional Papers in Prehistory No. 20, 268 p.

Allen, J., and White, J.P., 1989, The Lapita homeland: Some new data and an interpretation: Journal of the Polynesian Society, v. 98, p. 129–146.

Allen, M.S., 2004, Revisiting and revising Marquesan cultural history: New archaeological investigations at Anaho Bay, Nuku Hiva island: Journal of the Polynesian Society, v. 113, p. 143–196.

Ambrose, W.R., 1991, Lapita or not Lapita: The case of the Manus pots, in Allen, J., and Gosden, C., eds., Report of the Lapita homeland project: Canberra, Australian National University Occasional Papers in Prehistory No. 20, p. 103–112.

Ambrose, W.R., 1993, Pottery source materials: Source recognition in the Manus Islands, in Spriggs, M., et al., eds., A community of culture: The people and prehistory of the Pacific: Canberra, Australian National University Occasional Papers in Prehistory No. 21, p. 206–217.

Ambrose, W.R., 1997, Contradictions in Lapita pottery, a composite clone: Antiquity, v. 71, p. 525–538.

Ambrose, W.R., 1999, Curves, tines, scutes, and Lapita ware, in Galipaud, J.-C., and Lilley, I., eds., Le Pacifique de 5000 à 2000 avant le present: Paris, Éditions IRD (Institut de Recherche pour le Développement), p. 119–126.

Ambrose, W.R., and Gosden, C., 1991, Investigations on Boduna Island, in Allen, J., and Gosden, C., eds., Report of the Lapita Homeland Project: Canberra, Australian National University Occasional Papers in Prehistory No. 20, p. 182–188.

Amesbury, J.R., Moore, D.R., and Hunter-Anderson, R.L., 1996, Cultural adaptations and late Holocene sea level change in the Marianas: Recent excavations at Chalan Piao, Saipan: Indo-Pacific Prehistory Association Bulletin, v. 15, p. 53–69.

Anderson, A., 2001, Mobility models of Lapita migration, in Clark, G.R., et al., eds., The archaeology of Lapita dispersal in Oceania [Terra Australis 17]: Canberra, Australian National University Pandanus Books, p. 15–23.

Anderson, A., and Sinoto, Y., 2002, New radiocarbon ages of colonization sites in East Polynesia: Asian Perspectives, v. 41, p. 242–257.

Anderson, A., Bedford, S., Clark, G., Lilley, I., Sand, C., Summerhayes, G., and Torrence, R., 2001, An inventory of Lapita sites containing dentate-stamped pottery, in Clark, G.R., et al., eds., The archaeology of Lapita dispersal in Oceania [Terra Australis 17]: Canberra, Australian National University Pandanus Books, p. 1–13.

Anderson, A., Roberts, R., Dickinson, W., Clark, G., Burley, D., de Biran, A., Hope, G., and Nunn, P., 2006, Times of sand: Sedimentary history and archaeology of the Sigatoka dunes, Fiji: Geoarchaeology, v. 21, p. 131–154.

Anson, D., 1986, Lapita pottery of the Bismarck Archipelago and its affinities: Archaeology in Oceania, v. 21, p. 157–165.

Anson, D., 2000a, Excavations at Vunavaung (SDI), Rakival Village, Watom Island, Papua New Guinea: New Zealand Journal of Archaeology, v. 20 (1998), p. 95–118.

Anson, D., 2000b, Reber-Rakival dentate-stamped motifs: Documentation and comparative implications: New Zealand Journal of Archaeology, v. 20 (1998), p. 119–135.

Arnold, D.E., Neff, H., and Bishop, R.L., 1991, Compositional analysis and "sources" of pottery: An ethnoarchaeological approach: American Anthropologist, v. 93, p. 70–90, doi: 10.1525/aa.1991.93.1.02a00040.

Aronson, K.F., 1999, A compositional analysis of ceramics from the Qaranicagi rockshelter: Implications for exchange [M.A. thesis]: Honolulu, University of Hawaii, 114 p.

Ash, R.P., Carney, J.N., and MacFarlane, A., 1978, Geology of Efate and offshore islands: Port Vila, New Hebrides Condominium Geological Survey Regional Report, 49 p.

Ash, R.P., Carney, J.N., and MacFarlane, A., 1980, Geology of the northern Banks Islands: Port Vila, New Hebrides Condominium Geological Survey Regional Report, 52 p.

Ashley, P.M., and Flood, R.H., 1981, Low-K tholeiites and high-K igneous rocks from Woodlark Island, Papua New Guinea: Geological Society of Australia Journal, v. 28, p. 227–240.

Athens, J.S., 1990a, Nan Madol pottery: Pohnpei: Micronesica, v. 2, Supplement, p. 17–32.

Athens, J.S., 1990b, Kosrae: Pottery, clay, and early settlement: Micronesica, v. 2, Supplement, p. 171–186.

Athens, J.S., and Ward, J.V., 2001, Paleoenvironmental evidence for early human settlement in Palau: The Ngerchau core, in Stevenson, C.M., Lee, G., and Morin, F.J., eds., Pacific 2000: Proceedings of the Fifth International Conference on Easter Island and the Pacific: Los Osos, California, Easter Island Foundation, p. 165–178.

Audley-Charles, M.G., Carter, D.J., Barber, A.J., Norvick, M.S., and Tjokrosapoetro, S., 1979, Reinterpretation of the geology of Seram: Implications for the Banda arcs and northern Australia: Geological Society [London] Journal, v. 136, p. 547–568.

Austin, J.A., Jr., Taylor, F.W., and Cagle, C.D., 1989, Seismic stratigraphy of the central Tonga Ridge: Marine and Petroleum Geology, v. 6, p. 71–92, doi: 10.1016/0264-8172(89)90077-9.

Auzende, J.-M., Pelletier, B., and Lafoy, Y., 1994, Twin active spreading ridges in the North Fiji Basin (southwest Pacific): Geology, v. 22, p. 63–66, doi: 10.1130/0091-7613(1994)022<0063:TASRIT>2.3.CO;2.

Ayres, W.S., Wozniak, J.A., Robbins, G., and Suafo'a, E., 2001, Archaeology of American Samoa: Maloata, Malaeimi, and Malaeloa, in Stevenson, C.M., et al., eds., Pacific 2000: Proceedings of the Fifth International Conference on Easter Island and the Pacific: Los Osos, California, Easter Island Foundation, p. 227–235.

Babcock, T., 1977, A re-analysis of pottery from fortified sites on Taveuni, Fiji: Archaeology and Physical Anthropology in Oceania, v. 12, p. 112–134.

Bain, T.H.C., Mackenzie, D.E., and Ryburn, R.J., 1975, Geology of the Kubor anticline, central highlands of Papua New Guinea: Canberra, Australian Bureau of Mineral Resources, Geology, and Geophysics Bulletin 155, 106 p.

Baldwin, S.L., Lister, G.S., Foster, D.A., and MacDougall, I., 1993, Thermochronologic constraints on the tectonic evolution of active metamorphic core complexes, D'Entrecasteaux Islands, Papua New Guinea: Tectonics, v. 12, p. 611–628.

Band, R.A., 1968, The geology of southern Viti Levu and Mbengga: Suva, Fiji Department of Geological Surveys Bulletin 15, 49 p.

Bautista, B.C., Bautista, M.L.P., Oike, K., Wu, F.T., and Punongbayan, R.S., 2001, A new insight on the geometry of subducting slabs in northern Luzon, Philippines: Tectonophysics, v. 339, p. 279–310, doi: 10.1016/S0040-1951(01)00120-2.

Bedford, S., 1999, Lapita and post-Lapita ceramic sequences from Erromango, southern Vanuatu, in Galipaud, J.-C., and Lilley, I., eds., Le Pacifique de 5000 à 2000 avant le present: Paris, Éditions IRD (Institut de Recherche pour le Développement), p. 127–137.

Bedford, S., 2000, Results from excavations at the Mangaasi type site: A re-assessment of the ceramic sequence and its implications for Melanesian prehistory: Indo-Pacific Prehistory Association Bulletin, v. 19, p. 159–166.

Bedford, S., 2001, Ceramics from Malekula, northern Vanuatu: The two ends of a potential 3000-year sequence, in Clark, G.R., et al., eds., The archaeology of Lapita dispersal in Oceania [Terra Australis 17]: Canberra, Australian National University Pandanus Books, p. 105–114.

Bedford, S., 2003, The timing and nature of Lapita colonisation in Vanuatu: The haze begins to clear, in Sand, C., ed., Pacific archaeology: Assessments and prospects: Noumea, Cahiers de l'Archéologie en Nouvelle-Calédonie, v. 15, p. 147–158.

Bedford, S., and Clark, G., 2001, The rise and rise of the incised and applied relief tradition: A review and reassessment, in Clark, G.R., et al., eds., The archaeology of Lapita dispersal in Oceania [Terra Australis 17]: Canberra, Australian National University Pandanus Books, p. 61–74.

Bedford, S., and Spriggs, M., 2000, Crossing the Pwanmwou: Preliminary report on recent excavations adjacent to and south west of Mangaasi, Efate, Vanuatu: Archaeology in Oceania, v. 35, p. 120–126.

Bedford, S., Spriggs, M., Wilson, M., and Regenvanu, R., 1998, The Australian National University–National Museum of Vanuatu archaeology project: A preliminary report on the establishment of cultural sequences and rock art research: Asian Perspectives, v. 37, p. 165–193.

Bellon, H., Marcelot, G., Lefévre, C., and Maillet, P., 1984, Le volcanisme de l'île d'Erromango (République de Vanuatu): Calendrier de l'activité (données ^{40}K/^{39}K): Comptes Rendus de l'Académie des Sciences de Paris, ser. II, v. 299, p. 257–262.

Bellwood, P., 1995, Austronesian prehistory in southeast Asia: Homeland, expansion, and transformation, in Bellwood, P., et al., eds., The Austronesians: Historical and comparative perspectives: Canberra, Australian National University, p. 96–111.

Benes, V., Scott, S.D., and Binns, R.A., 1994, Tectonics of rift propagation into a continental margin: Western Woodlark Basin, Papua New Guinea: Journal of Geophysical Research, v. 99, p. 4439–4455, doi: 10.1029/93JB02878.

Bentley, R.A., 2000, Provenience analysis of pottery from Fijian hillforts: Preliminary implications for exchange within the archipelago: Archaeology in Oceania, v. 35, p. 82–91.

Best, S., 1984, Lakeba: The prehistory of a Fijian island [Ph.D. thesis]: Auckland, University of Auckland, 684 p.

Best, S., 1988, Tokelau archaeology: A preliminary report of an initial survey and excavations: Indo-Pacific Prehistory Bulletin, v. 8, p. 104–118.

Best, S., 2002, Lapita: A view from the east: Auckland, New Zealand Archaeological Association Monograph 54, 113 p.

Best, S., Sheppard, P., Green, R., and Parker, R., 1992, Necromancing the stone: Archaeologists and adzes in Samoa: Journal of the Polynesian Society, v. 101, p. 45–85.

Bickler, S.H., 1997, Early pottery exchange along the south coast of Papua New Guinea: Archaeology in Oceania, v. 32, p. 151–162.

Birks, L., 1973, Archaeological excavations at Sigatoka dune site, Fiji: Suva, Fiji Museum Bulletin 1, 176 p.

Bishop, R.L., Rands, R.L., and Holley, G.R., 1982, Ceramic compositional analysis in archaeological perspective, in Schiffer, M.B., ed., Advances in archaeological method and theory, vol. 5: New York, Academic Press, p. 275–330.

Blake, D.H., and Ewart, A., 1974, Petrography and geochemistry of the Cape Hoskins volcanoes, New Britain, Papua New Guinea: Geological Society of Australia Journal, v. 21, p. 319–331.

Blake, D.H., and Miezitis, Y., 1967, Geology of Bougainville and Buka Islands, New Guinea: Australian Bureau of Mineral Resources, Geology, and Geophysics Bulletin 93 [PNG Bulletin 1], 56 p.

Blust, R., 1996, Austronesian culture history: The window of language, in Goodenough, W.H., ed., Prehistoric settlement of the Pacific: American Philosophical Society Transactions, v. 86, p. 28–35.

Bowin, C., Purdy, G.M., Johnston, C., Shor, G., Lawver, L., Hartono, H.M.S., and Jexek, P., 1980, Arc-continent collision in Banda Sea region: American Association of Petroleum Geologists Bulletin, v. 64, p. 868–915.

Bracey, D.R., 1975, Reconnaissance geophysical survey of the Caroline Basin: Geological Society of America Bulletin, v. 86, p. 775–784, doi: 10.1130/0016-7606(1975)86<775:RGSOTC>2.0.CO;2.

Brousse, R., and Sevin, M.-A., 1978, Pétrologie de l'ile de Ua Huka dans le Pacifique central: Cahiers du Pacifique, v. 21, p. 203–213.

Brousse, R., Guille, G., and Maury, R.C., 1978a, Volcanisme et pétrologie de l'île de Nuku-Hiva dans les îles Marquises (Pacifique central): Cahiers du Pacifique, v. 21, p. 145–187.

Brousse, R., Guille, G., and Gibert, J.-P., 1978b, Volcanisme et pétrologie de l'île de Hiva-Oa dans les îles Marquises (Pacifique central): Cahiers du Pacifique, v. 21, p. 189–202.

Bryan, W.B., Stice, G.D., and Ewart, A., 1972, Geology, petrography, and geochemistry of the volcanic islands of Tonga: Journal of Geophysical Research, v. 77, p. 1566–1585.

Bulmer, S., 1999, Revisiting red slip: The Laloki style pottery of southern Papua and its possible relationship to Lapita, in Galipaud, J.-C., and Lilley, I., eds., Le Pacifique de 5000 à 2000 avant le présent: Paris, Éditions IRD (Institut de Recherche pour le Développement), p. 543–577.

Burley, D.V., 1994, Settlement pattern and Tongan prehistory: Reconsiderations from Ha'apai: Journal of the Polynesian Society, v. 103, p. 379–411.

Burley, D.V., 1998, Tongan archaeology and the Tongan past, 2850–150 B.P.: Journal of World Prehistory, v. 12, p. 337–392, doi: 10.1023/A:1022322303769.

Burley, D.V., 1999, Lapita settlement to the east: New data and changing perspectives from Ha'apai (Tonga) prehistory, in Galipaud, J.-C., and Lilley, I., eds., Le Pacifique de 5000 à 2000 avant le présent: Paris, Éditions IRD (Institut de Recherche pour le Développement), p. 189–200.

Burley, D.V., 2003, Dynamic landscapes and episodic occupations: Archaeological interpretation and implications in the prehistory of the Sigatoka sand dunes, in Sand, C., ed., Pacific archaeology: Assessments and prospects: Noumea, Cahiers de l'Archéologie en Nouvelle-Calédonie, v. 25, p. 307–315.

Burley, D.V., and Clark, J.T., 2003, The archaeology of Fiji/western Polynesia in the post-Lapita era, in Sand, C., ed., Pacific archaeology: Assessments

and prospects: Noumea, Cahiers de l'Archéologie en Nouvelle-Calédonie, v. 15, p. 235–254.
Burley, D.V., and Dickinson, W.R., 2001, Origin and significance of a founding settlement in Polynesia: National Academy of Sciences Proceedings, v. 98, p. 11,829–11,831, doi: 10.1073/pnas.181335398.
Burley, D.V., and Dickinson, W.R., 2004, Late Lapita occupation and ceramic assemblage at the Sigatoka sand dune site, Fiji, and their place in Oceanic prehistory: Archaeology in Oceania, v. 39, p. 12–25.
Burley, D.V., Nelson, E., and Shutler, R., 1995, Rethinking Lapita chronology in Ha'apai: Archaeology in Oceania, v. 30, p. 132–134.
Burley, D.V., Nelson, D.E., and Shutler, R., Jr., 1999, A radiocarbon chronology for the eastern Lapita frontier in Tonga: Archaeology in Oceania, v. 34, p. 59–70.
Burley, D.V., Dickinson, W.R., Barton, A., and Shutler, R., Jr., 2001, Lapita on the periphery: New data on old problems in the Kingdom of Tonga: Archaeology in Oceania, v. 36, p. 89–104.
Burley, D.V., Steadman, D.W., and Anderson, A.J., 2004, The volcanic outlier of 'Ata in Tongan prehistory: Reconsideration of its role and settlement chronology: New Zealand Journal of Archaeology, v. 25 (2003), p. 89–106.
Butler, B.M., 1994, Early prehistoric settlement in the Mariana Islands: New evidence from Saipan: Man and Culture in Oceania, v. 10, p. 15–38.
Carney, J.N., 1986, Geology of Maewo: Port Vila, Vanuatu Department of Geology, Mines, and Rural Water Supply General Report, 58 p.
Carney, J.N., and MacFarlane, A., 1979, Geology of Tanna, Aneityum, Futuna and Aniwa: Port Vila, New Hebrides Condominium Geological Survey Regional Report, 71 p.
Carney, J.N., and MacFarlane, A., 1982, Geological evidence bearing on the Miocene to Recent structural evolution of the New Hebrides arc: Tectonophysics, v. 87, p. 147–175, doi: 10.1016/0040-1951(82)90225-6.
Carney, J.N., MacFarlane, A., and Mallick, D.I.J., 1985, The Vanuatu island arc: An outline of the stratigraphy, structure, and petrology, in Nairn, A.E.M., et al., eds., The ocean basins and margins, vol. 7A: The Pacific Ocean: New York, Plenum Press, p. 683–718.
Carter, M.J., 2004, North of the Cape and south of the Fly: The archaeology of settlement and subsistence in the Murray Islands, eastern Torres Strait [Ph.D. thesis]: Townsville, James Cook University, 513 p.
Charlton, T.R., 2000, Tertiary evolution of the eastern Indonesian collision complex: Journal of Asian Earth Sciences, v. 18, p. 603–631, doi: 10.1016/S1367-9120(99)00049-8.
Charlton, T.R., 2004, The petroleum potential of inversion anticlines in the Banda arc: American Association of Petroleum Geologists Bulletin, v. 88, p. 565–585.
Chiu, S., 2003a, Social and economic meanings of Lapita pottery: A New Caledonian case, in Sand, C., ed., Pacific archaeology: Assessments and prospects: Noumea, Cahiers de l'Archéologie en Nouvelle-Calédonie, v. 15, p. 159–183.
Chiu, S., 2003b, The socio-economic functions of Lapita ceramic production and exchange: A case study from Site WKO013A, Koné, New Caledonia [Ph.D. thesis]: Berkeley, University of California, 440 p.
Chiu, S., 2005, Meanings of a Lapita face: Materialized social memory in ancient house societies: Taiwan Journal of Archaeology, v. 3, p. 1–47.
Chivas, A.R., Andrew, A.S., Sinha, A.K., and O'Neil, J.R., 1982, Geochemistry of a Pliocene-Pleistocene oceanic-arc plutonic complex, Guadalcanal: Nature, v. 300, p. 139–143, doi: 10.1038/300139a0.
Clark, G.R., 1999, Post-Lapita Fiji: Cultural transformation in the mid-sequence [Ph.D. thesis]: Canberra, Australian National University, 278 p.
Clark, G., 2003, Shards of meaning: Archaeology and the Melanesia-Polynesia divide: Journal of Pacific History, v. 38, p. 197–215, doi: 10.1080/0022334032000120530.
Clark, G., 2004, Radiocarbon dates from the Ulong site in Palau and implications for western Micronesian prehistory: Archaeology in Oceania, v. 39, p. 26–33.
Clark, G., and Anderson, A., 1999, The age of Lapita settlement in Fiji: Archaeology in Oceania, v. 34, p. 31–39.
Clark, G., and Anderson, A., 2001a, The pattern of Lapita settlement in Fiji: Archaeology in Oceania, v. 36, p. 77–88.
Clark, G., and Anderson, A., 2001b, The age of the Yanuca Lapita site, Viti Levu, Fiji: New Zealand Journal of Archaeology, v. 22 (2000), p. 15–30.
Clark, G., and Wright, D., 2003, The colonisation of Palau: Preliminary results from Angaur and Ulong, in Sand, C., ed., Pacific archaeology: Assessments and prospects: Noumea, Cahiers de l'Archéologie en Nouvelle-Calédonie, v. 15, p. 85–94.

Clark, G., Anderson, A., and Matararaba, S., 2001, The Lapita site at Votua, northern Lau Islands, Fiji: Archaeology in Oceania, v. 36, p. 134–145.
Clark, J.T., 1993, Radiocarbon dates from American Samoa: Radiocarbon, v. 35, p. 323–330.
Clark, J.T., 1996, Samoan prehistory in review, in Davidson, J., et al., eds., Oceanic culture history: Essays in honour of Roger Green: Dunedin, New Zealand Journal of Archaeology Special Publication, p. 445–460.
Clark, J.T., and Michlovic, M.G., 1996, An early settlement in the Polynesian homeland: Excavations at 'Aoa Valley, Tutuila Island, American Samoa: Journal of Field Archaeology, v. 23, p. 151–167.
Clark, J.T., Wright, E., and Herdrich, D.J., 1997, Interactions within and beyond the Samoan archipelago: Evidence from basaltic rock geochemistry, in Weisler, M.I., ed., Prehistoric long-distance interaction in Oceania: An interdisciplinary approach: Auckland, New Zealand Archaeological Association Monograph 21, p. 68–84.
Clark, J.T., Cole, A.O., and Nunn, P.D., 1999, Environmental change and human prehistory on Totoya Island, Fiji, in Galipaud, J.-C., and Lilley, I., eds., Le Pacifique de 5000 à 2000 avant le present: Paris, Éditions IRD (Institut de Recherche pour le Développement), p. 227–240.
Clay, R.B., 1974, Archaeological reconnaissance in central New Ireland: Archaeology and Physical Anthropology in Oceania, v. 9, p. 1–17.
Cloud, P.E., Jr., Schmidt, R.G., and Burke, E.W., 1956, Geology of Saipan, Mariana Islands, Part 1: General geology: U.S. Geological Survey Professional Paper 280-A, 126 p.
Clough, R., 1992, Firing temperatures and the analysis of oceanic ceramics: A study of Lapita ceramics from Reef/Santa Cruz, Solomon Islands, in Galipaud, J.-C., ed., Poterie Lapita et peuplement: Noumea, ORSTOM (Office de Recherche Scientifique et Technologique Outre-Mer), p. 177–192.
Cluzel, D., and Meffre, S., 2002, L'unité de la Boghen (Nouvelle-Calédonie, Pacifique sud-ouest): Un complexe d'accretion Jurassiques; données radiochronologiques preliminaires U-Pb sur les zircons détriques: Comptes Rendus Geoscience de l'Académie des Sciences de Paris, v. 334, p. 867–874.
Cluzel, D., Aitchison, J., Clarke, G., Meffre, S., and Picard, C., 1994, Point de vue sur l'évolution tectonique et géodynamique de la Nouvelle-Calédonie (Pacifique, France): Comptes Rendus de l'Académie des Sciences de Paris, ser. II, v. 319, p. 683–690.
Cluzel, D., Picard, C., Aitchison, J.C., Laporte, C., Meffre, S., and Parat, F., 1997, La nappe de Poya (ex-formation des basaltes) de Nouvelle-Calédonie (Pacifique sud-ouest): Un plateau océanique Campanien-Paléocene supérieur obducté à l'Éocene supérieur: Comptes Rendus de l'Académie des Sciences de Paris, ser. II, v. 324, p. 443–451.
Cluzel, D., Aitchison, J.C., and Picard, C., 2001, Tectonic accretion and underplating of mafic terranes in the later Eocene intraoceanic forearc of New Caledonia (southwest Pacific): Geodynamic implications: Tectonophysics, v. 340, p. 23–59, doi: 10.1016/S0040-1951(01)00148-2.
Cochrane, E.E., 2002, Explaining the prehistory of ceramic technology on Waya Island, Fiji: Archaeology in Oceania, v. 37, p. 37–50.
Cochrane, E.E., and Neff, H., 2006, Investigating compositional diversity among Fijian ceramics with laser-ablation inductively coupled plasma–mass spectrometry (LA-ICP-MS): Implications for interaction studies on geologically similar islands: Journal of Archaeological Science, v. 33, p. 378–390.
Cole, J.W., Gill, J.B., and Woodhall, D., 1985, Petrologic history of the Lau Ridge, Fiji, in Scholl, D.W., and Vallier, T.L., eds., Geology and offshore resources of Pacific island arcs—Tonga region: Houston, Circum-Pacific Council for Energy and Mineral Resources Earth Science Series, v. 2, p. 379–414.
Cole, W.S., Todd, R., and Johnson, C.G., 1960, Conflicting age determinations suggested by foraminifera on Yap, Caroline Islands: American Paleontology Bulletin, v. 41, p. 77–112.
Colley, H., and Ash, R.P., 1971, The geology of Erromango: Port Vila, New Hebrides Condominium Geological Survey Regional Report, 111 p.
Colley, H., and Greenbaum, D., 1980, The mineral deposits and metallogenesis of the Fiji platform: Economic Geology and the Bulletin of the Society of Economic Geologists, v. 75, p. 807–829.
Colley, H., and Hindle, W.H., 1984, Volcano-tectonic evolution of Fiji and adjoining marginal basins, in Kokelaar, B.P., and Howells, A.M.F., eds., Marginal basin geology: Geological Society [London] Special Publication 16, p. 151–162.
Colley, H., and Rice, C.M., 1975, A Kuroko-type ore deposit in Fiji: Economic Geology and the Bulletin of the Society of Economic Geologists, v. 70, p. 1373–1386.

Colley, H., and Rice, C.M., 1978, Kuroko-type deposits in Vanua Levu, Fiji: New Zealand Journal of Geology and Geophysics, v. 21, p. 277–285.

Colley, H., and Warden, A.J., 1974, Petrology of the New Hebrides: Geological Society of America Bulletin, v. 85, p. 1635–1646, doi: 10.1130/0016-7606(1974)85<1635:POTNH>2.0.CO;2.

Collot, J.Y., Malahoff, A., Récy, J., Latham, G., and Missègue, F., 1987, Overthrust emplacement of New Caledonia ophiolite: Geophysical evidence: Tectonics, v. 6, p. 215–232.

Conte, E., and Anderson, A., 2003, Radiocarbon ages for two sites on Ua Huka, Marquesas Islands: Asian Perspectives, v. 42, p. 155–160.

Cooper, P.A., and Taylor, B., 1985, Polarity reversal in the Solomon Islands arc: Nature, v. 314, p. 428–430, doi: 10.1038/314428a0.

Corwin, C.G., Rogers, C.L., and Elmquist, P.O., 1956, Military geology of Palau Islands: Tokyo, Intelligence Division, Office of the Engineer, Headquarters U.S. Army Far East, 285 p.

Coulon, C., Maillet, P., and Maury, R.C., 1979, Contribution à l'étude du volcanisme de l'arc des Nouvelles-Hébrides: Données pétrologique sur les laves d'île d'Efaté: Société Géologique de France Bulletin, ser. 7, v. 21, p. 619–629.

Coulson, F.I.E., 1971, The geology of western Vanua Levu: Suva, Fiji Geological Survey Department Bulletin 17, 49 p.

Coulson, F.I.E., 1976, Geology of the Lomaiviti and Moala island groups: Suva, Fiji Mineral Resources Division Bulletin 2, 162 p.

Coulson, F.I., 1985, Solomon Islands, in Nairn, A.E.M., et al., eds., The ocean basins and margins, vol. 7A: The Pacific Ocean: New York, Plenum Press, p. 607–682.

Coulson, F.I., and Vedder, J.G., 1986, Geology of the central and western Solomon Islands, in Vedder, J.G., et al., eds., Geology and offshore resources of Pacific island arcs—central and western Solomon Islands: Houston, Circum-Pacific Council for Energy and Mineral Resources Earth Science Series, v. 4, p. 59–87.

Craib, J.L., 1981, Settlement on Ulithi atoll, western Caroline Islands: Asian Perspectives, v. 24, p. 47–55.

Craib, J.L., 1999, Colonisation of Mariana Islands: New evidence and implications for human movements in the western Pacific, in Galipaud, J.-C., and Lilley, I., eds., Le Pacifique de 5000 à 2000 avant le present: Paris, Éditions IRD (Institut de Recherche pour le Développement), p. 477–485.

Critelli, S., and Ingersoll, R.V., 1995, Interpretation of neovolcanic versus paleovolcanic sand grains: An example from Miocene deep-marine sandstone of the Topanga Group (southern California): Sedimentology, v. 42, p. 783–804.

Cronin, S.J., and Neall, V.E., 2001, Holocene volcanic geology, volcanic hazard, and risk on Taveuni, Fiji: New Zealand Journal of Geology and Geophysics, v. 44, p. 417–437.

Crook, K.A.W., and Taylor, B., 1994, Structural and Quaternary tectonic history of the Woodlark triple junction region, Solomon Islands: Marine Geophysical Researches, v. 16, p. 65–89, doi: 10.1007/BF01812446.

Cullen, A.B., and Pigott, J.D., 1989, Post-Jurassic tectonic evolution of Papua New Guinea: Tectonophysics, v. 162, p. 291–302, doi: 10.1016/0040-1951(89)90250-3.

Cunningham, J.K., and Anscombe, K.J., 1985, Geology of 'Eua and other islands, Kingdom of Tonga, in Scholl, D.W., and Vallier, T.L., eds., Geology and offshore resources of Pacific island arcs—Tonga region: Houston, Circum-Pacific Council for Energy and Mineral Resources Earth Science Series, v. 2, p. 221–257.

Davenport, W., 1972, Preliminary excavations on Santa Ana Island, eastern Solomon Islands: Archaeology and Physical Anthropology in Oceania, v. 7, p. 165–183.

David, B., McNiven, I., Mitchell, R., Orr, M., Haberle, S., Brady, L., and Crouch, J., 2004, Badu 15 and the Papuan-Austronesian settlement of Torres Strait: Archaeology in Oceania, v. 39, p. 67–78.

Davidson, J., 1971, Preliminary report on an archaeological survey of the Vava'u Group, Tonga, in Fraser, R., ed., Cook bicentenary expedition to the southwest Pacific: Royal Society of New Zealand Bulletin 8, p. 29–40.

Davidson, J., and Leach, F., 1993, The chronology of the Natunuku site, Fiji: New Zealand Journal of Archaeology, v. 15, p. 99–105.

Davidson, J., Hinds, E., Holdaway, S., and Leach, F., 1990, The Lapita site of Natunuku, Fiji: New Zealand Journal of Archaeology, v. 12, p. 121–155.

Davies, H.L., and Jaques, A.L., 1984, Emplacement of ophiolite in Papua New Guinea, in Gass, I.G., et al., eds., Ophiolites and oceanic lithosphere: Geological Society [London] Special Publication 13, p. 341–349.

Davies, H.L., and Smith, I.E.M., 1971, Geology of eastern Papua: Geological Society of America Bulletin, v. 82, p. 3299–3312.

Davies, H.L., and Warren, R.G., 1992, Eclogites of the D'Entrecasteaux Islands: Contributions to Mineralogy and Petrology, v. 112, p. 463–474, doi: 10.1007/BF00310778.

Descantes, C., 1998, Integrating archaeology and ethnohistory: The development of exchange between Yap and Ulithi, western Caroline Islands [Ph.D. thesis]: Eugene, University of Oregon, 337 p.

Descantes, C., Neff, H., Glascock, M.D., and Dickinson, W.R., 2001, Chemical characterization of Micronesian ceramics through instrumental neutron activation analysis: A preliminary provenance study: Journal of Archaeological Science, v. 28, p. 1185–1190, doi: 10.1006/jasc.2000.0635.

Descantes, C.H., Intoh, M., Neff, H., and Glascock, M.D., 2004, Chemical characterization of Yapese clays and ceramics by instrumental neutron activation analysis: Journal of Radioanalytical and Nuclear Chemistry, v. 262, p. 83–91, doi: 10.1023/B:JRNC.0000040857.23051.a3.

Dickinson, P., and Sykes, M., 1965, Kuro manufacture in Yavulo village: Suva, Fiji Museum Records, v. 1, p. 69–72.

Dickinson, W.R., 1968a, Singatoka dune sands, Viti Levu, Fiji: Sedimentary Geology, v. 2, p. 115–124, doi: 10.1016/0037-0738(68)90031-6.

Dickinson, W.R., 1968b, Sedimentation of volcaniclastic strata of the Pliocene Koroimavua Group in northwest Viti Levu, Fiji: American Journal of Science, v. 266, p. 440–453.

Dickinson, W.R., 1969, Temper sands in prehistoric potsherds from Vailele and Falefa, in Green, R.C., and Davidson, J.M., eds., Archaeology in Western Samoa, Volume I: Auckland, Auckland Institute and Museum Bulletin 6, p. 271–273.

Dickinson, W.R., 1970, Interpreting detrital modes of graywacke and arkose: Journal of Sedimentary Petrology, v. 40, p. 695–707.

Dickinson, W.R., 1971a, Petrography of some temper sands in prehistoric pottery from Viti Levu, Fiji: Suva, Fiji Museum Records, v. 1, p. 107–121.

Dickinson, W.R., 1971b, Temper sands in Lapita-style potsherds on Malo: Journal of the Polynesian Society, v. 80, p. 244–246.

Dickinson, W.R., 1972a, Temper sand in prehistoric potsherds, Efate and vicinity, central New Hebrides, in Garanger, J., Archéologie des Nouvelles-Hébrides: Paris, Musée de l'Homme Publications de la Société des Océanistes No. 30, p. 111–112.

Dickinson, W.R., 1972b, Sand tempers in prehistoric potsherds from the Shortland Islands [Appendix IV], in Irwin, G.J., An archaeological survey of the Shortland Islands [M.A. thesis]: Auckland, University of Auckland, p. 235–238.

Dickinson, W.R., 1972c, Dissected erosion surfaces in northwest Viti Levu, Fiji: Zeitschrift für Geomorphologie, v. 16, p. 252–267.

Dickinson, W.R., 1973a, Reconstructions of past arc-trench systems from petrotectonic assemblages in the island arcs of the western Pacific, in Coleman, P.J., ed., The western Pacific: Island arcs, marginal seas, geochemistry: New York, Crane Russak, p. 569–601.

Dickinson, W.R., 1973b, Sand temper in prehistoric potsherds from the Sigatoka dunes, Viti Levu, Fiji [Appendix I], in Birks, L., Archaeological excavations at Sigatoka dune site, Fiji: Suva, Fiji Museum Bulletin 1, p. 69–72.

Dickinson, W.R., 1973c, Petrographic report on pottery sherds and fired clay samples from the Paubake survey area, south Bougainville: Chicago, Field Museum of Natural History Reports of the Bougainville Archaeological Survey No. 8, 7 p.

Dickinson, W.R., 1973d, Temper sands in potsherds from Anuta, in Yen, D.E., and Gordon, J., eds., Anuta: A Polynesian outlier in the Solomon Islands: Honolulu, Bishop Museum Pacific Anthropological Records No. 21, p. 109–111.

Dickinson, W.R., 1974a, Temper sands in sherds from Mulifanua and comparison with similar tempers at Vailele and Sasoa'a (Falefa), in Green, R.C., and Davidson, J.M., eds., Archaeology in Western Samoa, Volume II: Auckland, Auckland Institute and Museum Bulletin 7, p. 179–180.

Dickinson, W.R., 1974b, Sand tempers in sherds from Niuatoputapu and elsewhere in Tonga [Appendix 2], in Rogers, G., Archaeological discoveries on Niuatoputapu Island, Tonga: Journal of the Polynesian Society, v. 83, p. 342–345.

Dickinson, W.R., 1975a, Potash-depth (K-h) relations in continental margin and intraoceanic magmatic arcs: Geology, v. 3, p. 53–56, doi: 10.1130/0091-7613(1975)3<53:PKRICM>2.0.CO;2.

Dickinson, W.R., 1975b, Petrographic report on supplementary sherds from Paubake, Bougainville: Chicago, Field Museum of Natural History Reports of the Bougainville Archaeological Survey No. 6, 3 p.

Dickinson, W.R., 1975c, Petrographic report on sand tempers in sherds of Buka and Kieta traditions from Teop Island off Bougainville: Chicago, Field

Museum of Natural History Reports of the Bougainville Archaeological Survey No. 7, 8 p.

Dickinson, W.R., 1976a, Temper sands in sherds from Futuna, Alofi, and Uvea (Horne and Wallis Islands): Journal of the Polynesian Society, v. 85, p. 64–69.

Dickinson, W.R., 1976b, Mineralogy and petrology of sand tempers in sherds from the Ferry Berth site, Paradise site, and Jane's Camp, in Jennings, J.D., et al., eds., Excavations on Upolu, Western Samoa: Honolulu, Bishop Museum Pacific Anthropological Records No. 25, p. 99–103.

Dickinson, W.R., 1978, Sand tempers in southeast Solomons sherds from Feru rockshelter on Santa Ana and Lapita sites in the Santa Cruz Group: Auckland, University of Auckland Oceanic Prehistory Records No. 7, 13 p.

Dickinson, W.R., 1980a, Sand tempers in Lesu sherds from New Ireland [Appendix 2], in White, J.P., and Downie, J.E., Excavations at Lesu, New Ireland: Asian Perspectives, v. 23, p. 217–218.

Dickinson, W.R., 1980b, Sand tempers in prehistoric potsherds from Yanutha Island off Viti Levu, Fiji [Appendix], in Hunt, T.L., Towards Fiji's past: Archaeological research on southwestern Viti Levu [M.A. thesis]: Auckland, University of Auckland, p. 215–217.

Dickinson, W.R., 1981, Sand temper in a sherd from Kapingamarangi [Appendix 6], in Leach, F., and Ward, G., Archaeology on Kapingamarangi atoll, a Polynesian outlier in the eastern Caroline Islands: Dunedin, University of Otago Studies in Prehistoric Anthropology, v. 16, p. 134.

Dickinson, W.R., 1982a, Temper sands from prehistoric potsherds excavated at Pemrang site on Yap and from nearby Ngulu atoll: Indo-Pacific Prehistory Association Bulletin, v. 3, p. 115–117.

Dickinson, W.R., 1982b, Sand tempers in sherds from Tikopia [Appendix], in Kirch, P.V., and Yen, D.E., Tikopia: Honolulu, Bishop Museum Bulletin 238, p. 370–372.

Dickinson, W.R., 1982c, Compositions of sandstones in circum-Pacific subduction complexes and fore-arc basins: American Association of Petroleum Geologists Bulletin, v. 66, p. 121–137.

Dickinson, W.R., 1984, Indigenous and exotic temper sands in prehistoric potsherds from the central Caroline Islands, in Sinoto, Y.H., ed., Caroline Islands archaeology: Investigations on Fefan, Faraulep, Woleai, and Lamotrek: Honolulu, Bishop Museum Pacific Anthropological Records No. 35, p. 131–135.

Dickinson, W.R., 1985, Interpreting provenance relations from detrital modes of sandstones, in Zuffa, G.G., ed., Provenance of arenites: Dordrecht, Reidel, p. 333–361.

Dickinson, W.R., 1987, Temper sands in some Tongan Lapita sherds [Appendix 6], in Poulsen, J., The prehistory of the Tongan islands [Terra Australis 12]: Canberra, Australian National University, p. 278.

Dickinson, W.R., 1988, Temper sands in sherds from Niuatoputapu excavations in Tonga [Appendix B], in Kirch, P.V., Niuatoputapu, the prehistory of a Polynesian chiefdom: Seattle, Thomas Burke Memorial Washington State Museum Monograph 5, p. 274–277.

Dickinson, W.R., 1993, Sand temper in prehistoric potsherds from the To'aga site, in Kirch, P.V., and Hunt, T.L., eds., The To'aga site: Three millennia of Polynesian occupation in the Manu'a Islands, American Samoa: Berkeley, University of California Archaeological Research Facility Contribution No. 51, p. 151–156.

Dickinson, W.R., 1994, Natural beach placer analogous to prehistoric island tempers: Journal of the Polynesian Society, v. 103, p. 217–219.

Dickinson, W.R., 1995, Temper sand in prehistoric sherds from Kosrae, in Athens, J.S., Landscape archaeology: Prehistoric settlement, subsistence, and environment of Kosrae, eastern Caroline Islands, Micronesia: Honolulu, International Archaeological Research Institute, p. 271–276.

Dickinson, W.R., 1998a, Petrographic temper provinces of prehistoric pottery in Oceania: Sydney, Australian Museum Records, v. 50, p. 263–276.

Dickinson, W.R., 1998b, Geomorphology and geodynamics of the Cook-Austral island-seamount chain in the South Pacific Ocean: Implications for hotspots and plumes: International Geology Review, v. 40, p. 1039–1075.

Dickinson, W.R., 2000a, Petrography of temper sands in prehistoric Watom sherds and comparison with other temper suites of the Bismarck Archipelago: New Zealand Journal of Archaeology, v. 20 (1998), p. 161–182.

Dickinson, W.R., 2000b, Hydro-isostatic and tectonic influences on emergent Holocene paleoshorelines in the Mariana Islands, western Pacific Ocean: Journal of Coastal Research, v. 16, p. 725–746.

Dickinson, W.R., 2001a, Petrography and geologic provenance of sand tempers in prehistoric potsherds from Fiji and Vanuatu: South Pacific: Geoarchaeology, v. 16, p. 275–322, doi: 10.1002/1520-6548(200103)16:3<275::AID-GEA1005>3.0.CO;2-E.

Dickinson, W.R., 2001b, Paleoshoreline record of relative Holocene sea levels on Pacific islands: Earth-Science Reviews, v. 55, p. 191–234.

Dickinson, W.R., 2002a, The Basin and Range province as a composite extensional domain: International Geology Review, v. 44, p. 1–28.

Dickinson, W.R., 2002b, Petrologic character and geologic sources of sand tempers in prehistoric New Caledonian pottery, in Bedford, S., et al., eds., Fifty years in the field: Essays in honour and celebration of Richard Shutler Jr.'s archaeological career: Auckland, New Zealand Archaeological Association Monograph 25, p. 165–172.

Dickinson, W.R., 2003, Impact of mid-Holocene hydro-isostatic highstand in regional sea level on habitability of islands in Pacific Oceania: Journal of Coastal Research, v. 19, p. 489–502.

Dickinson, W.R., 2004a, Petrography of Polynesian Plainware from 'Ata: New Zealand Journal of Archaeology, v. 25 (2003), p. 107–111.

Dickinson, W.R., 2004b, Nature of sand tempers in sherds from the Murray Islands east of Torres Strait [Appendix L], in Carter, M.J., North of the Cape and south of the Fly: The archaeology of settlement and subsistence in the Murray Islands, eastern Torres Strait [Ph.D. thesis]: Townsville, James Cook University, p. 1–9.

Dickinson, W.R., 2005, Petrography of temper sands in prehistoric potsherds from the Aru Islands, south of West Irian near the shelf edge of the Arafura Sea [Appendix 6.1], in O'Connor, S., Spriggs, M., and Veth, P., The archaeology of the Aru Islands, eastern Indonesia [Terra Australis 22]: Canberra, Pandanus Books, p. 115–124.

Dickinson, W.R., 2006a, Sand tempers in Mussau sherds: Evidence for ceramic transfer from multiple unspecified localities within the Bismarck Archipelago, in Kirch, P.V., ed., Archaeological investigations in the Mussau Islands, Papua New Guinea, 1985–1988, vol. III: Berkeley, University of California Archaeological Research Facility Contribution No. 60 (in press).

Dickinson, W.R., 2006b, Petrography of manuports and control samples from Mussau, in Kirch, P.V., ed., Archaeological investigations in the Mussau Islands, Papua New Guinea, 1985–1988, vol. III: Berkeley, University of California Archaeological Research Facility Contribution No. 60 (in press).

Dickinson, W.R., 2006c, Petrography of coastal sands and prehistoric sherd tempers from the northwest coast of Papua New Guinea, in Terrell, J.E., and Schechter, eds., Archaeological investigations on the Sepik coast of Papua New Guinea: Chicago, Field Museum Press Fieldiana: Anthropology (in press).

Dickinson, W.R., and Green, R.C., 1974, Temper sands in A.D. 1595 Spanish ware from the Solomon Islands: Journal of the Polynesian Society, v. 83, p. 293–300.

Dickinson, W.R., and Green, R.C., 1998, Geoarchaeological context of Holocene subsidence at the ferry berth Lapita site, Mulifanua, Upolu, Samoa: Geoarchaeology, v. 13, p. 239–263, doi: 10.1002/(SICI)1520-6548(199802)13:3<239::AID-GEA1>3.0.CO;2-O.

Dickinson, W.R., and Rich, E.I., 1972, Petrologic intervals and petrofacies in the Great Valley sequence, Sacramento Valley, California: Geological Society of America Bulletin, v. 83, p. 3007–3024.

Dickinson, W.R., and Shutler, R., Jr., 1968, Insular sand tempers of prehistoric pottery from the southwest Pacific, in Yawata, I., and Sinoto, Y.H., eds., Prehistoric culture in Oceania: Honolulu, Bishop Museum Press, p. 29–37.

Dickinson, W.R., and Shutler, R., Jr., 1971, Temper sands in prehistoric pottery of the Pacific islands: Archaeology and Physical Anthropology in Oceania, v. 6, p. 191–203.

Dickinson, W.R., and Shutler, R., Jr., 1974, Probable Fijian origin of quartzose tempers in prehistoric pottery from Tonga and the Marquesas: Science, v. 185, p. 454–457.

Dickinson, W.R., and Shutler, R., Jr., 1979, Petrography of sand tempers in Pacific islands potsherds: Geological Society of America Bulletin, v. 90, Part I [summary], p. 993–995; Part II [microfiche], No. 11: Card, v. 1, p. 1644–1701.

Dickinson, W.R., and Shutler, R., Jr., 2000, Implications of petrographic temper analysis for Oceanian prehistory: Journal of World Prehistory, v. 14, p. 203–266, doi: 10.1023/A:1026557609805.

Dickinson, W.R., Rickard, M.J., Coulson, F.I., Smith, J.G., and Lawrence, R.L., 1968, Late Cenozoic shoshonitic lavas in northwestern Viti Levu, Fiji: Nature, v. 219, p. 148.

Dickinson, W.R., Helmold, K.P., and Stein, J.S., 1979, Mesozoic lithic sandstones in central Oregon: Journal of Sedimentary Petrology, v. 49, p. 501–516.

Dickinson, W.R., Takayama, J., Snow, E.A., and Shutler, R., Jr., 1990, Sand temper of probable Fijian origin in prehistoric potsherds from Tuvalu: Antiquity, v. 64, p. 307–312.

Dickinson, W.R., Burley, D.V., and Shutler, R., Jr., 1994, Impact of hydro-isostatic Holocene sea-level change on the geologic context of island archaeological sites, northern Ha'apai Group: Kingdom of Tonga: Geoarchaeology, v. 9, p. 85–111.

Dickinson, W.R., Shutler, R., Jr., Shortland, R., Burley, D.V., and Dye, T.S., 1996, Sand tempers in indigenous Lapita and Lapitoid Polynesian Plainware and imported protohistoric Fijian pottery of Ha'apai (Tonga) and the question of Lapita tradeware: Archaeology in Oceania, v. 31, p. 87–98.

Dickinson, W.R., Burley, D.V., Nunn, P.D., Anderson, A., Hope, G., De Biran, A., Burke, C., and Matararaba, S., 1998a, Geomorphic and archaeological landscapes of the Sigatoka dune site, Viti Levu, Fiji: Interdisciplinary investigations: Asian Perspectives, v. 37, p. 1–31.

Dickinson, W.R., Rolett, B.V., Sinoto, Y.H., Rosenthal, M.E., and Shutler, R., Jr., 1998b, Temper sands in exotic Marquesan pottery and the significance of their Fijian origin: Journal de la Société des Océanistes, v. 107, p. 119–133.

Dickinson, W.R., Sinoto, Y.H., Shutler, R., Jr., Shutler, M.E., Garanger, J., and Teska, T.M., 1999a, Japanese Jomon sherds in artifact collections from Mele Plain on Efate in Vanuatu: Archaeology in Oceania, v. 34, p. 15–24.

Dickinson, W.R., Burley, D.V., and Shutler, R., Jr., 1999b, Holocene paleoshoreline record in Tonga: Geomorphic features and archaeological implications: Journal of Coastal Research, v. 15, p. 682–700.

Dickinson, W.R., Butler, B.M., Moore, D.R., and Swift, M., 2001, Geologic sources and geographic distribution of sand tempers in prehistoric potsherds from the Mariana Islands: Geoarchaeology, v. 16, p. 827–854, doi: 10.1002/gea.1023.

Divis, A.F., 1975, Structure and metamorphism of the Yap Islands, western Caroline Islands: Geological Society of America Abstracts with Programs, v. 7, p. 1053–1054.

Dixon, T.H., Batiza, R., Futa, K., and Martin, D., 1984, Petrochemistry, age, and isotopic composition of alkali basalts from Ponape Island, western Pacific: Chemical Geology, v. 43, p. 1–28, doi: 10.1016/0009-2541(84)90138-4.

Dodson, J.R., and Intoh, M., 1999, Prehistory and palaeoecology of Yap, Federated States of Micronesia: Quaternary International, v. 59, p. 17–26, doi: 10.1016/S1040-6182(98)00068-8.

Dumont d'Urville, M.J., 1832, Notice sur les îles du Grand Océan et sur l'origine des peuples qui les habitent: Bulletin de la Société de Géographie, v. 17, p. 1–21.

Duncan, R.A., 1985, Radiometric ages from volcanic rocks along the New Hebrides–Samoa lineament, in Brocher, T.M., ed., Investigations of the northern Melanesian borderland: Houston, Circum-Pacific Council for Energy and Mineral Resources Earth Science Series, v. 3, p. 67–76.

Dunkley, P.N., 1983, Volcanism and the evolution of the ensimatic Solomon Islands arc, in Shimozuru, D., and Yokoyama, I., eds., Arc volcanism: Physics and tectonics: Tokyo, Terra Scientific, p. 225–241.

Dunkley, P.N., 1986, Geology of the New Georgia Group, western Solomon Islands: British Geological Survey Overseas Directorate Report MP/86/6, 83 p.

Dupuy, C., Dostal, J., Marcelot, G., Bougault, H., Joron, J.L., and Treuil, M., 1982, Geochemistry of basalts from central and southern New Hebrides island arc: Implications for their source rock composition: Earth and Planetary Science Letters, v. 60, p. 207–225, doi: 10.1016/0012-821X(82)90004-8.

Dye, T.S., 1987, Social and cultural change in the prehistory of the ancestral Polynesian homeland [Ph.D. thesis]: New Haven, Yale University, 303 p.

Dye, T.S., and Dickinson, W.R., 1996, Sources of sand tempers in prehistoric Tongan pottery: Geoarchaeology, v. 11, p. 141–164.

Eissen, J.P., Crawford, A.J., Cotton, J., Meffre, S., Bellon, H., and Delaune, M., 1998, Geochemistry and tectonic significance of basalts in the Poya terrane, New Caledonia: Tectonophysics, v. 284, p. 203–219, doi: 10.1016/S0040-1951(97)00183-2.

Ewart, A., 1976, A petrological study of the younger Tongan andesites and dacites, and the olivine tholeiites of Niuafo'ou Island, S.W. Pacific: Contributions to Mineralogy and Petrology, v. 58, p. 1–21, doi: 10.1007/BF00384740.

Ewart, A., and Bryan, W.B., 1972, Petrography and geochemistry of igneous rocks from Eua, Tongan islands: Geological Society of America Bulletin, v. 83, p. 3281–3298.

Ewart, A., and Bryan, W.B., 1973, The petrology and geochemistry of the Tongan islands, in Coleman, P.J., ed., The western Pacific: Island arcs, marginal seas, geochemistry: New York, Crane Russak, p. 503–522.

Ewart, A., and Hawkesworth, C.J., 1987, The Pleistocene-Recent Tonga-Kermadec arc lavas: Interpretation of new isotopic and rare earth data in terms of a depleted mantle source model: Journal of Petrology, v. 28, p. 495–530.

Ewart, A., Bryan, W.B., and Gill, J.B., 1973, Mineralogy and geochemistry of the younger volcanic islands of Tonga: Journal of Petrology, v. 14, p. 429–465.

Ewart, A., Brothers, R.N., and Mateen, A., 1977, An outline of the geology and geochemistry, and the possible geotectonic evolution, of the volcanic rocks of the Tonga–Kermadec–New Zealand island arc: Journal of Volcanology and Geothermal Research, v. 2, p. 205–250, doi: 10.1016/0377-0273(77)90001-4.

Exon, N.F., and Marlow, M.S., 1988, Tripartite study of the New Ireland–Manus region: An introduction, in Marlow, M.S., et al., eds., Geology and offshore resources of Pacific island arcs—New Ireland and Manus region, Papua New Guinea: Houston, Circum-Pacific Council for Energy and Mineral Resources Earth Science Series, v. 9, p. 1–9.

Falvey, D.A., 1978, Analysis of paleomagnetic data from the New Hebrides: Australian Society of Exploration Geophysicists Bulletin, v. 9, p. 117–123.

Falvey, D.A., Colwell, J.B., Coleman, P.J., Greene, H.G., Vedder, J.G., and Bruns, T.R., 1991, Petroleum prospectivity of Pacific island arcs: Solomon Islands and Vanuatu: Australasian Petroleum Exploration Association Journal, v. 31, p. 191–212.

Felgate, M.W., 2001, A Roviana ceramic sequence and the prehistory of Near Oceania: Work in progress, in Clark, G.J., et al., eds., The archaeology of Lapita dispersal in Oceania [Terra Australis 17]: Canberra, Australian National University Pandanus Books, p. 39–60.

Felgate, M.W., and Dickinson, W.R., 2001, Late-Lapita and post-Lapita pottery transfers: Evidence from intertidal-zone sites of Roviana Lagoon, Western Province, Solomon Islands, in Jones, M., and Sheppard, P., eds., Australasian connections and new directions [Proceedings of the 7th Australasian Archaeometry Conference]: Auckland, University of Auckland Research in Anthropology and Linguistics No. 5, p. 105–122.

Finney, B., 1996, Colonizing an island world, in Goodenough, W.H., ed., Prehistoric settlement of the Pacific: American Philosophical Society Transactions, v. 86, p. 71–116.

Fitzpatrick, S.M., 2003, Early human burials in the western Pacific: Evidence for a c.3000 year old occupation on Palau: Antiquity, v. 77, p. 719–731.

Fitzpatrick, S.M., Dickinson, W.R., and Clark, G., 2003, Ceramic petrography and cultural interaction in Palau, Micronesia: Journal of Archaeological Science, v. 30, p. 1175–1184, doi: 10.1016/S0305-4403(03)00014-1.

Fleck, R.J., and Vallier, T.L., 1985, Strontium isotopic measurements of igneous rocks from Ata and Eua islands, Kingdom of Tonga, in Scholl, D.W., and Vallier, T.L., eds., Geology and offshore resources of Pacific island arcs—Tonga region: Houston, Circum-Pacific Council for Energy and Mineral Resources Earth Science Series, v. 2, p. 213–219.

Francis, G., 1988, Stratigraphy of Manus Island, western New Ireland basin, Papua New Guinea, in Marlow, M.S., et al., eds., Geology and offshore resources of Pacific island arcs—New Ireland and Manus region, Papua New Guinea: Houston, Circum-Pacific Council for Energy and Mineral Resources Earth Science Series, v. 9, p. 31–40.

Francis, G., Lock, J., and Okuda, Y., 1987, Seismic stratigraphy and structure of the area to the southeast of the Trobriand platform: Geo-Marine Letters, v. 7, p. 121–128, doi: 10.1007/BF02238041.

Freestone, I.C., 1982, Applications and potential of electron probe micro-analysis in technological and provenance investigations of ancient ceramics: Archaeometry, v. 24, p. 99–116.

Freestone, I.C., 1991, Extending ceramic petrology, in Middleton, A., and Freestone, I.C., eds., Recent developments in ceramic petrology: London, British Museum Occasional Paper 81, p. 399–410.

Fryer, P., 1974, Petrology of some volcanic rocks from the northern Fiji plateau: Geological Society of America Bulletin, v. 85, p. 1717–1720, doi: 10.1130/0016-7606(1974)85<1717:POSVRF>2.0.CO;2.

Fryer, P., 1995, Geology of the Mariana Trough, in Taylor, B., ed., Backarc basins: Tectonics and magmatism: New York, Plenum Press, p. 237–279.

Fujimura, K., and Alkire, W.H., 1984, Archaeological test excavations on Faraulep, Woleai, and Lamotrek in the Caroline Islands of Micronesia, in Sinoto, Y.H., ed., Caroline Islands archaeology: Investigations on Fefan, Faraulep, Woleai, and Lamotrek: Honolulu, Bishop Museum Pacific Anthropological Records No. 35, p. 65–129.

Gaina, C., Müller, D.R., Royer, J.Y., Stock, J., Hardebeck, J., and Symonds, P., 1998, The tectonic history of the Tasman Sea: A puzzle with 13 pieces: Journal of Geophysical Research, v. 103, p. 12,413–12,433, doi: 10.1029/98JB00386.

Galipaud, J.-C., 1990, The physico-chemical analysis of ancient pottery from New Caledonia, *in* Spriggs, M., ed., Lapita design, form, and composition: Canberra, Australian National University Occasional Papers in Prehistory No. 19, p. 134–142.

Galipaud, J.-C., 1996, New Caledonia: Some recent archaeological perspectives, *in* Davidson, J.M., et al., eds., Oceanic culture history: Essays in honour of Roger Green: Dunedin, New Zealand Archaeological Association Special Publication, p. 297–305.

Galipaud, J.-C., 1998, Recherches archaéologique aux îles Torres: Journal de la Société des Océanistes, v. 107, p. 159–168.

Galipaud, J.-C., 2001, Le peuplement initial de Pohnpei: Journal de la Société des Océanistes, v. 112, p. 49–60.

Garanger, J., 1966, Recherches archéologiques aux Nouvelles-Hébrides: L' Homme, v. 6, p. 59–81.

Garanger, J., 1971, Incised and applied-relief pottery, its chronology and development in southeastern Melanesia, and extra-areal comparisons, *in* Green, R.C., and Kelly, M., eds., Studies in Oceanic culture history, vol. 2: Honolulu, Bishop Museum Pacific Anthropological Records No. 12, p. 53–66.

Garanger, J., 1972, Archéologie des Nouvelles Hébrides: Contribution à la connaissance des îles du Centre: Paris, Publications de la Société des Océanistes No. 30, 156 p.

Garling, S., 2003, Tanga takes to the stage: Another model "transitional" site? New evidence and a contribution to the "incised and applied relief tradition" in New Ireland, *in* Sand, C., ed., Pacific archaeology: Assessments and prospects: Noumea, Cahiers de l'Archéologie en Nouvelle-Calédonie, v. 15, p. 213–233.

Garzanti, E., Andò, S., and Scutellà, M., 2000, Actualistic ophiolite provenance: The Cyprus case: Journal of Geology, v. 108, p. 199–218, doi: 10.1086/314391.

Garzanti, E., Verzoli, G., and Andò, S., 2002, Modern sand from obducted ophiolite belts (Sultanate of Oman and United Arab Emirates): Journal of Geology, v. 110, p. 371–391, doi: 10.1086/340440.

Gerrard, C.M., 1991, Sedimentary petrology and the archaeologist: The study of ancient ceramics, *in* Morton, A.C., Todd, S.P., and Haughton, P.D.W., eds., Developments in sedimentary provenance studies: Geological Society of London Special Publication No. 57, p. 189–197.

Gifford, E.W., 1951, Archaeological excavations in Fiji: Berkeley, University of California Anthropological Records, v. 13, p. 189–288.

Gifford, E.W., and Gifford, D.S., 1959, Archaeological excavations in Yap: Berkeley, University of California Anthropological Records, v. 18, p. 149–224.

Gifford, E.W., and Shutler, R., Jr., 1956, Archaeological excavations in New Caledonia: Berkeley, University of California Anthropological Records, v. 18, p. 1–149.

Gill, J.B., 1976, Composition of Lau Basin and Ridge volcanic rocks: Implications for evolution of an interarc basin and remnant arc: Geological Society of America Bulletin, v. 87, p. 1384–1395, doi: 10.1130/0016-7606(1976)87<1384:CAAOLB>2.0.CO;2.

Gill, J.B., 1981, Orogenic andesites and plate tectonics: New York, Springer-Verlag, 390 p.

Gill, J.B., 1987, Early geochemical evolution of an oceanic island arc and back-arc: Fiji and the South Fiji Basin: Journal of Geology, v. 95, p. 589–615.

Gill, J.B., and Stork, A.L., 1979, Miocene low-K dacites and trondhjemites of Fiji, *in* Barker, F., ed., Trondhjemites, dacites, and related rocks: Amsterdam, Elsevier, p. 619–649.

Gill, J.B., and Whelan, P.W., 1989a, Early rifting of an oceanic island arc (Fiji) produced shoshonitic to tholeiitic basalts: Journal of Geophysical Research, v. 94, p. 4561–4578.

Gill, J.B., and Whelan, P.W., 1989b, Postsubduction ocean island alkali basalts in Fiji: Journal of Geophysical Research, v. 94, p. 4579–4588.

Gill, J.B., Stork, A.L., and Whelan, P.W., 1984, Volcanism accompanying back-arc basin development in the southwest Pacific: Tectonophysics, v. 102, p. 207–224, doi: 10.1016/0040-1951(84)90014-3.

Gladczenko, T.P., Coffin, M.F., and Eldholm, O., 1997, Crustal structure of the Ontong Java Plateau: Modeling of new gravity and existing seismic data: Journal of Geophysical Research, v. 102, p. 22,711–22,729, doi: 10.1029/97JB01636.

Goles, G.G., 2001, Cathodoluminescence of quartz grains as a provenance indicator: Proof-of-concept using sand-temper grains of prehistoric Fijian ceramics, *in* Stevenson, C.M., et al., eds., Pacific 2000: Proceedings of the Fifth International Conference on Easter Island and the Pacific: Los Osos, California, Easter Island Foundation, p. 51–58.

Golson, J., 1971, Lapita ware and its transformations, *in* Green, R.C., and Kelly, M., eds., Studies in Oceanic culture history, Volume 2: Honolulu, Bishop Museum Pacific Anthropological Records No. 12, p. 67–76.

Golson, J., 1991, Two sites at Lasigi, New Ireland, *in* Allen, J., and Gosden, C., eds., Report of the Lapita homeland project: Canberra, Australian National University Occasional Papers in Prehistory No. 20, p. 244–259.

Golson, J., 1992, The pottery from Lasigi, New Ireland in Galipaud, J.-C., ed., Poterie Lapita et peuplement: Noumea, ORSTOM (Office de Recherche Scientifique et Technologique Outre-Mer), p. 155–167.

Gosden, C., 1989, Prehistoric social landscapes of the Arawe Islands, west New Britain Province, Papua New Guinea: Archaeology in Oceania, v. 24, p. 45–58.

Gosden, C., 1991, Towards an understanding of the regional archaeological record from the Arawe Islands, west New Britain, Papua New Guinea, *in* Allen, J., and Gosden, C., eds., Report of the Lapita homeland project: Canberra, Australian National University Occasional Papers in Prehistory No. 20, p. 205–216.

Gosden, C., and Specht, J., 1991, Diversity, continuity and change in the Bismarck Archipelago, Papua New Guinea: Indo-Pacific Prehistory Association Bulletin, v. 11, p. 276–280.

Gosden, C., and Webb, J., 1994, The creation of a Papua New Guinea landscape: Archaeological and geomorphological evidence: Journal of Field Archaeology, v. 21, p. 29–51.

Gosden, C., Allen, J., Ambrose, W., Anson, D., Golson, J., Green, R., Kirch, P., Lilley, I., Specht, J., and Spriggs, M., 1989, Lapita sites of the Bismarck Archipelago: Antiquity, v. 63, p. 561–586.

Graham, S.A., Ingersoll, R.V., and Dickinson, W.R., 1976, Common provenance for lithic grains in Carboniferous sandstones from the Ouachita Mountains and the Black Warrior basin: Journal of Sedimentary Petrology, v. 46, p. 620–632.

Greene, H.G., Collot, J.-Y., Fisher, M.A., and Crawford, A.J., 1994, Neogene tectonic evolution of the New Hebrides island arc: A review incorporating ODP drilling results, *in* Greene, H.G., et al., eds., Vanuatu [Sites 827–833]: Proceedings of the Ocean Drilling Program, Scientific Results, Volume 134: College Station, Texas, Ocean Drilling Program, p. 19–46.

Green, R.C., 1974, A review of portable artifacts from Western Samoa, *in* Green, R.C., and Davidson, J.M., eds., Archaeology in Western Samoa, Volume II: Auckland, Auckland Institute and Museum Bulletin 7, p. 245–275.

Green, R.C., 1976, Lapita sites in the Santa Cruz Group, *in* Green, R.C., and Cresswell, M.M., eds., Southeast Solomons cultural history: A preliminary survey: Royal Society of New Zealand Bulletin 11, p. 245–265.

Green, R.C., 1979, Lapita, *in* Jennings, J.D., ed., The prehistory of Polynesia: Cambridge, Massachusetts, Harvard University Press, p. 27–60.

Green, R.C., 1991a, Near and Remote Oceania—Disestablishing "Melanesia" in culture history, *in* Pawley, A., ed., Man and a half: Essays in Pacific anthropology and ethnobotany in honour of Ralph Bulmer: Auckland, Polynesian Society, p. 491–502.

Green, R.C., 1991b, The Lapita cultural complex: Current evidence and proposed models: Indo-Pacific Prehistory Association Bulletin, v. 11, p. 275–305.

Green, R.C., 1991c, A reappraisal of the dating for some Lapita sites in the Reef–Santa Cruz Group of the southeast Solomons: Journal of the Polynesian Society, v. 100, p. 197–207.

Green, R.C., 1992, Definitions of the Lapita cultural complex and its nonceramic component, *in* Galipaud, J.C., ed., Poterie Lapita et peuplement: Noumea, ORSTOM (Office de Recherche Scientifique et Technologique Outre-Mer), p. 7–20.

Green, R.C., 1996, Prehistoric transfers of portable items during the Lapita horizon in Remote Oceania: A review: Indo-Pacific Prehistory Association Bulletin, v. 15, p. 119–130.

Green, R.C., 1997, Linguistic, biological, and cultural origins of the initial inhabitants of Remote Oceania: New Zealand Journal of Archaeology, v. 17, p. 5–27.

Green, R.C., 2000a, An introduction to investigations on Watom Island, Papua New Guinea: New Zealand Journal of Archaeology, v. 20 (1998), p. 5–27.

Green, R.C., 2000b, Lapita and the cultural model for intrusion, integration, and innovation, *in* Anderson, A., and Murray, T., eds., Australian archaeologist: Collected papers in honor of Jim Allen: Canberra, Coombs Academic Publishing, p. 372–392.

Green, R.C., 2002, A retrospective view of settlement pattern studies in Samoa, *in* Ladefoged, T.N., and Graves, M.W., eds., Pacific landscapes: Archaeological approaches: Los Osos, California, Easter Island Foundation, p. 125–152.

Green, R.C., 2003, The Lapita horizon and traditions—Signature for one set of oceanic migrations, *in* Sand, C., ed., Pacific archaeology: Assessments and prospects: Noumea, Cahiers de l'Archéologie en Nouvelle-Calédonie, v. 15, p. 95–120.

Green, R.C., and Anson, D., 1991, The Reber-Rakival site on Watom: Implications of the 1985 excavations at the SAC and SDI localities, *in* Allen, J., and Gosden, C., eds., Report of the Lapita homeland project: Canberra, Australian National University Occasional Papers in Prehistory No. 20, p. 170–181.

Green, R.C., and Anson, D., 2000a, Excavations at Kainipirina (SAC), Watom Island, Papua New Guinea: New Zealand Journal of Archaeology, v. 20 (1998), p. 29–94.

Green, R.C., and Anson, D., 2000b, Archaeological investigations on Watom Island: Early work, outcomes of recent investigations, and future prospects: New Zealand Journal of Archaeology, v. 20 (1998), p. 283–197.

Green, R.C., and Kirch, P.V., 1997, Lapita exchange systems and their Polynesian transformations: Seeking explanatory models, *in* Weisler, M.I., ed., Prehistoric long-distance interaction in Oceania: An interdisciplinary approach: Auckland, New Zealand Archaeological Association Monograph 21, p. 19–37.

Greenbaum, D., Mallick, D.I.J., and Radford, N.W., 1975, Geology of the Torres Islands: Port Vila, New Hebrides Condominium Geological Survey Regional Report, 46 p.

Grzeszyk, A., Eissen, J.-P., Dupont, J., Lefévre, C., Maillet, P., and Monzier, M., 1987, Pétrographie et minéralogie des îles Futuna et Alofi, TOM de Wallis et Futuna (Pacifique sud-ouest): Comptes Rendus de l'Académie des Sciences de Paris, ser. II, v. 305, p. 93–98.

Grzeszyk, A., Lefévre, C., Monzier, M., Eissen, J.-P., Dupont, J., and Maillet, P., 1991, Mise en evidence d'un volcanisme transitionnel pliocéne supérieur sur Futuna et Alofi (SW Pacifique): Un nouveau témoin de l'évolution géodynamique nord-Tonga: Comptes Rendus de l'Académie des Sciences de Paris, ser. II, v. 312, p. 713–720.

Guillon, J.-H., 1975, Les massifs péridotiques de Nouvelle-Calédonie: Type d'appareil ultrabasique stratiforme de chaine Récente: Paris, ORSTOM (Office de Recherche Scientifique et Technologique Outre-Mer) Mémoire 76, 120 p.

Hackman, B.D., 1980, The geology of Guadalcanal, Solomon Islands: British Geological Survey Overseas Memoir 6, 107 p.

Hall, R., 1996, Reconstructing Cenozoic southeast Asia, *in* Hall, R., and Blondell, D., eds., Tectonic evolution of southeast Asia: Geological Society [London] Special Publication 106, p. 153–184.

Hall, R., 2002, Cenozoic geological and plate tectonic evolution of SE Asia and the SW Pacific: Computer-based reconstructions, model and animations: Journal of Asian Earth Sciences, v. 20, p. 353–431, doi: 10.1016/S1367-9120(01)00069-4.

Hall, R., Audley-Charles, M.G., Banner, F.T., Hidayat, S., and Tobing, S.L., 1988a, Basement rocks of the Halmahera region, eastern Indonesia: A Late Cretaceous–Early Tertiary arc and forearc: Geological Society [London] Journal, v. 145, p. 65–84.

Hall, R., Audley-Charles, M.G., Banner, F.T., Hidayat, S., and Tobing, S.L., 1988b, Late Palaeogene–Quaternary geology of Halmahera, eastern Indonesia: Initiation of a volcanic island arc: Geological Society [London] Journal, v. 145, p. 577–590.

Hamilton, W., 1979, Tectonics of the Indonesian region: U.S. Geological Survey Professional Paper 1078, 345 p.

Hart, S.R., Coetzee, M., Workman, R.K., Blustajn, J., Johnson, K.T.M., Sinton, J.M., Steinberger, B., and Hawkins, J.W., 2004, Genesis of the western Samoa seamount province: Age, geochemical fingerprint and tectonics: Earth and Planetary Science Letters, v. 227, p. 37–56, doi: 10.1016/j.epsl.2004.08.005.

Haston, R., Fuller, M., and Schmidtke, E., 1988, Paleomagnetic results from Palau, west Caroline Islands: Geology, v. 16, p. 654–657, doi: 10.1130/0091-7613(1988)016<0654:PRFPWC>2.3.CO;2.

Hathway, B., 1993, The Nadi basin: Neogene strike-slip faulting and sedimentation in a fragmented arc, western Viti Levu, Fiji: Geological Society [London] Journal, v. 150, p. 563–581.

Hathway, B., 1994, Sedimentation and volcanism in an Oligocene-Miocene intra-oceanic arc and fore-arc, southwestern Viti Levu, Fiji: Geological Society [London] Journal, v. 151, p. 499–514.

Hathway, B., 1995, Deposition and diagenesis of Miocene arc-fringing platform and debris-apron carbonates, southwestern Viti Levu, Fiji: Sedimentary Geology, v. 94, p. 187–208, doi: 10.1016/0037-0738(94)00086-A.

Hathway, B., and Colley, H., 1994, Eocene to Miocene geology of southwest Viti Levu, Fiji, *in* Stevenson, A.J., et al., eds., Geology and submarine resources of the Fiji-Lau-Tonga region: Suva, South Pacific Applied Geoscience Commission (SOPAC) Technical Bulletin 8, p. 153–169.

Hawkins, J.W., Jr., 1995, The geology of the Lau Basin, *in* Taylor, B., ed., Backarc basins: Tectonics and magmatism: New York, Plenum Press, p. 61–138.

Hawkins, J.W., 2003, Geology of supra-subduction zones—Implications for the origin of ophiolites, *in* Dilek, Y., and Newcomb, S., eds., Ophiolite concept and the evolution of geological thought: Geological Society of America Special Paper 373, p. 227–268.

Hawkins, J., and Batiza, R., 1977, Metamorphic rocks of the Yap arc-trench system: Earth and Planetary Science Letters, v. 37, p. 216–229, doi: 10.1016/0012-821X(77)90166-2.

Hawkins, J.W., Jr., and Falvey, D.A., 1985, Petrology of andesitic dikes and flows from 'Eua, Tonga, *in* Scholl, D.W., and Vallier, T.L., eds., Geology and offshore resources of Pacific island arcs—Tonga region: Houston, Circum-Pacific Council for Energy and Mineral Resources Earth Science Series, v. 2, p. 269–279.

Hawkins, J.W., Jr., and Natland, J.H., 1975, Nephelinites and basanites of the Samoan linear volcanic chain: Their possible tectonic significance: Earth and Planetary Science Letters, v. 24, p. 427–439, doi: 10.1016/0012-821X(75)90150-8.

Hedrick, J.D., 1971, Lapita style pottery from Malo Island: Journal of the Polynesian Society, v. 80, p. 5–19.

Hegarty, K.A., and Weissel, J.K., 1988, Complexities in the development of the Caroline plate region, western equatorial Pacific, *in* Nairn, A.E.M., et al., eds., The ocean basins and margins, vol. 7B: New York, Plenum Press, p. 277–301.

Hegarty, K.A., Weissel, J.K., and Hayes, D.E., 1983, Convergence at the Caroline-Pacific plate boundary: Collision and subduction, *in* Hayes, D.E., ed., The tectonic and geologic evolution of southeast Asian seas and islands, Part 2: American Geophysical Union Geophysical Monograph 27, p. 326–348.

Heming, R.F., 1974, Geology and petrology of the Rabaul Caldera, Papua New Guinea: Geological Society of America Bulletin, v. 85, p. 1253–1264, doi: 10.1130/0016-7606(1974)85<1253:GAPORC>2.0.CO;2.

Heming, R.F., 1977, Mineralogy and proposed *P-T* paths of basaltic lavas from Rabaul Caldera, Papua New Guinea: Contributions to Mineralogy and Petrology, v. 61, p. 15–33, doi: 10.1007/BF00375943.

Heming, R.F., 1979, Undersaturated lavas from Ambitle Island, Papua New Guinea: Lithos, v. 12, p. 173–186, doi: 10.1016/0024-4937(79)90002-1.

Herzer, R.H., and Exon, N.F., 1985, Structure and basin analysis of the southern Tonga forearc, *in* Scholl, D.W., and Vallier, T.L., eds., Geology and offshore resources of Pacific island arcs—Tonga region: Houston, Circum-Pacific Council for Energy and Mineral Resources Earth Science Series, v. 2, p. 55–73.

Hill, E.J., and Baldwin, S.L., 1993, Exhumation of high-pressure metamorphic rocks during crustal extension in the D'Entrecasteaux region, Papua New Guinea: Journal of Metamorphic Petrology, v. 11, p. 261–277.

Hill, E.J., Baldwin, S.L., and Lister, G.S., 1992, Unroofing of active metamorphic core complexes in the D'Entrecasteaux Islands, Papua New Guinea: Geology, v. 20, p. 907–910, doi: 10.1130/0091-7613(1992)020<0907:UOAMCC>2.3.CO;2.

Hill, K.C., and Raza, A., 1999, Arc-continent collision in Papua New Guinea: Constraints from fission track thermochronology: Tectonics, v. 18, p. 950–966, doi: 10.1029/1999TC900043.

Hilyard, D., and Rogerson, R., 1989, Revised stratigraphy of Bougainville and Buka islands, Papua New Guinea, *in* Vedder, J.G., and Bruns, T.R., eds., Geology and offshore resources of Pacific island arcs—Solomon Islands and Bougainville, Papua New Guinea region: Houston, Circum-Pacific Council for Energy and Mineral Resources Earth Science Series, v. 12, p. 87–92.

Hindle, W.H., 1976, The geology of west-central Vanua Levu: Suva, Fiji Mineral Resources Division Bulletin 1, 76 p.

Hindle, W.H., and Colley, H., 1981, An oceanic volcano in an island arc setting—Seatura volcano, Fiji: Geological Magazine, v. 118, p. 1–14.

Hine, R., and Mason, D.R., 1978, Intrusive rocks associated with porphyry copper mineralization, New Britain, Papua New Guinea: Economic Geology and the Bulletin of the Society of Economic Geologists, v. 73, p. 749–760.

Hinschberger, F., Malod, J.-A., Réhault, J.-P., Villeneuve, M., Royer, J.I., and Burhanuddin, S., 2005, Late Cenozoic geodynamic evolution of eastern Indonesia: Tectonophysics, v. 404, p. 91–118.

Hirst, J.A., 1965, Geology of east and northeast Viti Levu: Suva, Fiji Geological Survey Bulletin 12, 51 p.

Hodges, H.W.M., 1963, The examination of ceramic materials in thin section, in Pyddoke, E., ed., The scientist and archaeology: New York, Roy, p. 101–110.

Hohnen, P.D., 1978, The geology of New Ireland: Australian Bureau of Mineral Resources Bulletin 194 [PNG Bulletin 12], 39 p.

Honthaas, C., Réhault, J.-P., Maury, R.C., Bellon, H., Hémond, C., Malod, J.-A., Cornée, J.-J., Villeneuve, M., Cotton, J., Burhunuddin, S., Gillou, H., and Arnaud, N., 1998, A Neogene back-arc origin for the Banda Sea basins: Geochemical and geochronological constraints from the Banda ridges (east Indonesia): Tectonophysics, v. 298, p. 297–317, doi: 10.1016/S0040-1951(98)00190-5.

Honza, E., Davies, H.L., Keene, J.B., and Tiffin, D.I., 1987, Plate boundaries and evolution of the Solomon Sea region: Geo-Marine Letters, v. 7, p. 161–168, doi: 10.1007/BF02238046.

Hu, J.-C., Yu, S.-B., Chu, H.-T., and Angelier, J., 2002, Transition tectonics of northern Taiwan induced by convergence and trench retreat, in Byrne, T.B., and Liu, C.-S., eds., Geology and geophysics of an arc-continent collision, Taiwan: Geological Society of America Special Paper 358, p. 147–160.

Hubbard, N.J., 1971, Some chemical features of lavas from the Manu'a Islands, Samoa: Pacific Science, v. 25, p. 178–187.

Hughes, G.W., 1978, The relationship between volcanic island genesis and the Indo-Australian Pacific plate margins in the eastern outer islands, Solomon Islands, southwest Pacific: Journal of Physics of the Earth, v. 26, Supplement, p. S123–S138.

Hughes, G.W., and Turner, C.C., 1977, Upraised Pacific ocean floor, southern Malaita, Solomon Islands: Geological Society of America Bulletin, v. 88, p. 412–424, doi: 10.1130/0016-7606(1977)88<412:UPOFSM>2.0.CO;2.

Hughes, G.W., Craig, P.M., and Dennis, R.S., 1981, Geology of the eastern outer islands: Honiara, Solomon Islands Geological Survey Division Bulletin 4, 108 p.

Hunt, T.L., 1980, Toward Fiji's past: Archaeological research on southwestern Viti Levu [M.A. thesis]: Auckland, University of Auckland, 235 p.

Hunt, T.L., 1986, Conceptual and substantive issues in Fijian prehistory, in Kirch, P.V., ed., Island societies: Archaeological approaches to evolution and transformation: Cambridge, Cambridge University Press, p. 20–32.

Hunt, T.L., 1987, Patterns of human interaction and evolutionary divergence in the Fiji Islands: Journal of the Polynesian Society, v. 96, p. 299–334.

Hunt, T.L., 1988, Lapita ceramic technological and compositional studies: A critical review, in Kirch, P.V., and Hunt, T.L., eds., Archaeology of the Lapita cultural complex: A critical review: Seattle, Thomas Burke Memorial Washington State Museum Research Report 5, p. 49–59.

Hunt, T.L., and Erkelens, C.L., 1993, The To'aga ceramics, in Kirch, P.V., and Hunt, T.L., eds., The To'aga site: Three millennia of Polynesian occupation in the Manu'a Islands, American Samoa: Berkeley, University of California Archaeological Research Facility Contribution No. 51, p. 123–149.

Hunt, T.L., and Graves, M.W., 1990, Some methodological issues of exchange in Oceanic prehistory: Asian Perspectives, v. 29, p. 107–117.

Hunt, T.L., and Kirch, P.V., 1988, An archaeological survey of the Manu'a Islands, American Samoa: Journal of the Polynesian Society, v. 97, p. 153–183.

Hunt, T.L., and Kirch, P.V., 1997, The historical ecology of Ofu Island, American Samoa, 3000 B.P. to the present, in Kirch, P.V., and Hunt, T.L., eds., Historical ecology in the Pacific islands: New Haven, Yale University Press, p. 105–121.

Huntley, D.J., Dickinson, W.R., and Shutler, R., Jr., 1983, Petrographic studies and thermoluminescence dating of some potsherds from Maré and Ouvea, Loyalty Islands: Archaeology in Oceania, v. 18, p. 106–108.

Ibbotson, P., 1961, The geology of Ovalau, Moturiki, and Naingani: Suva, Fiji Geological Survey Department Bulletin 9, 7 p.

Ibbotson, P., 1962, The geology of the Tavua area, Viti Levu: Suva, Fiji Geological Survey Department Bulletin 8, 22 p.

Ibbotson, P., 1967, Petrology of the Tertiary caldera, Tavua goldfield: Suva, Fiji Geological Survey Department Memoir 3, 59 p.

Ibbotson, P., 1969, The geology of east-central Vanua Levu: Suva, Fiji Department of Geological Surveys Bulletin 16, 44 p.

Ielsch, G., Caroff, M., Barsczus, H.G., Maury, R.C., Guillou, H., Guille, G., and Cotten, J., 1998, Géochemie des basaltes de l'île de Ua Huka (archipel des Marquises): Variation du taux de fusion partielle et hétérogénéité de la source mantellique: Comptes Rendus de l'Académie des Sciences de Paris, v. 326, p. 413–420.

Ingersoll, R.V., Bullard, T.F., Ford, R.L., Grimm, J.P., Pickle, J.D., and Sares, S.W., 1984, The effect of grain size on detrital modes: A test of the Gazzi-Dickinson point-counting method: Journal of Sedimentary Petrology, v. 54, p. 103–116.

Intoh, M., 1981, Reconnaissance archaeological research on Ngulu atoll in the western Caroline Islands: Asian Perspectives, v. 24, p. 69–80.

Intoh, M., 1990, Ceramic environment and technology: A case study in the Yap Islands in Micronesia: Man and Culture in Oceania, v. 6, p. 35–52.

Intoh, M., 1992a, Pottery traditions in Micronesia, in Galipaud, J.C., ed., Poterie Lapita et peuplement: Noumea, ORSTOM (Office de Recherche Scientifique et Technologique Outre-Mer), p. 67–82.

Intoh, M., 1992b, Why were pots imported to Ngulu atoll? A consideration of subsistence strategy: Journal of the Polynesian Society, v. 101, p. 159–168.

Intoh, M., 1996, Multi-regional contacts of prehistoric Fais islanders in Micronesia: Indo-Pacific Prehistory Association Bulletin, v. 15, p. 111–117.

Intoh, M., 1997, Human dispersals into Micronesia: Anthropological Science, v. 105, p. 15–28.

Intoh, M., 1999, Cultural contacts between Micronesia and Melanesia, in Galipaud, J.-C., and Lilley, I., eds., Le Pacifique de 5000 à 2000 avant le présent: Paris, Éditions IRD (Institut de Recherche pour le Développement), p. 407–422.

Intoh, M., and Dickinson, W.R., 2002, Prehistoric pottery movements in western Micronesia: Technological and petrological study of potsherds from Fais Island, in Bedford, S., et al., eds., Fifty years in the field: Essays in honour and celebration of Richard Shutler Jr.'s archaeological career: Auckland, New Zealand Archaeological Association Monograph 25, p. 123–134.

Intoh, M., and Leach, F., 1985, Archaeological investigations in the Yap Islands, Micronesia: First millennium B.C. to the present day: Oxford, BAR International Series 277, 200 p.

Irwin, G., 1991, Themes in the prehistory of coastal Papua and the Massim, in Pawley, A., ed., Man and a half: Essays in Pacific anthropology and ethnobiology in honour of Ralph Bulmer: Auckland, Polynesian Society, p. 503–510.

Irwin, G., 1992, The prehistoric exploration and colonisation of the Pacific: Cambridge, UK, Cambridge University Press, 240 p.

Irwin, G., 1993, Voyaging, in Spriggs, M., et al., eds., A community of culture: The people and prehistory of the Pacific: Canberra, Australian National University Occasional Papers in Prehistory No. 21, p. 73–87.

Irwin, G., Bellwood, P., Nitihaminoto, G., Tanudirjo, D., and Siswanto, L., 1999, Prehistoric relations between island southeast Asia and Oceania: Recent archaeological investigations in the northern Moluccas, in Galipaud, J.-C., and Lilley, I., eds., Le Pacifique de 5000 à 2000 avant le présent: Paris, Éditions IRD (Institut de Recherche pour le Développement), p. 363–374.

Jaques, A.L., 1976, High-K_2O island-arc volcanic rocks from the Finisterre and Adelbert Ranges, northern Papua New Guinea: Geological Society of America Bulletin, v. 87, p. 861–867, doi: 10.1130/0016-7606(1976)87<861:HIVRFT>2.0.CO;2.

Jaques, A.L., 1981, Quaternary volcanism on Manus and M'Buke islands, in Johnson, R.W., ed., Cooke-Ravian volume of volcanological papers: Port Moresby, Papua New Guinea Geological Survey Memoir 10, p. 213–219.

Jaques, A.L., and Robinson, G.P., 1977, The continent/island arc collision in northern Papua New Guinea: BMR Journal of Australian Geology and Geophysics, v. 2, p. 289–303.

Jennings, J.D., and Holmer, R.N., 1980, Archaeological investigations in western Samoa: Honolulu, Bishop Museum Pacific Anthropological Records No. 32, 155 p.

Johnson, C.G., Alvis, R.J., and Hetzler, R.L., 1960, Military geology of Yap Islands, Caroline Islands: Tokyo, Intelligence Division, Office of the Engineer, Headquarters U.S. Army Far East, 164 p.

Johnson, R.W., 1976, Late Cenozoic volcanism and plate tectonics at the southern margin of the Bismarck Sea, Papua New Guinea, in Johnson, R.W., ed., Volcanism in Australasia: Amsterdam, Elsevier, p. 101–116.

Johnson, R.W., 1979, Geotectonics and volcanism in Papua New Guinea: A review of the late Cainozoic: BMR Journal of Australian Geology and Geophysics, v. 4, p. 181–207.

Johnson, R.W., 1982, Papua New Guinea, in Thorpe, R.S., ed., Andesites: Orogenic andesites and related rocks: New York, Wiley, p. 225–244.

Johnson, R.W., and Arculus, R.W., 1978, Volcanic rocks of the Witu Islands, Papua New Guinea: The origin of magmas above the deepest part of the New Britain Benioff zone: Bulletin Volcanologique, v. 41, p. 609–655.

Johnson, R.W., and Smith, I.S.E., 1974, Volcanoes and rocks of St. Andrew Strait, Papua New Guinea: Geological Society of Australia Journal, v. 21, p. 333–351.

Johnson, R.W., Davies, R.A., and White, A.J.R., 1972, Ulawun volcano, New Britain: Australian Bureau of Mineral Resources, Geology, and Geophysics Bulletin 142, 42 p.

Johnson, R.W., Wallace, D.A., and Ellis, D.J., 1976, Feldspathoid-bearing potassic rocks and associated types from volcanic islands off the coast of New Ireland, Papua New Guinea: A preliminary account of geology and petrology, in Johnson, R.W., ed., Volcanism in Australasia: Amsterdam, Elsevier, p. 297–316.

Johnson, R.W., Smith, I.S.E., and Taylor, S.R., 1978, Hot-spot volcanism in St. Andrew Strait, Papua New Guinea: BMR Journal of Australian Geology and Geophysics, v. 3, p. 55–69.

Johnson, R.W., Macnab, R.P., Ryburn, R.J., Cooke, R.J.S., and Chappell, B.W., 1983, Bamus volcano, Papua New Guinea: Dormant neighbor of Ulawun and magnesian andesite locality: Geologische Rundschau, v. 72, p. 207–237, doi: 10.1007/BF01765907.

Johnson, R.W., Jaques, A.L., Hickey, R.L., McKee, C.O., and Chappell, B.W., 1985, Manam Island, Papua New Guinea: Petrology and geochemistry of a low-TiO_2 basaltic island-arc volcano: Journal of Petrology, v. 26, p. 283–323.

Johnstone, R.D., 1978, 'Ata, the most southerly island of Tonga: Royal Society of New Zealand Bulletin 17 [Lau-Tonga, 1977], p. 153–164.

Joseph, L.J., and Finlayson, E.J., 1991, A revised stratigraphy of Muyua (Woodlark Island), in Rogerson, R., ed., PNG geology, exploration, and mining conference proceedings: Parkville (Victoria, Australia), Australasian Institute of Mining and Metallurgy, p. 26–33.

Kaeppler, A., 1973a, Pottery sherds from Tungua, Ha'apai, and remarks on pottery and social structure in Tonga: Journal of the Polynesian Society, v. 82, p. 218–222.

Kaeppler, A., 1973b, Pottery sherds from Ha'apai, Tonga: Journal of the Polynesian Society, v. 82, p. 414.

Kaplan, S., 1976, Ethnological and biogeographical significance of pottery sherds from Nissan Island, Papua New Guinea: Chicago, Field Museum Press Fieldiana: Anthropology, v. 66, p. 35–89.

Kay, R.M.A., 1984, Analysis of archaeological material from Naigani [M.A. thesis]: Auckland, University of Auckland, 196 p.

Kear, D., 1967, Geological notes on Western Samoa: New Zealand Journal of Geology and Geophysics, v. 10, p. 1446–1451.

Kear, D., and Wood, B.L., 1959, The geology and hydrology of Western Samoa: New Zealand Geological Survey Bulletin n.s. 63, 90 p.

Keating, B.H., Mattey, D.P., Haughton, J.J., and Helsley, C.E., 1984a, Age and origin of Truk atoll, eastern Caroline Islands: Geochemical, radiometric-age, and paleomagnetic evidence: Geological Society of America Bulletin, v. 95, p. 350–356, doi: 10.1130/0016-7606(1984)95<350: AAOOTA>2.0.CO;2.

Keating, B.H., Mattey, D.P., Helsley, C.E., Naughton, J.J., and Epp, D., 1984b, Evidence for a hot spot origin of the Caroline Islands: Journal of Geophysical Research, v. 89, p. 9937–9948.

Kennedy, A.K., Hart, S.R., and Frey, F.A., 1990, Composition and isotopic constraints on the petrogenesis of alkaline arc lavas: Lihir Island, Papua New Guinea: Journal of Geophysical Research, v. 85, p. 6929–6942.

Kennedy, J., 1981, Lapita colonization of the Admiralty Islands?: Science, v. 213, p. 757–759.

Kennedy, J., 1982, Archaeology in the Admiralty Islands: Indo-Pacific Prehistory Association Bulletin, v. 3, p. 22–35.

Kennett, D.J., Sakai, S., Neff, H., Gossett, R., and Larson, D., 2002, Compositional characterization of prehistoric ceramics: A new approach: Journal of Archaeological Science, v. 29, p. 443–455, doi: 10.1006/jasc.2001.0737.

Kennett, D.J., Anderson, A.J., Cruz, M.J., Clark, G.R., and Summerhayes, G.R., 2004, Geochemical characterization of Lapita pottery via inductively coupled plasma–mass spectrometry (ICP-MS): Archaeometry, v. 46, p. 35–46, doi: 10.1111/j.1475-4754.2004.00142.x.

Key, C.O., 1969, Mineralogy and petrology of the Buka and Sohano pottery [Appendix 6], in Specht, J., Prehistoric and modern pottery industries of Buka Island [Ph.D. thesis]: Canberra, Australian National University, p. 368–372.

Key, C.O., 1987, The petrology of some Tongatapu pottery [Appendix 5], in Poulsen, J., The prehistory of the Tongan islands [Terra Australis 12]: Canberra, Australian National University, p. 274–277.

Kirch, P.V., 1976a, Early Anutan settlement and the position of Anuta in the prehistory of the southwest Pacific, in Green, R.C., and Cresswell, M.M., eds., Southeast Solomon Islands cultural history: A preliminary survey: Royal Society of New Zealand Bulletin 11, p. 225–244.

Kirch, P.V., 1976b, Ethno-archaeological investigations in Futuna and Uvea (western Polynesia): A preliminary report: Journal of the Polynesian Society, v. 85, p. 27–64.

Kirch, P.V., 1978, The Lapitoid period in west Polynesia: Excavations and survey in Niuatoputapu, Tonga: Journal of Field Archaeology, v. 5, p. 1–13.

Kirch, P.V., 1981, Lapitoid settlements of Futuna and Alofi, western Polynesia: Archaeology in Oceania, v. 16, p. 127–143.

Kirch, P.V., 1982a, Mangaasi-style ceramics from Tikopia and Vanikoro and their implications for east Melanesian prehistory: Indo-Pacific Prehistory Association Bulletin, v. 3, p. 67–76.

Kirch, P.V., 1982b, A revision of the Anuta sequence: Journal of the Polynesian Society, v. 91, p. 245–254.

Kirch, P.V., 1983, An archaeological exploration of Vanikoro, Santa Cruz Islands, eastern Melanesia: New Zealand Journal of Archaeology, v. 5, p. 69–113.

Kirch, P.V., 1984, The Polynesian outliers: Continuity, change, and replacement: Journal of Pacific History, v. 19, p. 224–238.

Kirch, P.V., 1986, Exchange systems and inter-island contact in the transformation of an island society: The Tikopia case, in Kirch, P.V., ed., Island societies: Archaeological approaches to evolution and transformation: Cambridge, UK, Cambridge University Press, p. 33–41.

Kirch, P.V., 1987, Lapita and oceanic cultural origins: Excavations in the Mussau Islands, Bismarck Archipelago, 1985: Journal of Field Archaeology, v. 14, p. 163–180.

Kirch, P.V., 1988a, Radiocarbon dates from the Mussau Islands and the Lapita colonization of the southwestern Pacific: Radiocarbon, v. 30, p. 161–169.

Kirch, P.V., 1988b, The Talepakemalai Lapita site and oceanic prehistory: National Geographic Research, v. 4, p. 328–342.

Kirch, P.V., 1988c, Niuatoputapu, the prehistory of a Polynesian chiefdom: Seattle, Thomas Burke Memorial Washington State Museum Monograph 5, 287 p.

Kirch, P.V., 1993a, The To'aga site: Modelling the morphodynamics of the land-sea interface, in Kirch, P.V., and Hunt, T.L., eds., The To'aga site: Three millennia of Polynesian occupation in the Manu'a Islands, American Samoa: Berkeley, University of California Archaeological Research Facility Contribution No. 51, p. 31–42.

Kirch, P.V., 1993b, Radiocarbon chronology of the To'aga site, in Kirch, P.V., and Hunt, T.L., eds., The To'aga site: Three millennia of Polynesian occupation in the Manu'a Islands, American Samoa: Berkeley, University of California Archaeological Research Facility Contribution No. 51, p. 85–92.

Kirch, P.V., 1995, The Lapita culture of western Melanesia in the context of Austronesian origins and dispersal, in Li, P.J., et al., eds., Austronesian studies relating to Taiwan: Taipei, Symposium Series of the Institute of History and Philology (Academica Sinica) No. 3, p. 255–294.

Kirch, P.V., 1996, Lapita and its aftermath: The Austronesian settlement of Oceania, in Goodenough, W.H., ed., Prehistoric settlement of the Pacific: American Philosophical Society Transactions, v. 86, p. 57–70.

Kirch, P.V., 1997, The Lapita peoples: Ancestors of the oceanic world: Cambridge, UK, Blackwell, 353 p.

Kirch, P.V., 2000, On the road of the winds: An archaeological history of the Pacific islands before European contact: Berkeley, University of California Press, 424 p.

Kirch, P.V., 2001a, Three Lapita villages: Excavations at Talepakemalai (ECA), Etakosarai (ECB), and Etakengaroasa (EHB), Eloaua and Emanus islands, in Kirch, P.V., ed., Lapita and its transformations in Near Oceania: Archaeological investigations in the Mussau Islands, Papua New Guinea, 1985–1988, vol. 1: Introduction, excavations, chronology: Berkeley, University of California Archaeological Research Facility Contribution No. 59, p. 68–145.

Kirch, P.V., 2001b, A radiocarbon chronology for the Mussau Islands, in Kirch, P.V., ed., Lapita and its transformations in Near Oceania: Archaeological investigations in the Mussau Islands, Papua New Guinea, 1985–1988, vol. 1: Introduction, excavations, chronology: Berkeley, University of California Archaeological Research Facility Contribution No. 59, p. 196–236.

Kirch, P.V., and Catterall, C., 2001, The Mussau Islands: Natural and cultural environments, in Kirch, P.V., ed., Lapita and its transformations in Near Oceania: Archaeological investigations in the Mussau Islands, Papua New Guinea, 1985–1988, vol. 1: Introduction, excavations, chronology: Berkeley, University of California Archaeological Research Facility

Contribution No. 59, p. 28–56.

Kirch, P.V., and Green, R.C., 2001, Hawaiki, ancestral Polynesia: An essay in historical anthropology: Cambridge, UK, Cambridge University Press, 375 p.

Kirch, P.V., and Hunt, T.L., 1988, The spatial and temporal boundaries of Lapita, in Kirch, P.V., and Hunt, T.L., eds., The Lapita cultural complex: A critical review: Seattle, Thomas Burke Memorial Washington State Museum Research Report 5, p. 9–31.

Kirch, P.V., and Rosendahl, P.H., 1973, Archaeological investigation of Anuta, in Yen, D.E., and Gordon, J., eds., Anuta: A Polynesian outlier in the Solomon Islands: Honolulu, Bishop Museum Pacific Anthropological Records No. 21, p. 25–108.

Kirch, P.V., and Yen, D.E., 1982, Tikopia: The prehistory and ecology of a Polynesian outlier: Honolulu, Bishop Museum Bulletin 238, 396 p.

Kirch, P.V., Dickinson, W.R., and Hunt, T.L., 1988, Polynesian Plainware sherds from Hivaoa and their implications for early Marquesan prehistory: New Zealand Journal of Archaeology, v. 10, p. 101–107.

Kirch, P.V., Hunt, T.L., Nagaoka, L., and Tyler, J., 1990, An ancestral Polynesian occupation site at To'aga, Ofu Island, American Samoa: Archaeology in Oceania, v. 25, p. 1–15.

Kirch, P.V., Hunt, T.L., Weisler, M., Butler, V., and Allen, M.S., 1991, Mussau Islands prehistory: Results of the 1985–1986 excavations, in Allen, J., and Gosden, C., eds., Report of the Lapita homeland project: Canberra, Australian National University Occasional Papers in Prehistory No. 20, p. 144–163.

Kroenke, L.W., 1984, Cenozoic tectonic development of the southwest Pacific: Suva, South Pacific Applied Geoscience Commission (SOPAC) Technical Bulletin 6, 122 p.

Kumar, R., Nunn, P.D., Katayama, K., Oda, H., Matararaba, S., and Osborne, T., 2003, An early human settlement site in Fiji: The first dates from the 2002 excavations at Naitabale, Moturiki Island: Institute of Applied Earth Sciences, University of the South Pacific, Technical Report 2003/06, 22 p.

Kumar, R., Nunn, P.D., and Dickinson, W.R., 2004a, The emerging pattern of earliest Lapita settlement in Fiji: Four new Lapita sites on Viti Levu Island: Archaeology in New Zealand, v. 47, p. 108–117.

Kumar, R., Nunn, P.D., Katayama, K., Oda, H., Matararaba, S., and Osborne, T., 2004b, The earliest-known humans in Fiji and their pottery: The first dates from the 2002 excavations at Naitabale (Naturuku), Moturiki Island: South Pacific Journal of Natural Science, v. 22, p. 15–21.

Kurashina, H., and Clayshulte, R.W., 1983, Site formation processes and cultural sequence at Tarague, Guam: Indo-Pacific Prehistory Association Bulletin, v. 4, p. 114–122.

Ladefoged, T.N., Wall, J., Black, P., and Dickinson, W.R., 1998, Exotic and indigenous ceramic sherds from the island of Rotuma: Journal of the Polynesian Society, v. 107, p. 301–311.

Laniz, R.V., Stevens, R.E., and Norman, M.B., 1964, Staining of plagioclase feldspar and other minerals: U.S. Geological Survey Professional Paper 501-B, p. B152–B153.

Le Dez, A., Maury, R.C., Vidal, P., Bellon, H., Cotton, J., and Brousse, R., 1996, Geology and geochemistry of Nuku Hiva, Marquesas: Temporal trends in a large Polynesian shield volcano: Société Géologique de France Bulletin, v. 167, p. 197–209.

Lee, T.-Y., and Lawver, L.A., 1995, Cenozoic plate reconstruction of southeast Asia: Tectonophysics, v. 251, p. 85–138, doi: 10.1016/0040-1951(95)00023-2.

Lilley, I., 1988a, Type X: Description and discussion of a prehistoric ceramic ware from northeastern Papua New Guinea: Indo-Pacific Prehistory Association Bulletin, v. 8, p. 90–100.

Lilley, I., 1988b, Prehistoric exchange across the Vitiaz Strait, Papua New Guinea: Current Anthropology, v. 29, p. 513–516, doi: 10.1086/203669.

Lilley, I., 1991a, Lapita and post-Lapita developments in the Vitiaz Strait–West New Britain area: Indo-Pacific Prehistory Association Bulletin, v. 11, p. 313–322.

Lilley, I., 1991b, Lapita sites in the Duke of York Islands, in Allen, J., and Gosden, C., eds., Report of the Lapita homeland project: Canberra, Australian National University Occasional Papers in Prehistory No. 20, p. 164–169.

Lindley, D., 1988, Early Cainozoic stratigraphy and structure of the Gazelle Peninsula, east New Britain: An example of extensional tectonics in the New Britain arc-trench complex: Australian Journal of Earth Sciences, v. 35, p. 231–244.

Louat, R., and Pelletier, B., 1989, Seismotectonics and present-day relative plate motions in the New Hebrides–North Fiji Basin region: Tectonophysics, v. 167, p. 41–55, doi: 10.1016/0040-1951(89)90293-X.

Lowder, G.G., and Carmichael, I.S.E., 1970, The volcanoes and caldera of Talasea, New Britain: Geology and petrology: Geological Society of America Bulletin, v. 81, p. 17–38.

Lus, W.Y., McDougall, I., and Davies, H.L., 2004, Age of the metamorphic sole of the Papuan ultramafic belt ophiolite, Papua New Guinea: Tectonophysics, v. 392, p. 85–101, doi: 10.1016/j.tecto.2004.04.009.

Macdonald, G.A., 1944, Petrography of the Samoan islands: Geological Society of America Bulletin, v. 55, p. 1333–1362.

Macdonald, G.A., 1945, Petrography of the Wallis Islands: Geological Society of America Bulletin, v. 56, p. 861–872.

Macdonald, G.A., 1968, A contribution to the petrology of Tutuila, American Samoa: Geologische Rundschau, v. 57, p. 821–837, doi: 10.1007/BF01845367.

MacFarlane, A., Carney, J.N., Crawford, A.J., and Greene, H.G., 1988, Vanuatu—A review of the onshore geology, in Greene, H.G., and Wong, F.L., eds., Geology and offshore resources of Pacific island arcs—Vanuatu region: Houston, Circum-Pacific Council for Energy and Mineral Resources Earth Science Series, v. 8, p. 45–91.

Mackenzie, D.E., 1976, Nature and origin of late Cainozoic volcanoes in western Papua New Guinea, in Johnson, R.W., ed., Volcanism in Australasia: Amsterdam, Elsevier, p. 221–238.

Madsen, J.A., and Lindley, I.D., 1994, Large-scale structures on Gazelle Peninsula, New Britain: Implications for the evolution of the New Britain arc: Australian Journal of Earth Sciences, v. 41, p. 561–569.

Maillet, P., Ruellan, E., Gérard, M., Person, A., Bellon, H., Cotton, J., Joron, J.-L., Nakada, S., and Price, C., 1995, Tectonics, magmatism, and evolution of the New Hebrides backarc troughs (southwest Pacific), in Taylor, B., ed., Backarc basins: Tectonics and magmatism: New York, Plenum Press, p. 177–235.

Mallick, D.I.J., and Greenbaum, D., 1977, Geology of southern Santo: Port Vila, New Hebrides Condominium Geological Survey Regional Report, 84 p.

Mallick, D.I.J., and Neef, G., 1974, Geology of Pentecost: Port Vila, New Hebrides Condominium Geological Survey Regional Report, 103 p.

Marcelot, G., Dupuy, C., Girod, M., and Maury, R.C., 1983, Petrology of Futuna Island lavas (New Hebrides): An example of calc-alkaline magmatism associated with the initial stages of back-arc spreading: Chemical Geology, v. 38, p. 23–37, doi: 10.1016/0009-2541(83)90043-8.

Marsaglia, K., and Tazaki, K., 1992, Diagenetic trends in Leg 126 sandstones, in Taylor, B., and Fujioka, K., eds., Proceedings of the Ocean Drilling Program, Scientific Results, Volume 126: College Station, Texas, Ocean Drilling Program, p. 125–138.

Martínez, F., and Taylor, B., 1996, Backarc spreading, rifting, and microplate rotation between transform faults in the Manus Basin: Marine Geophysical Researches, v. 18, p. 203–224, doi: 10.1007/BF00286078.

Martínez, F., Fryer, P., Baker, N.A., and Yamazaki, T., 1995, Evolution of back-arc rifting: Mariana Trough, 20°–24° N: Journal of Geophysical Research, v. 100, p. 3807–3827, doi: 10.1029/94JB02466.

Martínez, F., Goodliffe, A., and Taylor, B., 2001, Metamorphic core complex formation by density inversion and lower-crust extrusion: Nature, v. 411, p. 930–934, doi: 10.1038/35082042.

Matsuda, J. I., Zashu, S., and Ozima, M., 1977, Sr isotopic studies of volcanic rocks from island arcs in the western Pacific: Tectonophysics, v. 37, p. 141–151, doi: 10.1016/0040-1951(77)90044-0.

Mattey, D.P., 1982, The minor and trace element geochemistry of volcanic rocks from Truk, Ponape, and Kusaie, eastern Caroline Islands: The evolution of a young hot spot trace across old Pacific Ocean crust: Contributions to Mineralogy and Petrology, v. 80, p. 1–13, doi: 10.1007/BF00376730.

McCoy, P.C., and Cleghorn, P.L., 1988, Archaeological excavations on Santa Cruz (Nendö), southeast Solomon Islands: Summary report: Archaeology in Oceania, v. 23, p. 104–115.

McDougall, I., 1985, Age and evolution of the volcanoes of Tutuila, American Samoa: Pacific Science, v. 39, p. 311–320.

McKee, C.O., Cooke, R.J.S., and Wallace, D.A., 1976, 1974–1975 eruption of Karkar volcano, Papua New Guinea, in Johnson, R.W., ed., Volcanism in Australasia: Amsterdam, Elsevier, p. 173–190.

McNiven, I.J., Dickinson, W.R., David, B., Weisler, M., von Gnielenski, F., Carter, M., and Zoppi, U., 2006, Mask Cave: Red-slipped pottery and the Australian-Papuan settlement of Torres Strait: Archaeology in Oceania, v. 41 (in press).

Meffre, S., Aitchison, J.C., and Crawford, J.C., 1996, Geochemical evolution and tectonic significance of boninites and tholeiites from the Koh ophiolite, New Caledonia: Tectonics, v. 15, p. 67–83, doi: 10.1029/95TC02316.

Meijer, A., Reagan, M., Ellis, H., Shafiqullah, M., Suter, J., Damon, P., and

Kling, S., 1983, Chronology of events in the eastern Philippine Sea, in Hayes, D.E., ed., The tectonic and geologic evolution of southeast Asian seas and islands, Part 2: American Geophysical Union Geophysical Monograph 27, p. 349–359.

Middleton, A.F., Freestone, I.C., and Leese, M.N., 1985, Textural analysis of ceramic thin sections: Evaluation of grain sampling procedures: Archaeometry, v. 1, p. 64–74.

Milsom, J., 2003, Forearc ophiolites: A view from the western Pacific, in Dilek, Y., and Robinson, P.T., eds., Ophiolites in Earth history: Geological Society [London] Special Publication 218, p. 507–515.

Milsom, J., Masson, D., Nichols, G., Sikembang, N., Dwiyanto, B., Parson, L., and Kallagher, H., 1992, The Manokwari Trough and the western end of the New Guinea Trench: Tectonics, v. 11, p. 145–153.

Mitchell, A.H.G., 1966, Geology of south Malekula: Port Vila, New Hebrides Condominium Geological Survey Report 3, 41 p.

Mitchell, A.H.G., 1971, Geology of northern Malekula: Port Vila, New Hebrides Condominium Geological Survey Regional Report, 56 p.

Mitchell, A.H.G., and Warden, A.J., 1971, Geological evolution of the New Hebrides island arc: Geological Society [London] Journal, v. 127, p. 501–529.

Mitrovica, J.X., and Milne, G.A., 2002, On the origin of late Holocene sea-level highstands within equatorial ocean basins: Quaternary Science Reviews, v. 21, p. 2179–2190, doi: 10.1016/S0277-3791(02)00080-X.

Mitrovica, J.X., and Peltier, W.R., 1991, On postglacial geoid subsidence over the equatorial oceans: Journal of Geophysical Research, v. 96, p. 20,053–20,071.

Monjaret, M.C., Bellon, H., and Maillet, P., 1991, Magmatism of the troughs behind the New Hebrides island arc (RV Jean Charcot SEAPSO 2 cruise): K-Ar geochronology and petrology: Journal of Volcanology and Geothermal Research, v. 46, p. 265–280, doi: 10.1016/0377-0273(91)90088-H.

Monzier, M., Robin, C., and Eissen, J.-P., 1994, Kuwae (~1425 A.D.): The forgotten caldera: Journal of Volcanology and Geothermal Research, v. 59, p. 207–218, doi: 10.1016/0377-0273(94)90091-4.

Moore, D.R., and Hunter-Anderson, R.L., 1999, Pots and pans in the Intermediate Pre-Latte (2500–1600 BP) Mariana Islands, Micronesia, in Galipaud, J.-C., and Lilley, I., eds., Le Pacifique de 5000 à 2000 avant le present: Paris, Éditions IRD (Institut de Recherche pour le Développement), p. 487–503.

Morris, J.D., Jezek, P.A., Hart, S.R., and Gill, J.B., 1983, The Halmahera island arc, Molucca Sea collision zone, Indonesia: A geochemical survey, in Hayes, D.E., ed., The tectonic and geologic evolution of southeast Asian seas and islands, Part 2: American Geophysical Union Geophysical Monograph 27, p. 373–387.

Motteler, E.L., 1986, Pacific island names: Honolulu, Bishop Museum Miscellaneous Publication 34, 91 p.

Musgrave, R.J., 1990, Paleomagnetism and tectonics of Malaita, Solomon Islands: Tectonics, v. 9, p. 735–759.

Musgrave, R.J., and Firth, J.V., 1999, Magnitude and timing of New Hebrides arc rotation: Paleomagnetic evidence from Nendö, Solomon Islands: Journal of Geophysical Research, v. 104, p. 2841–2853, doi: 10.1029/1998JB900080.

Mutter, J.C., Mutter, C.Z., and Fang, J., 1996, Analogies to oceanic behavior in the continental breakup of the western Woodlark Basin: Nature, v. 380, p. 333–336, doi: 10.1038/380333a0.

Natland, J.H., 1980, The progression of volcanism in the Samoan linear volcanic chain: American Journal of Science, v. 280-A, p. 709–735.

Natland, J.H., and Turner, D.L., 1985, Age progression and petrological development of Samoan shield volcanoes: Evidence from K-Ar ages, lava compositions, and mineral studies, in Brocher, T.M., ed., Investigations of the northern Melanesian borderland: Houston, Circum-Pacific Council for Energy and Mineral Resources Earth Science Series, v. 3, p. 139–171.

Neal, C.R., Mahoney, J.J., Duncan, R.A., Jain, J.C., and Petterson, M.G., 1997, The origin, evolution, and ultimate fate of the Ontong Java Plateau, SW Pacific: Evidence from exposed plateau basement on Malaita, Solomon Islands, in Mahoney, J.J., and Coffin, M.F., eds., Large igneous provinces: American Geophysical Union Geophysical Monograph 100, p. 183–216.

Neff, H., Bishop, R.L., and Sayre, E.V., 1988, A simulation approach to the problem of tempering in compositional studies of archaeological ceramics: Journal of Archaeological Science, v. 15, p. 159–172.

Neff, H., Bishop, R.L., and Sayre, E.V., 1989, More observations on the problem of tempering in compositional studies of archaeological ceramics: Journal of Archaeological Science, v. 16, p. 57–69.

Nunn, P.D., 1999, Lapita pottery from Moturiki island, central Fiji: Archaeology in New Zealand, v. 42, p. 309–313.

Nunn, P.D., and Areki, F., 2004, A Lapita site on Ovalau Island, central Fiji Islands: Archaeology in New Zealand, v. 47, p. 215–219.

Nunn, P.D., and Matararaba, S., 2000, New finds of Lapita pottery in northeast Fiji: Archaeology in Oceania, v. 35, p. 92–93.

Nunn, P.D., Kumar, R., and Matararaba, S., 2003, Recent research relating to Lapita settlement in Fiji, in Sand, C., ed., Pacific archaeology: Assessments and prospects: Noumea, Cahiers de l'Archéologie en Nouvelle-Calédonie, v. 15, p. 183–186.

Nunn, P.D., Kumar, R., Matararaba, M., Ishimura, T., Seeto, J., Rayawa, S., Kuruyawa, S., Nasila, A., Oloni, B., Ram, A.R., Saunivalu, P., Singh, P., and Tegu, E., 2004, Early Lapita settlement site at Bourewa, southwest Viti Levu Island, Fiji: Archaeology in Oceania, v. 39, p. 139–143.

O'Day, S.J., O'Day, P., and Steadman, D.W., 2004, Defining the Lau context: Recent findings on Nayau, Lau Islands, Fiji: New Zealand Journal of Archaeology, v. 25 (2003), p. 31–56.

Ollier, C.D., and Pain, C.F., 1978, Geomorphology and tectonics of Woodlark Island, Papua New Guinea: Zeitschrift für Geomorphologie, v. 22, p. 1–20.

Ollier, C.D., and Pain, C.F., 1980, Actively rising surficial gneiss domes in Papua New Guinea: Geological Society of Australia Journal, v. 27, p. 35–44.

Osborne, D., 1979, Archaeological test excavations, Palau Islands, 1968–1969: Micronesica Supplement 1, 353 p.

Page, R.W., and Ryburn, R.J., 1977, K-Ar ages and geological relations of intrusive rocks in New Britain: Pacific Geology, v. 12, p. 99–105.

Palfreyman, W.D., and Cooke, R.J.S., 1976, Eruptive history of Manam volcano, Papua New Guinea, in Johnson, R.W., ed., Volcanism in Australasia: Amsterdam, Elsevier, p. 117–131.

Palmer, J.B., 1965, Excavations at Karobo, Viti Levu: New Zealand Archaeological Association Newsletter, v. 8, p. 26–33.

Paris, J.-P., 1981, Géologie de la Nouvelle-Calédonie: Un essai de synthèse: Orleans, Territoire de Nouvelle-Calédonie Bureau de Recherches Géologiques et Minière Mémoire 113, 278 p.

Parke, A.L., 2000, Coastal and inland Lapita sites in Vanua Levu, Fiji: Archaeology in Oceania, v. 35, p. 116–119.

Parrot, J.F., and Dugas, F., 1980, The disrupted ophiolite belt of the southwest Pacific: Evidence of an Eocene subduction zone: Tectonophysics, v. 66, p. 349–372, doi: 10.1016/0040-1951(80)90249-8.

Peacock, D.P.S., 1970, The scientific analysis of ancient ceramics: A review: World Archaeology, v. 1, p. 375–389.

Pelletier, B., Calmant, S., and Pillet, R., 1998, Current tectonics of the Tonga–New Hebrides region: Earth and Planetary Science Letters, v. 164, p. 263–276, doi: 10.1016/S0012-821X(98)00212-X.

Pelletier, B., Lagabrielle, Y., Cabioch, G., Calmant, M., Régnier, M., and Perrier, J., 2000, Transpression active le long de la frontière décrochante Pacifique-Australie: Les apports de la cartographie multifaisceaux autour des îles Futuna et Alofi (Pacifique sud-ouest): Comptes Rendus de l'Académie des Sciences de Paris Sciences de la Terre et de Planètes, v. 331, p. 127–132.

Pelletier, B., Lagabrielle, Y., Benoit, M., Cabioch, G., Calmant, S., Garel, E., and Guivel, C., 2001, Newly identified segments of the Pacific-Australia plate boundary along the North Fiji fracture zone: Earth and Planetary Science Letters, v. 193, p. 347–358, doi: 10.1016/S0012-821X(01)00522-2.

Petchey, F.J., 1995, The archaeology of Kudon: Archaeological analysis of Lapita ceramics from Mulifanua, Samoa and Sigatoka, Fiji [M.A. thesis]: Auckland, University of Auckland, 192 p.

Petchey, F.J., 2001, Radiocarbon determinations from the Mulifanua Lapita site, Upolu, Western Samoa: Radiocarbon, v. 43, p. 63–68.

Pétrequin, P., and Pétrequin, A.-M., 1999, La poterie en Nouvelle-Guineé: Savoir-faire et transmission des techniques: Journal de la Société des Océanistes, v. 108, p. 71–101.

Petterson, M.G., 2004, The geology of north and central Malaita, Solomon Islands: The thickest and most accessible part of the world's largest (Ontong Java) ocean plateau, in Fitton, J.G., et al., eds., Origin and evolution of the Ontong Java Plateau: Geological Society [London] Special Publication 229, p. 63–81.

Petterson, M.G., Neal, C.R., Mahoney, J.J., Kroenke, L.W., Saunders, A.D., Babbs, T.L., Duncan, R.A., Tolia, D., and McGrail, B., 1997, Structure and deformation of north and central Malaita, Solomon Islands: Tectonic implications for the Ontong Java Plateau–Solomon Islands arc collision, and for the fate of oceanic plateaus: Tectonophysics, v. 283, p. 1–33, doi: 10.1016/S0040-1951(97)00206-0.

Petterson, M.G., Babbs, T., Neal, C.R., Mahoney, J.J., Saunders, A.D., Duncan, R.A., Tolia, D., Magu, R., Qopoto, C., Mahoa, H., and Natogga, D., 1999,

Geological-tectonic framework of Solomon Islands, SW Pacific: Crustal accretion and growth within an intra-oceanic setting: Tectonophysics, v. 301, p. 35–60, doi: 10.1016/S0040-1951(98)00214-5.

Phear, S., Clark, G., and Anderson, A., 2003, A radiocarbon chronology for Palau, in Sand, C., ed., Pacific archaeology: Assessments and prospects: Noumea, Cahiers de l'Archéologie en Nouvelle-Calédonie, v. 15, p. 255–263.

Phinney, E.J., Mann, P., Coffin, M.F., and Shipley, T.H., 1999, Sequence stratigraphy, structure, and tectonic history of the southwestern Ontong Java Plateau adjacent to the North Solomon Trench and Solomon Islands arc: Journal of Geophysical Research, v. 104, p. 20,449–20,466, doi: 10.1029/1999JB900169.

Pigram, C.J., and Davies, H.L., 1987, Terranes and the accretion history of the New Guinea orogen: BMR Journal of Australian Geology and Geophysics, v. 10, p. 193–211.

Polach, H.A., 1972, Outlier archaeology: Bellona, a preliminary report on field work and radiocarbon dates, Part II: Radiocarbon dating of Sikumango midden: Sample ANU 608: Archaeology and Physical Anthropology in Oceania, v. 7, p. 206–214.

Poulsen, J., 1972, Outlier archaeology: Bellona, a preliminary report of field work and radiocarbon dates, Part I: Archaeology: Archaeology and Physical Anthropology in Oceania, v. 7, p. 184–205.

Quarles van Ufford, A.I., and Cloos, M., 2005, Cenozoic tectonics of New Guinea: American Association of Petroleum Geologists Bulletin, v. 89, p. 119–140.

Rainbird, P., 1999, Entangled biographies: Western Pacific ceramics and the tombs of Pohnpei: World Archaeology, v. 31, p. 214–224.

Ramsay, W.R.H., Crawford, A.J., and Foden, J.D., 1984, Field setting, mineralogy, chemistry, and genesis of arc picrites, New Georgia, Solomon Islands: Contributions to Mineralogy and Petrology, v. 88, p. 386–402, doi: 10.1007/BF00376763.

Raos, A.M., and Crawford, A.J., 2004, Basalts from the Efate island group, central segment of the Vanuatu arc, SW Pacific: Geochemistry and petrogenesis: Journal of Volcanology and Geothermal Research, v. 134, p. 35–56, doi: 10.1016/j.jvolgeores.2003.12.004.

Reagan, M., and Meijer, A., 1984, Geology and geochemistry of early arc-volcanic rocks from Guam: Geological Society of America Bulletin, v. 95, p. 701–713, doi: 10.1130/0016-7606(1984)95<701: GAGOEA>2.0.CO;2.

Reeve, R., 1989, Recent work on the prehistory of the western Solomons, Melanesia: Indo-Pacific Prehistory Association Bulletin, v. 9, p. 44–67.

Rice, P.M., 1987, Pottery analysis: A sourcebook: Chicago, University of Chicago Press, 559 p.

Rickard, M.J., 1966, Reconnaissance geology of Vanua Levu: Suva, Fiji Geological Survey Memoir 2, 81 p.

Rickard, M.J., 1970, The geology of northeastern Vanua Levu: Suva, Fiji Geological Survey Bulletin 14, 13 p.

Ridgway, J., 1987, Neogene displacements in the Solomon Islands arc: Tectonophysics, v. 133, p. 81–93, doi: 10.1016/0040-1951(87)90282-4.

Ridgway, J., and Coulson, F.I.E., 1987, The geology of Choiseul and the Shortland Islands, Solomon Islands: British Geological Survey Overseas Memoir 8, 134 p.

Robin, C., Monzier, M., and Eissen, J.-P., 1994, Formation of the mid-fifteenth century Kuwae caldera (Vanuatu) by an initial hydroclastic and subsequent ignimbritic eruption: Bulletin of Volcanology, v. 56, p. 170–183, doi: 10.1007/s004450050026.

Robinson, G.P., 1969, The geology of north Santo: Port Vila, New Hebrides Condominium Geological Survey Regional Report, 77 p.

Rodda, P., 1976, Geology of northern and central Viti Levu: Suva, Fiji Mineral Resources Division Bulletin 3, 159 p.

Rodda, P., 1994, Geology of Fiji, in Stevenson, A.J., et al., eds., Geology and submarine resources of the Tonga-Lau-Fiji region: Suva, South Pacific Applied Geoscience Commission (SOPAC) Technical Bulletin 8, p. 131–151.

Rodda, P., and Kroenke, L.W., 1984, Fiji: A fragmented arc, in Kroenke, L.W., ed., Cenozoic tectonic development of the southwest Pacific: Suva, South Pacific Applied Geoscience Commission (SOPAC) Technical Bulletin 6, p. 87–99.

Rodda, P., and Lum, J., 1990, Geological evolution and mineral deposits of Fiji: Geologische Jahrbuch, v. D92, p. 37–66.

Rogers, G., 1974, Archaeological discoveries on Niuatoputapu Island, Tonga: Journal of the Polynesian Society, v. 83, p. 308–348.

Rogerson, R.J., Hilyard, D.B., Finlayson, E.J., Johnson, R.W., and McKee, C.O., 1989, The geology and mineral resources of Bougainville and Buka Islands, Papua New Guinea: Port Moresby, Papua New Guinea Geological Survey Memoir 16, 217 p.

Rolett, B.V., Chen, W.-C., and Sinton, J.M., 2000, Taiwan, Neolithic seafaring and Austronesian origins: Antiquity, v. 74, p. 54–61.

Rosenthal, M.E., 1995, The archaeological excavation of an outrigger canoe at the Nasilai site, Rewa Delta, Viti Levu, Fiji: Asian Perspectives, v. 34, p. 91–118.

Rowland, M.J., and Best, S., 1980, Survey and excavation on the Kedekede hillfort, Lakeba island, Lau Group, Fiji: Archaeology and Physical Anthropology in Oceania, v. 15, p. 29–50.

Rytuba, J.J., and Miller, W.R., 1990, Geology and geochemistry of epithermal precious metal vein systems in the intra-oceanic island arcs of Palau and Yap, western Pacific: Journal of Geochemical Exploration, v. 35, p. 413–447, doi: 10.1016/0375-6742(90)90046-D.

Rytuba, J.J., McKee, E.H., and Cox, D.P., 1993, Geochronology and geochemistry of the Ladolan gold deposit, Lihir Island, and gold deposits and volcanoes of Tabar and Tatau, Papua New Guinea, in Scott, R.W., Jr., et al., eds., Advances related to United States and international mineral resources: Developing frameworks and exploration technologies: U.S. Geological Survey Bulletin 2039, p. 119–126.

Sand, C., 1990, The ceramic chronology of Futuna and Alofi: An overview, in Spriggs, M., ed., Lapita design, form and composition: Canberra, Australian National University Occasional Papers in Prehistory No. 19, p. 123–133.

Sand, C., 1993, Données archéoloqiques de géomorphologiques du site ancient d'Asipani (Futuna–Polynésie occidentale): Journal de la Société des Océanistes, v. 97, p. 117–144.

Sand, C., 1996, Recent developments in the study of New Caledonia's prehistory: Archaeology in Oceania, v. 31, p. 45–71.

Sand, C., 1997, The chronology of Lapita ware in New Caledonia: Antiquity, v. 71, p. 539–547.

Sand, C., 1998a, Archaeological report on localities WKO013A and WKO013B at the site of Lapita (Koné, New Caledonia): Journal of the Polynesian Society, v. 107, p. 7–33.

Sand, C., 1998b, Archaeological research on Uvea Island, western Polynesia: New Zealand Journal of Archaeology, v. 18 (1996), p. 91–123.

Sand, C., 1998c, Recent archaeological research in the Loyalty Islands of New Caledonia: Asian Perspectives, v. 37, p. 194–223.

Sand, C., 1999a, Lapita and non-Lapita ware in New Caledonia's first millennium of Austronesian settlement, in Galipaud, J.-C., and Lilley, I., eds., Le Pacifique de 5000 à 2000 avant le présent: Paris, Éditions IRD (Institut de Recherche pour le Développement), p. 139–159.

Sand, C., 1999b, The beginning of southern Melanesian prehistory: The St. Maurice–Vatcha Lapita site, New Caledonia: Journal of Field Archaeology, v. 26, p. 307–323.

Sand, C., 2000a, The specificities of the "Southern Lapita Province": The New Caledonia case: Archaeology in Oceania, v. 25, p. 20–33.

Sand, C., 2000b, La datation du premier peuplement de Wallis et Futuna: Contribution à la définition de la chronologie Lapita en Polynésie occidentale: Journal de la Société des Océanistes, v. 111, p. 165–172.

Sand, C., 2001a, Evolutions in the Lapita cultural complex: A view from the Southern Lapita province: Archaeology in Oceania, v. 36, p. 65–76.

Sand, C., 2001b, Ancestral oceanic art: The Lapita design system, in Stevenson, C.M., et al., eds., Pacific 2000: Proceedings of the Fifth International Conference on Easter Island and the Pacific: Los Osos, California, Easter Island Foundation, p. 275–280.

Sand, C., 2002, Site LPO023 of Kurin: Characteristics of a Lapita settlement in the Loyalty Islands (New Caledonia): Asian Perspectives, v. 41, p. 129–147.

Sand, C., and Valentin, F., 1998, Cikobia: Données archéologiques preliminaries sur une île fidjienne à la frontière entre Melanésié et Polynésie: Journal de la Société des Océanistes, v. 106, p. 65–74.

Sand, C., Ouetcho, A., Bole, J., and Baret, D., 2001, Evaluating the "Lapita smoke screen": Site SGO015 of Goro, an early Austronesian settlement on the southeast coast of New Caledonia's Grand Terre: New Zealand Journal of Archaeology, v. 22 (2000), p. 91–111.

Schmidt, R.G., 1957, Geology of Saipan, Mariana Islands, Part 2B: Petrology of the volcanic rocks: U.S. Geological Survey Professional Paper 280-B, p. 127–174.

Scholl, D.W., and Herzer, R.H., 1992, Geology and resource potential of the southern Tonga platform, in Watkins, J.S., et al., eds., Geology and geophysics of continental margins: American Association of Petroleum Geologists Memoir 53, p. 139–156.

Scholl, D.W., Vallier, T.L., and Packham, G.H., 1985, Framework geology and resource potential of the southern Tonga platform and adjacent terranes—

A synthesis, *in* Scholl, D.W., and Vallier, T.L., eds., Geology and offshore resources of Pacific island arcs—Tonga region: Houston, Circum-Pacific Council for Energy and Mineral Resources Earth Science Series, v. 2, p. 457–488.

Schubert, P., 1986, Petrographic modal analysis—A necessary complement to chemical analysis of ceramic coarse ware: Archaeometry, v. 28, p. 163–178.

Schuth, S., Rohrbach, A., Münker, C., and Ballhaus, C., Garbe-Schönberg, D., and Qopoto, C., 2004, Geochemical constraints on the petrogenesis of arc picrites and basalts, New Georgia Group, Solomon Islands: Contributions to Mineralogy and Petrology, v. 148, p. 288–304.

Seeley, J.B., and Searle, E.J., 1970, Geology of the Rakiraki district, Viti Levu: New Zealand Journal of Geology and Geophysics, v. 13, p. 52–71.

Setterfield, T.N., Eaton, P.C., Rose, W.J., and Sparks, R.S.J., 1991, The Tavua caldera, Fiji: A complex shoshonitic caldera formed by concurrent faulting and sagging: Geological Society [London] Journal, v. 148, p. 115–127.

Sharp, N.D., 1988, Style and substance: A reconsideration of the Lapita design system, *in* Kirch, P.V., and Hunt, T.L., eds., Archaeology of the Lapita cultural complex: A critical review: Seattle, Thomas Burke Memorial Washington State Museum Research Report 5, p. 61–81.

Shepard, A.O., 1965, Ceramics for the archaeologist: Washington, D.C., Carnegie Institution Publication 609, 414 p.

Sheppard, P.J., Felgate, M., Roga, K., Keopo, J., and Walter, R., 1999, A ceramic sequence from Roviana Lagoon (New Georgia, Solomon Islands), *in* Galipaud, J.-C., and Lilley, I., eds., Le Pacifique de 5000 à 2000 avant le present: Paris, Éditions IRD (Institut de Recherche pour le Développement), p. 313–322.

Shiraki, K., 1971, Metamorphic basement rocks of Yap Islands, western Pacific Ocean: Earth and Planetary Science Letters, v. 13, p. 167–174, doi: 10.1016/0012-821X(71)90120-8.

Shiraki, K., Kuroda, N., Maruyamna, S., and Urano, H., 1978, Evolution of the Tertiary volcanic rocks in the Izu-Mariana arc: Bulletin of Volcanology, v. 41, p. 548–562.

Shutler, R., Jr., 1999, The relationship of red-slipped and lime-impressed pottery of the southern Philippine Islands to that of Micronesia and the Lapita of Oceania, *in* Galipaud, J.-C., and Lilley, I., eds., Le Pacifique de 5000 à 2000 avant le present: Paris, Éditions IRD (Institut de Recherche pour le Développement), p. 521–529.

Shutler, R., Jr., Sinoto, Y.H., and Takayama, J., 1984, Preliminary excavations of Fefan Island sites, Truk Islands, *in* Sinoto, Y.H., ed., Caroline Islands archaeology: Investigations on Fefan, Faraulep, Woleai, and Lamotrek: Honolulu, Bishop Museum Pacific Anthropological Records No. 35, p. 1–64.

Shutler, R., Jr., Burley, D.V., Dickinson, W.R., Nelson, E., and Carlson, A.K., 1994, Early Lapita sites, the colonisation of Tonga and recent data from northern Ha'apai: Archaeology in Oceania, v. 29, p. 53–68.

Sibuet, J.C., and Hsu, S.-K., 2004, How was Taiwan created?: Tectonophysics, v. 379, p. 159–181, doi: 10.1016/j.tecto.2003.10.022.

Silver, E.A., and Moore, J.C., 1978, Molucca Sea collision zone, Indonesia: Journal of Geophysical Research, v. 83, p. 1681–1691.

Sinoto, Y.H., 1966, A tentative prehistoric cultural sequence in the northern Marquesas Islands, French Polynesia: Journal of the Polynesian Society, v. 75, p. 287–303.

Sinoto, Y.H., 1983, An analysis of Polynesian migrations based on the archaeological assessments: Journal de la Société des Océanistes, v. 39, p. 57–67.

Sinoto, Y.H., Shutler, R., Jr., Dickinson, W.R., Shutler, M.E., Garanger, J., and Teska, T.M., 1999, Was there a pre-Lapita, Japanese Jomon, cord-marked pottery occupation in Vanuatu? *in* Galipaud, J.-C., and Lilley, I., eds., Le Pacifique de 5000 à 2000 avant le present: Paris, Éditions IRD (Institut de Recherche pour le Développement), p. 505–519.

Smart, C.D., 1965, An outline of Kabara prehistory: New Zealand Archaeological Association Newsletter, v. 8, p. 43–52.

Smith, A., 1995, The need for Lapita: Explaining change in the late Holocene Pacific archaeological record: World Archaeology, v. 26, p. 366–379.

Smith, I.E.M., 1976, Peralkaline rhyolites from the D'Entrecasteaux Islands, Papua New Guinea, *in* Johnson, R.W., ed., Volcanism in Australasia: Amsterdam, Elsevier, p. 275–285.

Smith, I.E.M., and Compston, W., 1982, Strontium isotopes in Cenozoic volcanic rocks from southeastern Papua New Guinea: Lithos, v. 15, p. 199–206, doi: 10.1016/0024-4937(82)90011-1.

Smith, I.E.M., and Johnson, R.W., 1981, Contrasting rhyolite suites in the late Cenozoic of Papua New Guinea: Journal of Geophysical Research, v. 86, p. 10,257–10,272.

Smith, I.E.M., and Milsom, J.S., 1984, Late Cenozoic extension and volcanism in eastern Papua, *in* Kokelaar, B.P., and Howells, M.E., eds., Marginal basin geology: Geological Society [London] Special Publication 16, p. 163–171.

Smith, I.E.M., Chapell, B.W., Ward, G.K., and Freeman, R.S., 1977, Peralkaline rhyolites associated with andesitic arcs of the southwest Pacific: Earth and Planetary Science Letters, v. 37, p. 230–236, doi: 10.1016/0012-821X(77)90167-4.

Specht, J., 1968, Preliminary report on excavations on Watom Island: Journal of the Polynesian Society, v. 77, p. 117–134.

Specht, J., 1972, The pottery industry of Buka Island, T.P.N.G: Archaeology and Physical Anthropology in Oceania, v. 7, p. 125–144.

Specht, J., 1974, Lapita pottery at Talasea, west New Britain, Papua New Guinea: Antiquity, v. 48, p. 302–306.

Specht, J., 1991, Kreslo: A Lapita pottery site in southwest New Britain, *in* Allen, J., and Gosden, C., eds., Report of the Lapita homeland project: Canberra, Australian National University Occasional Papers in Prehistory No. 20, p. 189–204.

Specht, J., 2003, Watom Island and Lapita: Observations on the Reber-Rakival localities, *in* Sand, C., ed., Pacific archaeology: Assessments and prospects: Noumea, Cahiers de l'Archéologie en Nouvelle-Calédonie, v. 15, p. 121–135.

Specht, J., and Gosden, C., 1997, Dating Lapita pottery in the Bismarck Archipelago, Papua New Guinea: Asian Perspectives, v. 36, p. 175–199.

Specht, J., Fullagar, R., and Torrence, R., 1991, What was the significance of Lapita pottery at Talasea?: Indo-Pacific Prehistory Association Bulletin, v. 11, p. 281–294.

Specht, J., Lilley, I., and Dickinson, W.R., 2006, Type X pottery, Morobe Province, Papua New Guinea: Petrography and possible Micronesian relationships: Asian Perspectives, v. 45, p. 24–47.

Spriggs, M., 1986, Landscape, land use, and political transformation in southern Melanesia, *in* Kirch, P.V., ed., Island societies: Archaeological approaches to evolution and transformation: Cambridge, UK, Cambridge University Press, p. 6–19.

Spriggs, M., editor, 1990, Lapita design, form, and composition: Canberra, Australian National University Occasional Papers in Prehistory No. 19, 142 p.

Spriggs, M., 1991, Nissan, the island in the middle: Summary report on excavations at the north end of the Solomons and the south end of the Bismarcks, *in* Allen, J., and Gosden, C., eds., Report of the Lapita homeland project: Canberra, Australian National University Occasional Papers in Prehistory No. 20, p. 222–243.

Spriggs, M., 1995, The Lapita culture and Austronesian prehistory in Oceania, *in* Bellwood, P., et al., eds., The Austronesians: Historical and comparative perspectives: Canberra, Australian National University, p. 112–133.

Spriggs, M., 1996, What is southeast Asian about Lapita? *in* Akazawa, T., and Szathmáry, E.J.E., eds., Prehistoric Mongoloid dispersals: Oxford, Oxford University Press, p. 324–348.

Spriggs, M., 1997a, Landscape catastrophe and landscape enhancement: Are either or both true in the Pacific? *in* Kirch, P.V., and Hunt, T.L., eds., Historical ecology in the Pacific islands: New Haven, Yale University Press, p. 80–104.

Spriggs, M., 1997b, The island Melanesians: Oxford, Blackwell, 326 p.

Spriggs, M., 1999, The stratigraphy of the Ponamla site, northwest Erromango, Vanuatu: Evidence for 2700 year old stone structures, *in* Galipaud, J.-C., and Lilley, I., eds., Le Pacifique de 5000 à 2000 avant le present: Paris, Éditions IRD (Institut de Recherche pour le Développement), p. 323–331.

Spriggs, M., 2000, The Solomon Islands as bridge and barrier in the settlement of the Pacific, *in* Anderson, A., and Murray, T., eds., Australian archaeologist: Collected papers in honour of Jim Allen: Canberra, Coombs Academic Publishing, p. 348–364.

Spriggs, M., 2003, Post-Lapita evolutions in island Melanesia, *in* Sand, C., ed., Pacific archaeology: Assessments and prospects: Noumea, Cahiers de l'Archéologie en Nouvelle-Calédonie, v. 15, p. 205–212.

Spriggs, M., and Anderson, A.J., 1993, Late colonization of East Polynesia: Antiquity, v. 67, p. 200–217.

Spriggs, M., and Bedford, S., 2001, Arapus: A Lapita site at Mangaasi in central Vanuatu? *in* Clark, G.J., et al, eds., The archaeology of Lapita dispersal in Oceania [Terra Australis 17]: Canberra, Australian National University Pandanus Books, p. 93–104.

Spriggs, M., and Bickler, S., 1989, Archaeological research on Erromango: Recent data on southern Melanesian prehistory: Indo-Pacific Prehistory

Association Bulletin, v. 9, p. 68–91.

Stanton, R.L., and Bell, J.D., 1969, Volcanic and associated rocks of the New Georgia Group, British Solomon Islands Protectorate: Overseas Geology and Mineral Resources, v. 10, p. 113–145.

Stark, J.T., 1963, Petrology of the volcanic rocks of Guam: U.S. Geological Survey Professional Paper 403-C, p. C1–C32.

Stark, J.T., and Hay, R.L., 1963, Geology and petrography of volcanic rocks of the Truk Islands, east Caroline Islands: U.S. Geological Survey Professional Paper 409, 41 p.

Stearns, H.T., 1944, Geology of the Samoan islands: Geological Society of America Bulletin, v. 55, p. 1279–1332.

Stearns, H.T., 1945, Geology of the Wallis Islands: Geological Society of America Bulletin, v. 56, p. 849–860.

Stearns, H.T., 1971, Geologic setting of an Eocene fossil deposit on Eua Island, Tonga: Geological Society of America Bulletin, v. 82, p. 2541–2552.

Stevenson, J., and Dodson, J.R., 1995, Palaeoenvironmental evidence for human settlement of New Caledonia: Archaeology in Oceania, v. 30, p. 36–41.

Stewart, W.D., and Sandy, M.J., 1988, Geology of New Ireland and Djaul islands, northeastern Papua New Guinea, in Marlow, M.S., et al., eds., Geology and offshore resources of Pacific island arcs—New Ireland and Manus region, Papua New Guinea: Houston, Circum-Pacific Council for Energy and Mineral Resources Earth Science Series, v. 9, p. 13–30.

Stice, G.D., 1968, Petrography of the Manu'a Islands, Samoa: Contributions to Mineralogy and Petrology, v. 19, p. 343–357, doi: 10.1007/BF00389417.

Stice, G.D., and McCoy, F.W., Jr., 1968, The geology of the Manu'a Islands, Samoa: Pacific Science, v. 22, p. 427–457.

Stoltman, J.B., 1989, A quantitative approach to the petrographic analysis of ceramic thin sections: American Antiquity, v. 54, p. 147–160.

Summerhayes, G.R., 1997, Losing your temper: The effect of mineral inclusions on pottery analysis: Archaeology in Oceania, v. 32, p. 108–117.

Summerhayes, G.R., 2000a, Lapita interaction [Terra Australis 15]: Canberra, Australian National University, 244 p.

Summerhayes, G.R., 2000b, What's in a pot? in Anderson, A., and Murray, T., eds., Australian archaeologist: Collected papers in honour of Jim Allen: Canberra, Coombs Academic Publishing, p. 291–307.

Summerhayes, G.R., 2000c, Recent archaeological investigations in the Bismarck Archipelago, Anir–New Ireland Province, Papua New Guinea: Indo-Pacific Prehistory Association Bulletin, v. 19, p. 167–174.

Summerhayes, G.R., 2001a, Defining the chronology of Lapita in the Bismarck Archipelago, in Clark, G.J., et al., eds., The archaeology of Lapita dispersal in Oceania [Terra Australis 17]: Canberra, Australian National University Pandanus Books, p. 25–38.

Summerhayes, G.R., 2001b, Lapita in the far west: Recent developments: Archaeology in Oceania, v. 36, p. 53–63.

Summerhayes, G.R., 2001c, Far Western, Western, and Eastern Lapita: A reevaluation: Asian Perspectives, v. 39, p. 109–138.

Summerhayes, G.R., and Scales, I., 2005, New Lapita pottery finds from Kolombangara, western Solomon Islands: Archaeology in Oceania, v. 40, p. 14–20.

Takayama, J., 1981, Early pottery and population movements in Micronesian prehistory: Asian Perspectives, v. 24, p. 1–10.

Tappin, D.R., 1993, The Tonga frontal-arc basin, in Balance, P.F., ed., South Pacific sedimentary basins: Amsterdam, Elsevier, p. 157–176.

Taylor, B., Goodliffe, A., Martínez, F., and Hey, R., 1995, Continental rifting and initial seafloor spreading in the Woodlark Basin: Nature, v. 374, p. 534–537, doi: 10.1038/374534a0.

Tejada, M.L.G., Mahoney, J.J., Duncan, R.A., and Hawkins, M.P., 1996, Age and geochemistry of basement and alkalic rocks of Malaita and Santa Isabel, Solomon Islands, southern margin of Ontong Java Plateau: Journal of Petrology, v. 37, p. 361–394.

Tejada, M.L.G., Mahoney, J.J., Neal, C.R., Duncan, R.A., and Petterson, M.G., 2002, Basement geochemistry and geochronology of central Malaita, Solomon Islands, with implications for the origin and evolution of the Ontong Java Plateau: Journal of Petrology, v. 43, p. 449–484, doi: 10.1093/petrology/43.3.449.

Terrell, J.E., and Welsh, R.L., 1990, Trade networks, areal integration, and diversity along the north coast of New Guinea: Asian Perspectives, v. 29, p. 155–165.

Terrell, J.E., and Welsh, R.L., 1997, Lapita and the temporal geography of prehistory: Antiquity, v. 71, p. 548–572.

Thomas, F.R., Nunn, P.D., Osborne, T., Kumar, R., Areki, F., Matararaba, S., Steadman, D., and Hope, G., 2004, Recent archaeological findings at Qaranilaca cave, Vanuabalavu island, Fiji: Archaeology in Oceania, v. 39, p. 42–49.

Thomas, N., 1989, The force of ethnology: Origins and significance of the Melanesia/Polynesia division: Current Anthropology, v. 30, p. 27–42, doi: 10.1086/203707.

Thomson, J.-A.R., and White, J.P., 2000, Localism of Lapita pottery in the Bismarck Archipelago, in Anderson, A., and Murray, T., eds., Australian archaeologist: Collected papers in honour of Jim Allen: Canberra, Coombs Academic Publishing, p. 308–322.

Thorpe, R.S., editor, 1982, Andesites: Orogenic andesites and related rocks: New York, Wiley, 724 p.

Torrence, R., 2001, Cultural landscapes on Garua Island, Papua New Guinea: Antiquity, v. 76, p. 766–776.

Torrence, R., and Stevenson, C.M., 2000, Beyond the beach: Changing Lapita landscapes on Garua Island, Papua New Guinea, in Anderson, A., and Murray, T., eds., Australian archaeologist: Collected papers in honour of Jim Allen: Canberra, Coombs Academic Publishing, p. 324–345.

Torrence, R., Pavlides, C., Jackson, P., and Webb, J., 2000, Volcanic disasters and cultural discontinuities in Holocene time in West New Britain, Papua New Guinea, in McGuire, W.G., et al., eds., The archaeology of geological catastrophes: Geological Society [London] Special Publication 171, p. 225–244.

Tracey, J.L., Jr., Schlanger, S.O., Stark, J.T., Doan, D.B., and May, B.G., 1964, General geology of Guam: U.S. Geological Survey Professional Paper 403-A, p. A1–A102.

Trail, D.S., 1967, Geology of Woodlark Island, Papua: Australian Bureau of Mineral Resources Report 115, 32 p.

Turner, C.C., and Ridgway, J., 1982, Tholeiitic, calc-alkaline, and (?)alkaline igneous rocks of the Shortland Islands, Solomon Islands: Tectonophysics, v. 87, p. 335–354, doi: 10.1016/0040-1951(82)90232-3.

Vallier, T.L., Stevenson, A.J., and Scholl, D.W., 1985, Petrology of igneous rocks from Ata Island, Kingdom of Tonga, in Scholl, D.W., and Vallier, T.L., eds., Geology and offshore resources of Pacific island arcs—Tonga region: Houston, Circum-Pacific Council for Energy and Mineral Resources Earth Science Series, v. 2, p. 301–316.

Van der Plas, L., and Tobi, A.C., 1965, A chart for determining the reliability of point counting results: American Journal of Science, v. 263, p. 87–90.

Van Wyck, N., and Williams, I.S., 2002, Age and provenance of basement metasediments from the Kubor and Bena Bena blocks, central highlands, Papua New Guinea: Constraints on the tectonic evolution of the northern Australian cratonic margin: Australian Journal of Earth Sciences, v. 49, p. 565–577, doi: 10.1046/j.1440-0952.2002.00938.x.

Verstappen, H.T., 1959, Geomorphology and crustal movements of the Aru Islands in relation to the Pleistocene drainage of the Sahul shelf: American Journal of Science, v. 257, p. 491–502.

von Gnielenski, F.E., Danaro, T.J., Wellman, P., and Pain, C.F., 2002, Torres Strait region, in Bain, J.H.C., and Draper, J.J., eds., North Queensland geology [Queensland Geology 9]: Australian Geological Survey Organization Bulletin 240, p. 159–164.

Vroon, P.Z., Van Bergen, M.J., White, W.M., and Varekamp, J.C., 1993, Sr-Nd-Pb isotope systematics of the Banda arc, Indonesia: Combined subduction and assimilation of continental material: Journal of Geophysical Research, v. 98, p. 22,349–22,366.

Walker, G.P.L., Heming, R.F., Speed, T.J., and Walker, R.F., 1981, Latest major eruption of Rabaul volcano, in Johnson, R.W., ed., Cooke-Ravian volume of volcanological papers: Port Moresby, Geological Survey of Papua New Guinea Memoir 10, p. 181–193.

Wallace, D.A., Johnson, R.W., Chappell, B.W., Arculus, R.J., Perfit, M.R., and Crick, I.H., 1983, Cainozoic volcanism of the Tabar, Lihir, Tanga, and Feni islands, Papua New Guinea: Australian Bureau of Mineral Resources, Geology, and Geophysics Report 243, 62 p. [BMR Microform MF197].

Walter, R., and Dickinson, W.R., 1989, A ceramic sherd from Ma'uke in the southern Cook Islands: Journal of the Polynesian Society, v. 98, p. 465–470.

Ward, G.K., 1979, Prehistoric settlement and economy in a tropical small island environment: The Banks Islands, insular Melanesia [Ph.D. thesis]: Canberra, Australian National University, 255 p.

Warden, A.J., 1967, The geology of the central islands: Port Vila, New Hebrides Condominium Geological Survey Report 5, 108 p.

Weissel, J.K., 1977, Evolution of the Lau Basin by the growth of small plates, in Talwani, T., and Pitman, W.C., III, eds., Island arcs, deep sea trenches, and back-arc basins: American Geophysical Union Maurice Ewing Series 1, p. 429–436.

Weissel, J.K., Taylor, B., and Karner, G.D., 1982, The opening of the Woodlark Basin, subduction of the Woodlark spreading system, and the evolution of northern Melanesia since mid-Pliocene time: Tectonophysics, v. 87,

p. 253–277, doi: 10.1016/0040-1951(82)90229-3.
Welch, D.J., 2001, Early upland expansion of Palauan settlement, in Stevenson, C.M., et al., eds., Pacific 2000: Proceedings of the Fifth International Conference on Easter Island and the Pacific: Los Osos, California, Easter Island Foundation, p. 179–184.
Wells, R.E., 1989a, The oceanic basalt basement of the Solomon Islands arc and its relationship to the Ontong Java Plateau—Insights from Cenozoic plate motion models, in Vedder, J.G., and Bruns, T.R., eds., Geology and offshore resources of Pacific island arcs—Solomon Islands and Bougainville, Papua New Guinea region: Houston, Circum-Pacific Council for Energy and Mineral Resources Earth Science Series, v. 12, p. 7–22.
Wells, R.E., 1989b, Origin of the oceanic basalt basement of the Solomon Islands arc and its relationship to the Ontong Java Plateau—Insights from Cenozoic plate motion models: Tectonophysics, v. 165, p. 219–235, doi: 10.1016/0040-1951(89)90048-6.
Whalen, J.B., 1985, Geochemistry of an island-arc plutonic suite: The Uasilau–Yau Yau intrusive complex, New Britain, P.N.G: Journal of Petrology, v. 26, p. 603–632.
Whalen, J.B., and McDougall, I., 1980, Geochronology of the Uasilau–Yau Yau porphyry copper deposit, New Britain, Papua New Guinea: Economic Geology and the Bulletin of the Society of Economic Geologists, v. 75, p. 566–571.
Wharton, M.R., Hathway, B., and Colley, A.H., 1995, Volcanism associated with extension in an Oligocene-Miocene arc, southwestern Viti Levu, Fiji, in Smellie, J.L., ed., Volcanism associated with extension at consuming plate margins: Geological Society [London] Special Publication 81, p. 95–114.
Whelan, P.M., Gill, J.B., Kollman, E., Duncan, R.A., and Drake, R.E., 1985, Radiometric dating of magmatic stages in Fiji, in Scholl, D.W., and Vallier, T.L., eds., Geology and offshore resources of Pacific island arcs—Tonga region: Houston, Circum-Pacific Council for Energy and Mineral Resources Earth Science Series, v. 2, p. 415–440.
Whitbread, I.K., 1986, The characterisation of argillaceous inclusions in ceramic thin sections: Archaeometry, v. 28, p. 79–88.
White, J.P., 1997, Archaeological survey in southern New Ireland: Journal de la Société des Océanistes, v. 105, p. 141–146.
White, J.P., and Downie, J.E., 1980, Excavations at Lesu, New Ireland: Asian Perspectives, v. 23, p. 193–220.
White, J.P., and Murray-Wallace, C.V., 1996, Site ENX (Fissoa) and the incised and applied pottery tradition in New Ireland, Papua New Guinea: Man and Culture in Oceania, v. 12, p. 31–46.
White, J.P., and Specht, J., 1971, Prehistoric pottery from Ambitle Island, Bismarck Archipelago: Asian Perspectives, v. 14, p. 88–94.
Whitford, D.J., and Jezek, P.A., 1979, Origin of late Cenozoic lavas from the Banda arc, Indonesia: Trace element and Sr isotopic evidence: Contributions to Mineralogy and Petrology, v. 68, p. 141–150, doi: 10.1007/BF00371896.
Wickler, S., 1990, Prehistoric Melanesian exchange: Recent evidence from the northern Solomon Islands: Asian Perspectives, v. 29, p. 135–154.
Wickler, S., 2001a, The prehistory of Buka: A stepping stone island in the northern Solomons [Terra Australis 16]: Canberra, Australian National University, 306 p.
Wickler, S., 2001b, The colonization of western Micronesia and early settlement in Palau, in Stevenson, C.M., et al., eds., Pacific 2000: Proceedings of the Fifth International Conference on Easter Island and the Pacific: Los Osos, California, Easter Island Foundation, p. 185–196.
Wickler, S., 2002, Terraces and villages: Transformations of the cultural landscape in Palau, in Ladefoged, T.N., and Graves, M.W., eds., Pacific landscapes: Archaeological approaches: Los Osos, California, Easter Island Foundation, p. 63–93.
Williams, D.F., 1983, Petrology of ceramics, in Kempe, D.R.C., and Harvey, A.P., eds., The petrology of archaeological artifacts: Oxford, Clarendon Press, p. 301–329.
Willmott, W.F., Whitaker, W.G., Palfreyman, W.D., and Trail, D.S., 1973, Igneous and metamorphic rocks of the Cape York Peninsula and Torres Strait: Canberra, Australian Bureau of Mineral Resources, Geology, and Geophysics Bulletin 135, 145 p.
Wolf, K.H., 1971, Textural and compositional transitional stages between various lithic grain types (with a comment on "interpreting detrital modes of graywacke and arkose"): Journal of Sedimentary Petrology, v. 41, p. 328–332.
Wood, C.P., Nairn, I.A., McKee, O.C., and Talai, B., 1995, Petrology of the Rabaul Caldera area, Papua New Guinea: Journal of Volcanology and Geothermal Research, v. 69, p. 285–302, doi: 10.1016/0377-0273(95)00034-8.
Woodhall, D., 1985a, Geology of the Lau Ridge, in Scholl, D.W., and Vallier, T.S., eds., Geology and offshore resources of Pacific island arcs—Tonga region: Houston, Circum-Pacific Council for Energy and Mineral Resources Earth Science Series, v. 2, p. 351–378.
Woodhall, D., 1985b, Geology of Taveuni, Qamea, Laucala, Cikobia, and adjacent islands: Suva, Fiji Mineral Resources Department, scale 1:50,000 [map intended to accompany an unpublished text also consulted in preliminary manuscript form].
Woodhall, D., 1987, Geology of Rotuma: Suva, Fiji Mineral Resources Department Bulletin 8, 40 p.
Woodrow, P.J., 1976, Geology of southeastern Vanua Levu: Suva, Fiji Mineral Resources Division Bulletin 4, 73 p.
Woodrow, P.J., 1980, Geology of Kadavu: Suva, Fiji Mineral Resources Division Bulletin 7, 31 p.
Wright, E., and White, W.M., 1986/87, The origin of Samoa: New evidence from Sr, Nd, and Pb isotopes: Earth and Planetary Science Letters, v. 81, p. 151–162, doi: 10.1016/0012-821X(87)90152-X.
Yagi, K., 1960, Petrochemistry of the alkalic rocks of Ponape Island, western Pacific Ocean: 21st International Geological Congress Reports, v. 13, p. 108–122.
Yan, C.Y., and Kroenke, L.W., 1993, A plate tectonic reconstruction of the southwest Pacific, 0–100 Ma: Proceedings of the Ocean Drilling Program, Scientific Results, Volume 130: College Station, Texas, Ocean Drilling Program, p. 697–709.
Zuffa, G.G., 1979, Hybrid arenites: Their composition and classification: Journal of Sedimentary Petrology, v. 50, p. 21–29.

MANUSCRIPT ACCEPTED BY THE SOCIETY 28 NOVEMBER 2005

APPENDIX 1. CATALOGUE OF PREHISTORIC OCEANIAN POTSHERDS EXAMINED PETROGRAPHICALLY IN THIN SECTION

The following catalogue of 2223 prehistoric potsherds from Pacific Oceania, listed alphabetically by island group, and secondarily by island within each group and site on each island, is provided to allow the specific archaeological provenience of each studied sherd suite to be traced by those who need more detailed information than is provided in the text (boldface numbers indicate total number of sherds examined from each island group). As listed, collections from the western (W) and eastern (E) Caroline Islands are tabulated separately, Bismarck Archipelago includes sherds examined for comparative purposes from the Huon Peninsula and the Aitape coastline of Papua New Guinea, the Solomon Islands include Bougainville and Buka lying politically within Papua New Guinea, Vanuatu includes the eastern Solomon outliers (politically part of the Solomon Islands) lying along or east of the northern extension of the New Hebrides island arc, the Lau Archipelago of eastern Fiji is distinguished from the remainder of Fiji, NE Melanesia includes islands along the northern Melanesian borderland between Fiji and Samoa, and Samoa includes American Samoa. N is the number of sherds from each collection examined personally in thin section, and all abbreviations (island, site, temper class, temper origin) are defined by endnotes, which also identify collectors or providers by the initials of the individual or institution from which each sherd collection (or set of thin sections) was received or borrowed for petrographic study.

APPENDIX TABLE A1

Island group	Island	Site(s)	N	Temper class	Temper origin(s)	Collector(s) or provider(s)
Aru **28**	-	Jur, NLi, Pap	18	THq [hyb]	In	SO'C, 2002
		Kar, Wan	10	AA	Ex (Ban)	SO'C, 2002
Bismarck **348**	Ani	Bal	8	PA	In	SG, 2004
		Mao	6	PA	In	WRA, 1971
	Bas	Kam	6	PA	In	GS, 2000
	Ara	Adw, Apo	9	AA	Ex (NBr)	CG-GS, 1997-1998
	Bal	N/A	2	AA-BAq	Ex (Man-Lou)	WRA, 1997
	Bod	N/A	2	AA	Ex (NBr)	JS, 1997
	DoY	several	6	DO [+hyb]	In	JRT-JPW, 1998
			8	AA	Ex (NBr)	JRT-JPW, 1998
	Elo	Tal	6	[cal]	In?	PVK, 2003
			25	BAq	Ex (Lou)	PVK, 1995-2003
			23	AA	Ex (Man)	PVK, 1995-2003
			3	PA	Ex (TLTF)	PVK, 1995-2003
	Ema	Eta	2	[cal]	In?	PVK, 2003
			2	AA	Ex (Man)	PVK, 2003
			2	BAq	Ex (Lou)	PVK, 2003
	Gar	N/A	2	AA	Ex (NBr)	RT, 1997
	Lou	N/A	4	BAq-AA	In, Ex (Man)	WRA, 1997
	Man	several	5	AA	In	JK, 1997
	MBu	N/A	3	AA-BAq	Ex (Man-Lou)	WRA, 1997
	NBr	Kre	15	AA-DO	In	GS, 1998
	NIr	Fis	10	AA	In	SG, 2004
		Las	3	AA	In	JGo, 1997
			1	AA	In	SG, 2005
			1	BAq	Ex (Lou)	SG, 2005
		Los	8	AA [+cal]	In	JPW, 1980
			3	AA	In	SG, 2005
	Nis	N/A	12	AA	Ex (Buk)	SK, 1974
			14	AA	Ex (NIr)	MSp, 1998
			2	AA	Ex (NBr)	MSp, 1998
	Pap	Ait	50	DO	In	JTe, 1997
		Huo	3	[com]	In	JS, 2002
			15	[gro]	In	JS, 2002
	Sia	Mai, Tua	10	[gro]	Ex (Huo)	IL, 2002
	Tan	Boe, Mad	33	PA	In	SG, 2004
			9	PA [+hyb]	Ex (TLTF)	SG, 2004
	Wat	Reb-Rak	31	AA [+hyb]	In	DA, 1997
			3	AA	Ex (Man)	DA, 1997
			1	DO	Ex (NBr)	DA, 1997

(*continued*)

APPENDIX TABLE A1 (*continued*)

Island group	Island	Site(s)	N	Temper class	Temper origin(s)	Collector(s) or provider(s)
Caroline (E) **76**	Fef	Sap	24	OB [hyb]	In	RS-JTa, 1977
			2	OB [hyb]	In	CD, 1997
	Kos	Lel	8	[cal]	In	JSA, 1993
			2	[cal]	In	CD, 1997
	Lam	Bol, Sab	6	THo [com]	Ex (Yap)	RS- JTa, 1977
	Poh	Nad	34	OB [com]	In	JSA-CD, 1980, 1997
Caroline (W) **300**	Ang	N/A	3	[gro=com]	Ex (Bab)	TR, 1970
			2	AA	Ex (Bab)	DO, 1970
	Bab	surface	3	[gro]	In	PVK, 1978
		several	16	[gro>com]	In	SP, 2003
		Air	7	[gro>com]	In	JSA-CD, 1996–1997
		Kor	1	AA	In	JSA, 1996
			3	[gro>com]	In	DO, 1970
			28	[com>gro]	In	SMF, 2000
		Mek	2	AA	In	DO, 1970
			12	[gro>com]	In	DO-JSA, 1970–1996
		Orr	11	AA [+hyb]	Ex (Bab)	SMF, 2001
			55	[gro>com]	Ex (Bab)	SMF, 2000
		Ngc	12	[com>gro]	In	SMF, 2001
		Ngr	1	AA	In	JSA, 1996
			4	[gro>com]	In	JSA, 1996
		Ngt	8	[gro>com]	In	JSA, 1996
		Ngw	4	[gro>com]	In	JSA, 1996
	Che	Ulo	1	AA	Ex (Bab)	DO, 1970
			3	[gro>com]	Ex (Bab)	DO, 1970
			5	[gro>com]	Ex (Bab)	GC, 2002
			8	AA [+hyb]	Ex (Bab)	GC, 2002
	Fai	N/A	15	THo [+hyb]	Ex (Yap)	MI, 1993
			4	AA-AAq	Ex (Yap)	MI, 1993
			16	[gro]	Ex (Yap>Pal)	MI, 1993
	Kay	N/A	4	[gro=com]	Ex (Bab)	TR, 2002
	Ngu	N/A	5	THo [+hyb]	Ex (Yap)	JTa, 1980
			3	AA-AAq	Ex (Yap)	JTa, 1980
			3	[gro]	Ex (Pal>Yap)	JTa, 1980
	Pel	N/A	4	[gro>com]	Ex (Bab)	TR, 2002
	Son	Fan	1	[gro>com]	Ex (Pal)	RH-A, 2002
	Uli	Mog	4	THo	Ex (Yap)	CD, 1997
			13	AA-AAq	Ex (Yap)	CD, 1997
			1	[gro]	Ex (Yap)	CD, 1997
	Yap	surface	2	THo	In	PVK, 1978
		several	6	THo	In	Im, 1977
			6	AA-AAq	In	Im, 1977

(*continued*)

APPENDIX TABLE A1 (continued)

Island group	Island	Site(s)	N	Temper class	Temper origin(s)	Collector(s) or provider(s)
Caroline (W) **300**		Gac	8	THo [+hyb]	In	CD, 1997
(continued)			7	AAq	In	CD, 1997
		Pem	4	THo [hyb]	In	JTa-CD, 1980–1997
			2	AAq	In	JTa-CD, 1980–1997
			3	[gro]	In	JTa, 1980
Cook (southern) **1**	Mau	Ano	1	AA	Ex (Ton)	RW, 1988
D'Entrecasteaux **2**	Goo	N/A	2	THq	In	WRA, 1997
Fiji (non-Lau) **329**	Beq	several	6	PA	In	RCG, 1997
			4	DO	Ex (Rew)	RCG, 1997
		Kul	4	PA [+hyb]	In-nl	GC, 1997
			4	PA	In	GC, 1997
	Cik	N/A	7	PA	Ex (VaL)	CS, 2001
			6	PA	Ex (VaL)	CS, 2001
			5	AAq	Ex (Udu)	CS, 2001
			1	DO	Ex (VaL)	CS, 2001
	Bul	N/A	2	AA	In	CH, 1977
	Kad	Num	4	AA	In	WRD, 1992
	Lac	Sep	1	PA	In	PDN, 2002
	Mot	Nab	22	PA	In	PDN, 2004
			3	AA	Ex (Kad)	PDN, 2004
			3	DO	Ex (ViL)	PDN, 2004
			2	PA	Ex (Ked)	PDN, 2004
			2	PA	Ex (Fij-Lau)	PDN, 2004
		Sau	1	PA	In	PDN, 2001
			1	PA	Ex (Ked)	PDN, 2002
	Nag	Mat	12	PA	Ex (Ova)	SBe, 1997
			2	PA	Ex (Ked)	SBe, 1997
			2	DO	Ex (ViL)	SBe, 1997
			6	PA	Ex (Fij-Lau)	SBe, 1997
	Nan	surface	2	PA	In	EH, 1967
	Ono	N/A	2	AA	In	CH, 1977
	Tav	N/A	2	PA	In	EH, 1967
	Tot	surface	2	PA	Ex (M/M)	JTC, 1994
			1	PA	Ex (Ked)	JTC, 1994
			1	DO	Ex (Rew)	JTC, 1994
		Lew	2	PA	In	JTC, 1994
	Uga	N/A	5	PA [hyb]	Ex (Beq)	GC, 1997–1999
			5	PA	Ex (Kul)	GC, 1998–1999
			3	PA	Ex (Vag)	GC, 1998–1999

(continued)

APPENDIX TABLE A1 (continued)

Island group	Island	Site(s)	N	Temper class	Temper origin(s)	Collector(s) or provider(s)
Fiji (non-Lau) 329			2	DO	Ex (ViL)	GC, 1998–1999
(continued)			2	AAq	Ex (Udu)	GC, 1998–1999
	VaL	Mal, Vau	2	PA	In	AP, 1997
	ViL	Bou	27	DO	In	PDN, 2005
		Kao	5	DO	In	GC, 1998
			5	AA	In-nl (Nvu)	GC, 1998
		Nas	20	DO	In	EH-fm, 1967–1996
			2	AA	Ex (Kad)	fm, 1996
			2	PA [hyb]	Ex (Fij)	fm, 1996
		Nat	16	PA	In	EH-GC, 1967–1997
			6	PA	In-nl	EH-GC, 1967–1997
			7	DO	In-nl	EH-GC, 1967–1997
		Nav	25	PA	In	Im-GC, 1966–1997
			2	DO	In-nl	Im-GC, 1966–1999
		Nvu	4	AA	In	GJI, 1997
			1	AA	Ex (Kad)	GJI, 1997
		Qaq	1	PA	In	PDN, 2002
		Sig	20	DO	In	LHB, 1966
			2	DO	In	GC, 1997–1999
			9	DO	In	DVB, 1998
			1	AA	Ex (Nvu)	DVB, 1998
		Vud	4	PA	In	Im, 1968
			4	DO	In-nl	Im, 1968
		Yad	1	DO	In	PDN, 2002
		Yan	19	DO [+hyb]	In	LHB, 1966–1967
			6	PA	In-nl	SBe-TLH, 1978
	Yas	surface	9	AA [hyb]	In	WRD, 1983
Lau (E. Fiji) 146	Aiw	several	9	AA	Ex (Lak)	SJO'D, 2003
	Kab	N/A	9	PA	Ex (Ked)	SBe, 1978
	Lak	several	90	AA	In	SBe, 1977
			6	PA	Ex (Ked)	SBe, 1977
			2	DO	Ex (ViL)	SBe, 1977
		Ulu	3	AA	In	SBe, 1979
			3	AA	Ex (Kad)	SBe, 1979
			1	DO	Ex (ViL)	SBe, 1979
			1	PA	Ex (Ked)	SBe, 1979
	Mag	Vot	6	PA [+hyb]	In	GC, 1997
	Nay	several	12	AA	Ex (Lak)	SJO'D, 2003
	VaB	surface	2	PA	Ex (Ked)	SBe, 1978
			2	PA	In	WRD, 1999
Loyalty 16	Mar	Kur	5	[cal]	In (?)	CS, 2001
		Wab	5	THq	Ex (NCa)	RS, 1980
	Ouv	Mou	6	THo	Ex (NCa)	RS, 1980

(continued)

APPENDIX TABLE A1 (*continued*)

Island group	Island	Site(s)	N	Temper class	Temper origin(s)	Collector(s) or provider(s)
Mariana **144**	Agu	several	12	AA-AAq	Ex (Sai)	JAS, 1998
	Ala	NWC	6	AA-AAq	Ex (Sai)	BMB, 1998
	Gua	Fou, Nom	14	AA	In	FMR-PM, 1999
		Pla, Taf	4	AA	In	FMR-PM, 1999
		Ipa, Tum	5	AA	In	WRD-DRM, 1998
		Lag	4	AA-AAq	In-Ex (Sai)	TR, 1999
		Ord, Wat	20	AA	In	DRM, 2001
		Tar	9	AA	In-nL	ERR-RO, 2002
		Tum	1	AAq	Ex (Sai)	DRM, 1998
		Wat	1	AAq	Ex (Sai)	DRM, 2001
	Pag	Apa, Reg, San	9	AA-AAq	Ex (Sai)	EW, 1998
	Roa	Asm	2	AA	Ex (Sai)	WRD, 1998
		Moc	8	AA-AAq	Ex (Sai-Gua)	JTa, 1977
		NEC	6	AA-AAq	Ex (Sai-Gua)	BMB, 1998
	Sai	Cha	8	AAq	In	FMR-LB, 1966–1998
		Saf	8	AA-AAq	In	MSw, 1998
		Una	4	AAq	In	BMB, 1998
		Unb	6	AA-AAq	In	MTC, 2005
	Tin	Cas	2	AAq	Ex (Sai)	DRM, 1998
		IBB	4	AA-AAq	Ex (Sai)	DRM, 1998
		Kah	4	AA-AAq	Ex (Sai)	BD, 1998
		Tag	5	AA-AAq	Ex (Sai)	FMR, 1966
		Unc	2	AAq	Ex (Sai)	DDeF, 1998
Marquesas **5**	Hiv	Ato	2	OB	In	PVK, 1987
	Nuk	Hom	1	OB	In	am, 1967
	UaH	Han	2	OB	In	YHS, 1967
Molucca **53**	Amb	several	10	AAq	In	MSp, 2002
			5	THq	Ex (Ser)	MSp, 2002
	Ban	Ay, BN	6	AA	In	PVL, 2004
			2	AA	Ex (UNK)	PVL, 2004
	Bur	N/A	6	THq	In	MSp, 2003
	Gor	several	9	THq	In	MSp, 2003
		Gir	6	AA	Ex (Ban)	MSp, 2003
	Hal	several	5	AA	In	SO'C, 2002
	Har	Oma	1	AAq	In	MSp, 2002
	Ser	N/A	1	THq	In	MSp, 2002
			1	AA	Ex (Ban)	MSp, 2002
			1	AAq	Ex (Lea)	MSp, 2002
Murray **4**	Dau	Orm	3	DO	Ex (NGH)	MC, 2002
	Mer	Zom	1	DO	EX (NGH)	MC, 2002

(*continued*)

APPENDIX TABLE A1 (*continued*)

Island group	Island	Site(s)	N	Temper class	Temper origin(s)	Collector(s) or provider(s)
NE Melanesia 29	Alo	Mam	2	BA-PA	In	PVK, 1975
	Fut	Nuu, Tai, Vel	16	BA-PA	In	PVK, 1975
	Rou	Mak	4	OB	In	TNL, 1996-97
			1	PA	Ex (VaL)	TNL, 1996
	Uve	several	6	OB	In	PVK, 1975
New Caledonia 54	GrT	several	12	THq-THo	In	RS, 1966-70
		Lap	16	THq-THo	In	RS, 1966
			15	THq-THo	In	SC, 2001
	IdP	Kap	2	THo	Ex (NCa)	RS, 1966
		Vac	5	THo	Ex (NCa)	RS, 1966
			4	THq-THo	Ex (NCa)	CS, 2001
Samoa 103	Aun	N/A	1	OB	In	JTC, 1994
	Ofu	Tog	30	OB	In	TLH, 1990
		Vat	2	OB	In	SBe, 1997
	Sav	Ple	2	OB	In	GC, 2005
	Tut	Aoa	13	OB	In	JTC, 1994
			3	OBt	In	JTC, 1994
		Mae	7	OB	In	ES, 1997
		Tat	3	OB	In	SBe, 1988
	Upo	Jan	8	OB	In	JDJ, 1975
			2	OBt	In	JDJ, 1975
		Mul	10	OB	In	RCG, 1973
			1	AAq	Ex (Udu)	FP, 1999
		Par	1	OB	In	JDJ, 1975
		Sas	4	OB	In	RCG, 1967
			2	OBt	In	RCG, 1967
		Val	10	OB	In	RCG, 1966-67
			4	OBt	In	RCG, 1966
Solomon 114	Bel	Sik	1	AA	Ex (NeG)	JP, 1972
	Bou	Mei	3	AA	In	MSp, 1997
		Pid	6	AA	In	MSp, 1997
		Pau	20	AA	In	JTe, 1972-1974
		Teo	12	AA	Ex (Buk)	SBk, 1974
			6	AA	Ex (Kie)	SBk, 1974
	Buk	several	9	AA	In	WRA, 1971, 1997
	Kol	N/A	3	DO	Ex (Cho)	DEY, 1972
	NeG	Rov	21	AA	In	MF, 2000
			7	DO [hyb]	Ex (UNK)	MF, 2000
	OnJ	Lua	2	AA	Ex (Buk)	WWH, 1973
		Pel	3	AA	Ex (Buk)	TB-S, 1995
	Sho	Alu	13	AA	In	GJI, 1971

(*continued*)

APPENDIX TABLE A1 (continued)

Island group	Island	Site(s)	N	Temper class	Temper origin(s)	Collector(s) or provider(s)
Solomon 114			2	AA	Ex (Bui)	GJI, 1971
(continued)		Mon	2	AA	In	GJI, 1971
	StA	Fer	3	AA [+hyb]	Ex (Mak)	RCG, 1972
			1	DO	Ex (Mak)	RCG, 1972
Solomon Sea 9	Muy	N/A	3	DO	In	SBi, 1997
	Tro	Kir	4	THq	Ex (Dee)	WRA, 1997
			2	AA	Ex (PP)	WRA, 1997
Tokelau 1	Atu	N/A	1	AA	Ex (Fij-Lau)	SBe, 1988
Tonga 97	Ata	surface	2	AA	In	DWS, 2002
	Eua	Ana	3	AA	In	DWS, 1995
	Foa	Fal	7	AA	In	DVB, 1995
	Hno	Puk	2	AA	In	DVB, 1995
	Hun	surface	1	AA	In	WRD, 1998
	Kap	surface	3	AA	In	JD, 1971
			6	AA	In	RS, 1992
			2	AA	In	DVB, 1995
	Kot	surface	1	AA	In	GR, 1972
	Lif	Hol	9	AA	In	DVB, 1995
	Mat	surface	1	AA	In	GR, 1972
	Niu	several	9	AA [+hyb]	In	GR, 1972
			18	AA [+hyb]	In	PVK, 1978
	Nom	surface	2	AA	In	GR-RS, 1972–1992
			2	AA	In	DVB, 1995
	Ttu	Maa, Mua	4	AA	In	RS, 1966
		Nuk	6	AA	IN	JP-DVB, 1975–1998
			4	AAq	Ex (Van)	JP-DVB, 1975–1998
		Pea	5	AA	In	DVB, 1998
		Vei	1	AA	In	WRA, 1997
	Tun	Fak	7	AA	In	YHS-GR, 1972
	Uih	Vap	2	AA	In	DVB, 1995
Torres Strait 3	Mab	Mas	3	DO	In	IMcN, 2003
Tuvalu 12	Nae	N/A	1	PA [hyb]	Ex (VaL/ViL)	AA, 1998
	Nau	Hah	5	PA [hyb]	Ex (VaL/ViL)	JTa, 1998
	Vai	Tem	5	PA	Ex (Nav)	JTa, 1985
			1	PA [hyb]	Ex (VaL/ViL)	JTa, 1985

(continued)

APPENDIX TABLE A1 (continued)

Island group	Island	Site(s)	N	Temper class	Temper origin(s)	Collector(s) or provider(s)
Vanuatu **347**	Anu	Rot	10	BA [+hyb]	In	DEY, 1972
			1	AA	Ex (Nen)	DEY, 1972
	Bak	Pak	4	AA	In	GW, 1997
			1	AA	Ex (San)	GW, 1997
	Efa	surface	1	AA	Ex (San)	RS, 1970
		several	9	AA	In	JGa, 1967
			16	AA	In	RS, 1970
		Mag	2	AA	In	MSp, 1997
		Mel	6	AA	In	YHS, 1993
			6	AA	In	MES, 1994
	Err	Ifo. Pon	10	AA	In	MSp, 1997
		Ifo	1	THq	Ex (NCa)	MSp, 1997
	Mlk	several	21	AA	In	RS, 1970
			10	AA	In	SBd, 1998
		Mau, Nvp	2	AA	In	SBd, 1997
		Ten	1	AA	Ex (San)	SBd, 2003
		Vao	20	AA	Ex (Mlk)	SBd, 2003
			1	AA	Ex (San)	SBd, 2003
			1	AA	Ex (She)	SBd, 2003
	Mlo	several	6	AA	Ex (San)	JCG, 1997
		Avn	14	AA	Ex (San)	JDH, 1969
			1	THq-THo	Ex (NCa)	JDH, 1969
	Nen	Gra, Tre	15	AA [+hyb]	In	PCMcC, 1979
		Nan	20	AA [+ hyb]	In	RCG, 1972
			1	AAq	Ex (UNK)	RCG, 1972
	Ree	Ngm, Ngn	33	AA [+hyb]	Ex (Nen)	RCG, 1972
		Ngn	4	AA	Ex (UNK)	RCG, 1972
		pLm	28	AA [+hyb]	Ex (Nen)	MD, 2000
	San	surface	2	THq	Ex (NCa)	RS, 1970
		several	6	AA	In	RS, 1970
		Wus	19	AA	In	JCG, 1997
	She	Ema, Ton	3	AA	In	JeG-RS, 1967–1970
	Tau	Tav	42	AA	In	BFL, 1978
	Tik	Kik	6	BA	In	PVK, 1979
		Sin	6	AA	Ex (Vak/Bak)	PVK, 1979
			3	AA	Ex (San)	PVK, 1979
	Tor	surface	5	AA	Ex (San)	JCG, 1997
	Vak	surface	3	AA	In	RD, 1984
		Emo	2	AA	In (?)	PVK, 1979
			2	AA	Ex (San)	PVK, 1979
		Lav	2	AA	In	JCG, 2000
			1	AA	Ex (San)	JCG, 2000
Volcano, **2**	KIJ	surface	2	AA	In	JTa, 2005

Islands: Agu—Aguijan; Aiw—Aiwa; Ala—Alamagan; Alo—Alofi; Amb—Ambon; Ang—Angaur (Palau); Ani—Anir (=Ambitle) of Feni Islands (TLTF chain); Anu—Anuta; Ara—Arawe Islands (off New Britain); Ata—'Ata; Atu—Atafu; Aun—Aunu'u; Bab—Babeldaob (Palau) and neighboring offshore islets (Arakabesan, Koror, Namakal, Ngurur); Bak—Banks Islands; Ban—Banda of Banda Archipelago; Bas—Babase of Feni Islands (TLTF chain); Bal—Baluan (Admiralty Group); Bel—Bellona; Beq—Beqa; Bod—Boduna; Bou—Bougainville; Buk—Buka; Bul—Bulia; Bur—Buru; Che—Chelbacheb Group (Palau); Cik—Cikobia; Dau—Dauar; DoY—Duke of York Islands; Efa—Efate; Elo—Eloaua (Mussau Group); Ema—Emananus (Mussau Group); Err—Erromango; Eua—'Eua; Fai—Fais; Fef—Fefan (Chuuk = Truk); Foa—Foa (Ha'apai Group); Fut—Futuna; Gar—Garua; Goo—Goodenough; Gor—Gorom (=Gorong); GrT—Grand Terre (main island); Gua—Guam; Hal—Halmahera; Har—Haruku; Hiv—Hiva Oa; Hno—Ha'ano (Ha'apai Group); Hun—Hunga (Vava'u Group); IdP—Île des Pins; Kab—Kabara; Kad—Kadavu; Kap—Kapa (Vava'u Group); Kay—Kayangel (Palau); KIJ—Kita Iwo Jima; Kol—Kolombangara (New Georgia Group); Kos—Kosrae (= Kusaie); Kot—Kotu (Ha'apai Group); Lac—Laucala; Lak—Lakeba; Lam—Lamotrek; Lif—Lifuka (Ha'apai Group); Lou—Lou (Admiralty Group); Mab—Mabuiag; Mag—Mago; Man—Manus (Admiralty Group); Mar—Maré; Mat—Matuku (Ha'apai Group); Mau—Mauke; MBu—M'Buke (Admiralty Group); Mer—Maer; Mlk—Malakula and adjacent offshore islets; Mlo—Malo; Mot—Moturiki; Muy—Muyua (=Woodlark); Nae—Nanumea; Nag—Naigani; Nan—Nananu; Nau—Nanumanga; Nay—Nayau; NBr—New Britain; NeG—New Georgia; Nen—Nendö; Ngu—Ngulu; NIr—New Ireland; Nis—Nissan; Niu—Niuatoputapu; Nom—Nomuka (Ha'apai Group); Nuk—Nuku Hiva; Ofu—Ofu (Manu'a Group); OnJ—Ontong Java atoll; Ono—Ono; Ouv—Ouvea; Pag—Pagan; Pap—Papua (New Guinea); Pel—Peleliu (Palau); Poh—Pohnpei (= Ponape); Ree—Reef Islands; Roa—Rota; Rou—Rotuma; Sai—Saipan; San—Santo; Sap—Saparua; Sav—Savai'i; Ser—Seram; Sia—Siassi Islands (off Umboi); She—Shepherd Islands; Sho—Shortland Islands; Son—Sonsorol Islands; StA—Santa Ana; Tan—Tanga Islands (TLTF chain); Tau—Taumako; Tav—Taveuni; Tik—Tikopia; Tin—Tinian; Tor—Torres Islands; Tot—Totoya; Tro—Trobriand Islands; Ttu—Tongatapu; Tun—Tungua (Ha'apai Group); Tut—Tutuila; UaH—Ua Huka; Uga—Ugaga; Uih—'Uiha (Ha'apai Group); Uli—Ulithi; Upo—Upolu; Uve—Uvea (= Wallis); Vai—Vaitupu; VaB—Vanua Balavu; Vak—Vanikolo; VaL—Vanua Levu; ViL—Viti Levu; Wat—Watom (off New Britain); Yap—Yap (four islets); Yas—Yasawa.

Sites (N/A—specific locale not available; several—from several excavated sites; surface—surface collection from unexcavated sites): Adw—Adwe; Air—Airai State; Ait—Aitape coastline (multiple sites); Alu—Alu (Shortland) Island; Ana—'Anatu; Ano—Anaio; Aoa—'Aoa Valley; Apa—Apansantate; Apo—Apalo; Asm—Asmatmos fishing cliff; Ato—Atuona Valley; Avn—Avnitare; Ay—Ay islet; Bal—Balbalankin; Boe—islet of Boeng; Bol—Bolipi; Bou—Bourewa; BN—islet of Banda Naira; Cas—Casino; Cha—Chalan Piao; Ema—Emae; Emo—Emo; Eta—Etapakengaroasa; Fak—Fakatafenga; Fal—Faleloa; Fan—Fana Island; Fer—Feru rockshelter; Fis—Fissoa; Fou—Fouha Bay; Gac—Gachpar; Gir—Giru Gaja; Gra—Graciosa Bay; Haa—Ha'atuatua; Hah—Hahula; Han—Hane; Hol—Holopeka; Hom—Ho'oumi; Huo—Huon Peninsula (multiple sites); IBB—IBB operations site; Ifo—Ifo; Ipa—Ipan Beach; Jan—Jane's Camp; Jur—Jurlay; Kao—Karobo; Kah—Kahet; Kam—Kamgot; Kap—Kapume; Kar—Karkur; Kik—Kiki; Kir—Kiriwina Island; Kor—islet of Koror; Kre—Kreslo; Kul—Kulu Bay; Kur—Kurin; Lag—Laguas; Lap—Lapita; Las—Lasigi; Lav—Lavaka; Lel—islet of Lelu; Lew—Lewaki Levu; Los—Lossu (Lesu); Lua—islet of Luangiua; Maa—Mangai'a; Mad—islet of Malendok; Mae—Malae'imi Valley; Mag—Mangaasi; Mai—Malai Island; Mak—Maka Bay; Mal—Malau (islet of Mali); Mam—Mamalua (same as Anatale = Anakele); Mao—Malekolon; Mas—Mask Cave; Mat—Matanamuani; Mau—Malua Bay; Mei—Meinakapa; Mel—Mele Plain; Mek—Melekeok; Moc—Mochong; Mog—Mogmog; Mon—Mono Island (Treasury Group); Mou—Mouly; Mua—Mu'a; Mul—Mulifanua; Nab—Naitabale; Nad—Nan Madol; Nan—Nanggu; Nas—Nasilai (Rewa Delta); Nat—Natunuku; Nav—Navatu; NEC—northeast coast; Ngc—Ngchesar State; Ngm—Ngambelipa; Ngn—Ngangaua; Ngr—Ngaraard State; Ngt—Ngatpang State; Ngw—Ngiwal State; NLi—Nabulei Lisa; Nom—Nomna Bay; Nuk—Nukuleka; Num—beach at mouth of Numbulevu Creek in cove east of Tiliva; Nuu—Nuku; Nvp—Navaprah; Nvu—Navua Delta; NWC—northwest coast; Oma—Oma; Orm—Ormi; Ord—Ordnance Annex; Orr—islet of Orrak; Pak—islet of Pakea; Pap—Papakula; Par—Paradise; Pau—Paubake (near Buin); Pea—Pea; Pel—islet of Pelau; Pem—Pemrang; Pid—Pidia; Pla—Pulantat; Ple—Pulemelei; pLm—post-Lapita mound; Pon—Ponamla; Puk—Pukotala; Qaq—Qaqaruku rockshelter; Reb-Rak—Reber-Rakival; Reg—Regusa; Rot—Rotoapi; Rov—Roviana Lagoon; Sab—Sabaig; Saf—Sabanan Fiang; San—Sanmeima; Sap—Sapota; Sas—Sasoa'a (Falefa Valley); Sau—Saulevu; Sep—Seputu; Sig—Sigatoka; Sik—Sikumango; Sin—Sinapupu; Taf—Talafofo; Tag—Taga; Tai—Tavai; Tal—Talepakemalai; Tar—Tarague; Tat—Tataga-matau; Tav—Tavatava; Tem—Temei; Ten—Tenmiel; Teo—islet of Teop; Tog—To'aga; Ton—Tongoa; Tre—islet of Trevanion; Tua—Tuam Island; Tum—Tumon Bay; Ulo—islet of Ulong; Ulu—Ulunikoro hillfort; Una—Unai Achugao; Unb—Unai Bapot; Unc—Unai Chulu; Vac—Vatcha; Val—Vailele; Vao—islet of Vao; Vap—Vaipuna; Vat—Va'ota; Vau—Vaturekuka; Vei—Veitongo; Vel—Vele; Vot—Votua; Vud—Vuda; Wab—Wabao; Wan—Wangil; Wat—Waterfront Annex; Wus—Wusi (and west coast sites); Yad—Yadua; Yan—Yanuca; Zom—Zomar.

Temper class: AA—andesitic arc (AAq—quartzose variants); BA—backarc (BAq—quartzose variants); DO—dissected orogen; OB—oceanic basalt (OBt—trachytic variants); PA—post-arc; TH—tectonic highland (THq—quartzose variants; THo—ophiolitic variants). Annotations in brackets: cal—dominantly calcareous; com—composite temper with admixed grog particles; gro—dominantly grog particles; hyb—hybrid sand temper with admixed calcareous grains.

Temper origin (In—indigenous [In-nl—nonlocal]; Ex—exotic, with site or island or island group of origin given in parenthesis): Bab—Babeldaob or neighboring offshore bedrock islets; Ban—Banda; Bak—Banks Islands; Beq—Beqa; Bui—Buin (Bougainville); Buk—Buka; Cho—Choiseul; Dee—D'Entrecasteaux Islands; Fij—Fiji (locality unknown); Gua—Guam; Huo—Huon Peninsula (Papua New Guinea); Kad—Kadavu; Ked—"kedekede" temper probably from Kanacea (Lau); Kie—Kieta (Bougainville); Kul—Kulu Bay on Beqa; Lak—Lakeba (Lau); Lau—Lau (locality unknown); Lou—Lou or neighboring rhyolitic islet (Admiralty Group); Lea—Lease Islands (northern Banda arc); Mak—Makira (=San Cristobal); Man—Manus (Admiralty Group); Mlk—Malakula; M/M—Moala or Matuku; Nav—Navatu on Viti Levu; NBr—New Britain; NCa—New Caledonia; NeG—New Georgia; Nen—Nendö; NGH—New Guinea Highlands (Papua New Guinea); NIr—New Ireland; Nvu—Navua Delta on Viti Levu; Ova—Ovalau (or Moturiki); Pal—Palau (Babeldaob or nearby bedrock islet); PP—Papuan Peninsula; Rew—Rewa Delta on Viti Levu (Fiji); Sai—Saipan; San—Santo; Ser—Seram; She—Shepherd Islands; TLTF—Tabar-Lihir-Tanga-Feni chain; Ton—Tonga; Udu—Udu Peninsula on Vanua Levu; UNK—unknown; Vag—Vaga Bay on Beqa; Vak—Vanikolo; VaL—Vanua Levu; Van—Vanuatu; ViL—Viti Levu; Yap—Yap.

Collector or provider, alphabetical by initial letter of last name (secondarily by initial letter of first name) of individuals (initials in capitals) or first word of museum name (in lowercase), with institutional affiliations (or domiciles) at present (where known), or as current when sherds or sherd thin sections were received for personal study:

am—American Museum of Natural History, New York, New York
AA—Atholl Anderson, Australian National University, Canberra
DA—Dimitri Anson, University of Otago (Dunedin, New Zealand)
JSA—J. Stephen Athens, International Archaeological Research Institute (Honolulu, Hawaii)
WRA—Wallace R. Ambrose, Australian National University (Canberra)
bm—Bernice P. Bishop Museum, Honolulu, Hawaii
BMB—Brian M. Butler, Southern Illinois University (Carbondale)
DB—Diana Brown, Neiafu, Vava'u Group, Tonga
DVB—David V. Burley, Simon Fraser University (Canada)
LB—Lon Bulgrin, Division of Historic Preservation (Saipan)
LHB—Lawrence and Helen Birks, Auckland, New Zealand
SBd—Stuart Bedford, University of Auckland (New Zealand)
SBe—Simon Best, University of Auckland (New Zealand)
SBi—Simon Bickler, University of Auckland (New Zealand)
SBk—Stephen Black, Field Museum of Natural History (Chicago, Illinois)
TB-S—Tim Bayliss-Smith, Cambridge University (UK)
GC—Geoffrey Clark, Australian National University (Canberra)
JTC—Jeffrey T. Clark, North Dakota State University (Fargo)
MC—Melissa Carter, James Cook University (Townsville, Australia)
MTC—Mike T. Carson, International Archaeological Research Institute (Honolulu, Hawaii)
SC—Scarlett Chiu, University of California (Berkeley)
BD—Boyd Dixon, International Archaeological Research Institute (Honolulu, Hawaii)
CD—Christophe Descantes, University of Oregon (Eugene, Oregon)
JD—Janet Davidson, University of Auckland (New Zealand)
MD—Moira Doherty, University of Auckland (New Zealand)
RD—Reece Discombe, Port Vila, Vanuatu
WRD—William R. Dickinson, University of Arizona (Tucson)
DDeF—David DeFant, Paul H. Rosendahl Incorporated (Guam)
MF—Matthew Felgate, University of Auckland (New Zealand)
SMF—Scott M. Fitzpatrick, North Carolina State University (Raleigh)
fm—Directors of the Fiji Museum, Suva (Fiji)
CG—Chris Gosden, Pitt Ropers Museum, Oxford University (UK)
JCG—Jean-Christophe Galipaud, ORSTOM, Noumea (New Caledonia)
JGa—Jean Garanger, Université de Paris (France)
JGo—Jack Golson, Australian National University (Canberra)
RCG—Roger C. Green, University of Auckland (New Zealand)
SG—Stephanie Garling, Australian National University (Canberra)
CH—Charles Hunt, Fiji Museum (Suva)
EH—Elizabeth Hinds (nee Shaw), University of Auckland (New Zealand)
JDH—John D. Hedrick, University of Pennsylvania (Philadelphia)
RH-A—Rosalind Hunter-Anderson, Micronesian Archaeological Research Services (Guam)
TLH—Terry L. Hunt, University of Hawaii (Honolulu)
WWH—William W. Howells, Harvard University (Cambridge, Massachusetts)
GJI—Geoffrey J. Irwin, University of Auckland (New Zealand)
MI—Michiko Intoh, Hokkaido Tokai University (Sapporo)
JDJ—Jesse D. Jennings, University of Utah (Salt Lake City)
JK—Jean Kennedy, Australian National University (Canberra)
PVK—Patrick V. Kirch, University of California (Berkeley)
SK—Susan Kaplan, Field Museum of Natural History (Chicago, Illinois)
BFL—B. Foss Leach, University of Otago (New Zealand)
IL—Ian Lilley, University of Queensland (Australia)
PVL—Peter V. Lape, Burke Museum, University of Washington (Seattle)
TNL—Thegn N. Ladefoged, University of Auckland (New Zealand)
lm—Lowie (now Hearst) Museum of Anthropology, University of California (Berkeley)
DRM—Darlene R. Moore, Micronesian Archaeological Research Services (Guam)
PM—Patricia Martz, California State University (Los Angeles)
PCMcC—Patrick C. McCoy, Bishop Museum (Honolulu, Hawaii)
IMcN—Ian McNiven, Monash University (Clayton, Victoria, Australia)
PDN—Patrick D. Nunn, University of the South Pacific (Suva)
DO—Douglas Osborne, California State University (Long Beach)
RO—Richard Olmo, International Archaeological Research Institute (Guam)
SO'C—Sue O'Connor, Australian National University (Canberra)

SJO'D—Sandra J. O'Day, University of Florida (Gainesville)
AP—Aubrey Parke, Australian National University (Canberra)
FP—Fiona Petchey, University of Auckland (New Zealand)
JP—Jens Poulsen, University of Aarhus (Denmark)
SP—Sarah Phear, Australian Natural University (Canberra)
ERR—Erwin R. Ray, Tempe, Arizona
FMR—Fred M. Reinman, Field Museum of Natural History (Chicago, Illinois)
GR—Garth Rogers, University of Auckland (New Zealand)
TR—Tim Rieth, International Archaeological Research Institute (Honolulu, Hawaii)
CS—Christophe Sand, New Caledonia Museum (Noumea)
DWS—David W. Steadman, University of Florida (Gainesville)
ES—Epi Suafo'a, National Park of Samoa (Pago Pago)
GS—Glenn Summerhayes, Australian National University (Canberra)
JAS—Jeannette A. Simons, Department of the Navy, Honolulu, Hawaii

JS—Jim Specht, Australian Museum (Sydney)
MES—Mary Elizabeth Shutler, Cypress College (California)
MSp—Matthew Spriggs, Australian National University (Canberra)
MSw—Marilyn Swift, Swift and Harper Archaeological Research Consulting (Saipan)
RS—Richard Shutler Jr., Simon Fraser University (Canada)
YHS—Yosihiko H. Sinoto, Bishop Museum (Honolulu, Hawaii)
JRT—Jo-Anne R. Thomson, Sydney University (Australia)
JTa—Jun Takayama, Tokai University (Tokyo)
JTe—John Terrell, Field Museum of Natural History (Chicago, Illinois)
RT—Robin Torrence, Sydney University (Australia)
EW—Eleanor Wells, Micronesian Archaeological Research Services (Guam)
GW—Graeme Ward, AIATSIS, Canberra (Australia)
JPW—J. Peter White, Sydney University (Australia)
RW—Richard Walter, University of Otago (New Zealand)
DEY—Douglas E. Yen, Bishop Museum (Honolulu, Hawaii)

APPENDIX 2. SHERD PHOTOMICROGRAPHS

For Figures A1–A30: PL—plane light; XN—crossed Nicol prisms; FoV—field of view (dimensions in mm).

Nonterrigenous and Nondetrital Tempers

Figure A1. Calcareous temper, Sapota site, Fefan Island, Chuuk (= Truk): XN, FoV 1.0 × 1.5 mm. Note extreme birefringence (high-order interference tints) of calcareous temper grains (rounded beach sand) set in dark clay paste (internal bioclastic texture not well shown under crossed Nicols).

Figure A2. Grog (broken-sherd) temper, Ngargasang site, Babeldaob Island, Palau (= Belau): PL, FoV 1.0 × 1.5 mm. Discontinuities in coloration of clay paste delimit edges of grog particles (similar texture but lighter or darker than background clay paste)

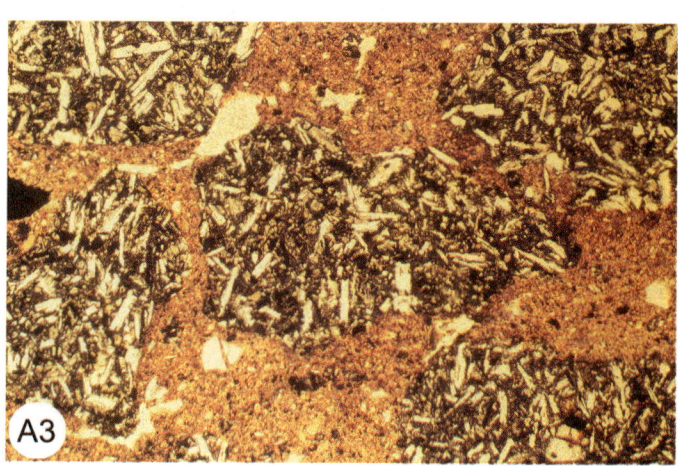

Figure A3. Crushed-rock temper, Sasoa'a site (Falefa Valley), Upolu Island, Samoa: PL, FoV 1.0 × 1.5 mm. Note identical lithology and superangular nature (ragged edges) of volcanic lithic fragments (internally polycrystalline and polymineralic) forming five temper grains set in silty brown clay paste.

Volcanic Lithic Fragments

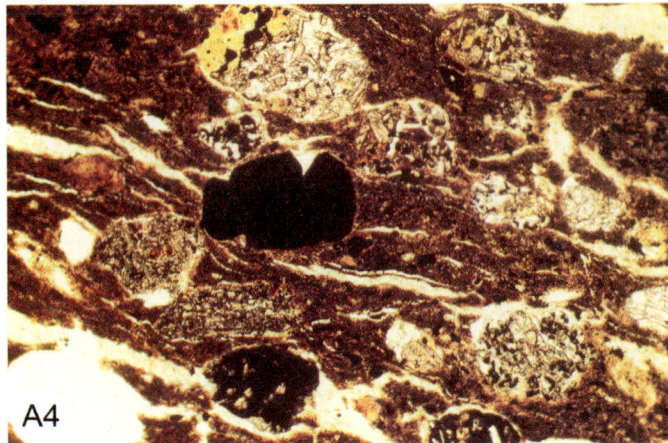

Figure A4. Oceanic basalt temper class, Mulifanua site, Upolu Island, Samoa: PL, FoV 1.0 × 1.5 mm. Note heterogeneity of multiple polycrystalline-polymineralic volcanic lithic fragments (dominantly lathwork but also sparse vitric grains) in beach sand temper. Contrast dark microlitic lithic fragments (semiopaque tachylitic groundmasses) at bottom with opaque ferromagnesian mineral grain near center.

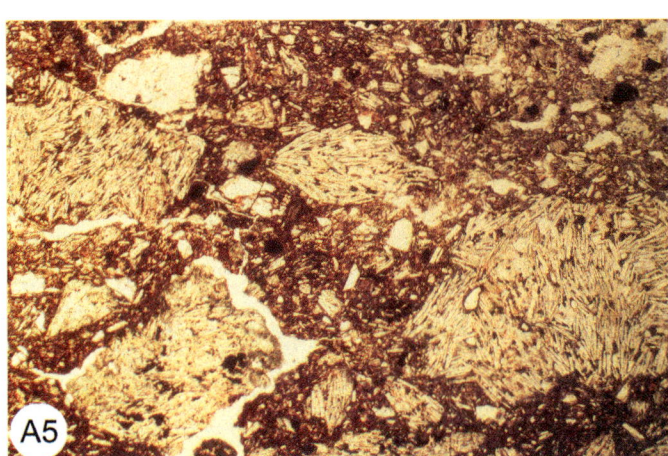

Figure A5. Oceanic basalt temper class (trachytic variant), Vailele site, Upolu Island, Samoa: PL, FoV 2 × 3 mm. Note microlitic internal texture of pilotaxitic volcanic lithic fragments (rich in flow-oriented feldspar microlites) forming poorly sorted alluvial or colluvial temper aggregate.

Figure A6. Andesitic arc temper class, Tongatapu Island, Kingdom of Tonga: PL, FoV 1.0 × 1.5 mm. Beach sand temper includes microlitic (hyalopilitic) volcanic lithic fragments with partly glassy groundmasses (three polycrystalline-polymineralic grains arranged diagonally across center of field of view), pyroxene mineral grains (pale green with high relief and distinct cleavage), and colorless plagioclase feldspar mineral grains (clear with low relief). Curvilinear and anastomosing white areas are cracks and pores in the sherd.

Figure A7. Andesitic arc temper class, Malakula sherd from offshore islet of Uripiv, Vanuatu: XN, FoV 1.0 × 1.5 mm. Temper sand is dominantly felsitic volcanic lithic fragments (internal polycrystalline quartz-feldspar mosaics) with green hornblende, twinned plagioclase, and opaque iron oxide grains also visible.

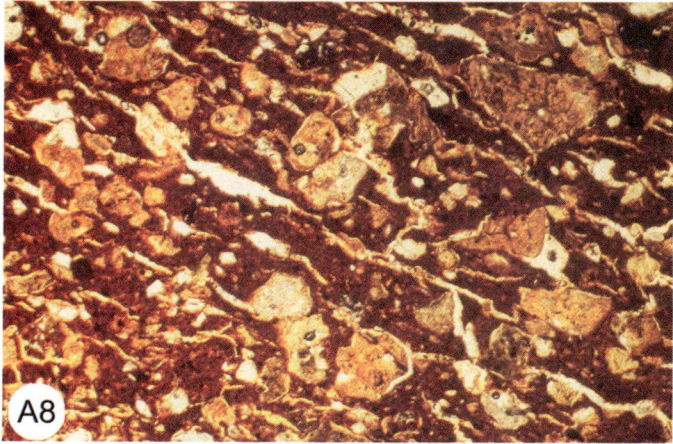

Figure A8. Andesitic arc temper class, Ipan site, Guam Island, Mariana Islands: PL, FoV 1.0 × 1.5 mm. Stream sand temper composed dominantly of variably rounded vitric volcanic lithic fragments (brown mafic volcanic glass).

Figure A9. Andesitic arc temper class, Niuatoputapu Island, Kingdom of Tonga: PL, FoV 0.4 × 0.6 mm. Visible temper grains are dominantly pale brown vitric volcanic lithic fragments with pumiceous (microvesicular) internal textures formed by elongate stretched vesicles (clear plagioclase feldspar mineral grain also present near center of field of view).

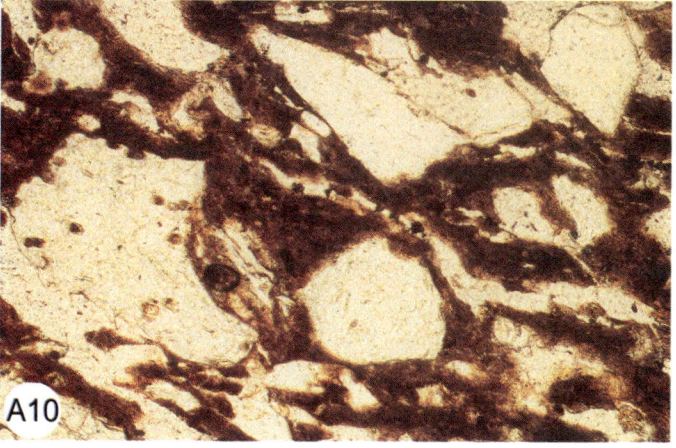

Figure A10. Andesitic arc temper class (quartzose variant), Hutumuri site, Ambon Island (Lease chain), Maluku (= Molucca Islands): PL, FoV 1.0 × 1.5 mm. Temper grains visible are dominantly internally heterogeneous vitric volcanic lithic fragments composed of pale felsic volcanic glass (nearly colorless in thin section), but two limpid volcanic quartz grains are also visible at upper center and upper right.

Nonvolcanic Lithic Fragments

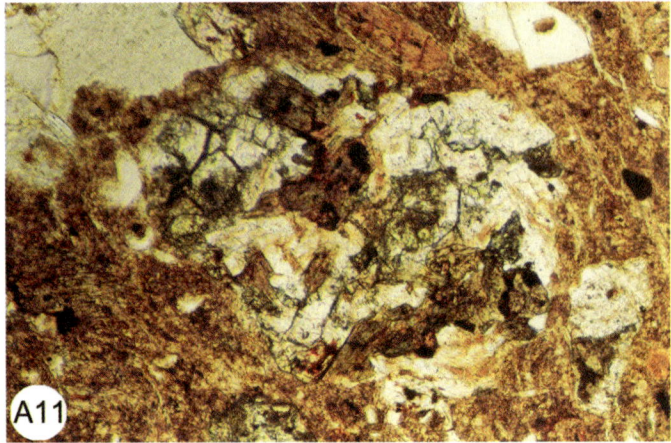

Figure A11. Postarc temper class, Navatu site, north coast of Viti Levu Island, Fiji: PL, FoV 0.4 × 0.6 mm. Microphanerite lithic fragment (nonvolcanic holocrystalline igneous internal texture of hypabyssal intrusive origin) centered in field of view.

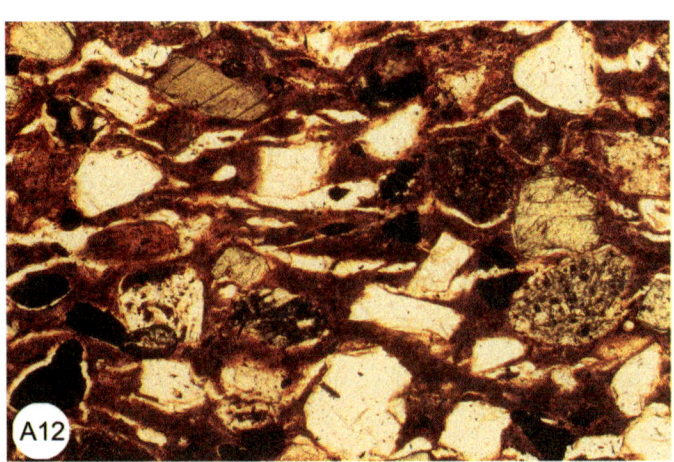

Figure A12. Dissected orogen temper class (heterogeneous sand derived from diverse bedrock sources), Nasilai site, Rewa Delta, Viti Levu Island, Fiji: PL, FoV 1.0 × 1.5 mm. Visible temper grains include colorless quartz and feldspar mineral grains (monocrystalline), tinted pyribole mineral grains (both pale green clinopyroxene and darker green hornblende), opaque iron oxide grains, and diverse polycrystalline-polymineralic lithic fragments.

Figure A13. Tectonic highland temper class (quartzose lithic variant), Lapita site near Koné, Grand Terre Island, New Caledonia: XN, FoV 1.0 × 1.5 mm. Temper grains are dominantly quartzose sedimentary-metasedimentary lithic fragments (monomineralic but polycrystalline), with a prominent monocrystalline quartz mineral grain visible in upper left. Aligned tiny mica flakes impart internal foliation to some lithic fragments.

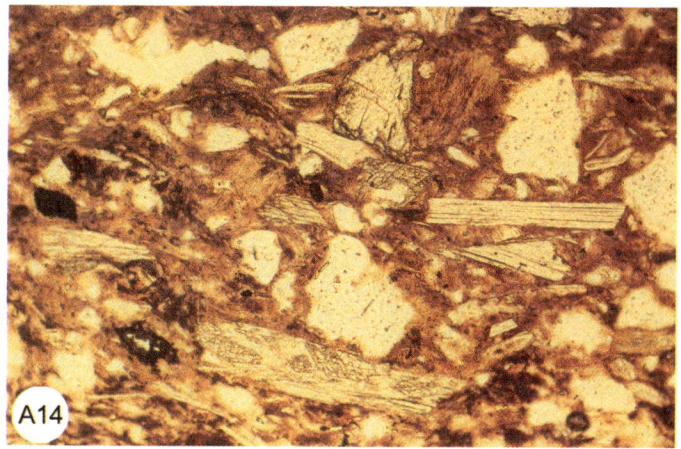

Figure A14. Tectonic highland temper class (gneissic source rocks), Goodenough Island, D'Entrecasteaux Islands, Papua New Guinea (Solomon Sea): PL, FoV 1.0 × 1.5 mm. Temper grains include colorless tabular muscovite mica flakes (example at center right), minor pale amphibole mineral grains with high relief (several examples in upper center), and varied quartzo-feldspathic mineral grains and metamorphic lithic fragments (the latter internally foliated) with low relief.

Nonplacer Temper Aggregates

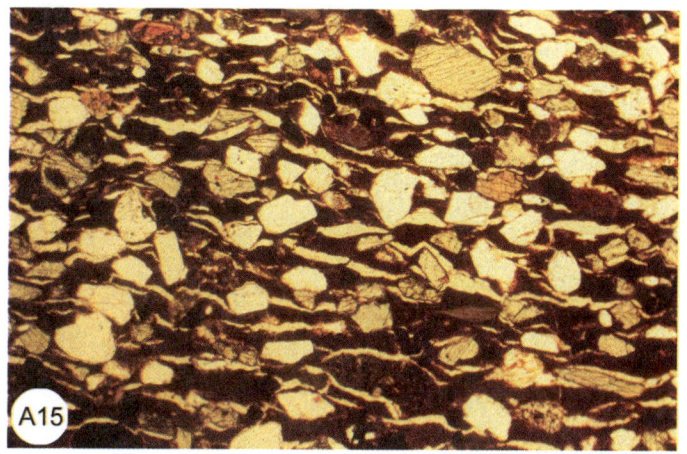

Figure A15. Andesitic arc temper class, Santo sherd from Avnitare site, islet of Malo, Vanuatu: PL, FoV 2 × 3 mm. Heterogeneous sand derived from volcanic rocks, with subequal proportions of colorless (clear) plagioclase and colored pyribole mineral grains (with both dominant pale green clinopyroxene and subordinate green-brown hornblende present).

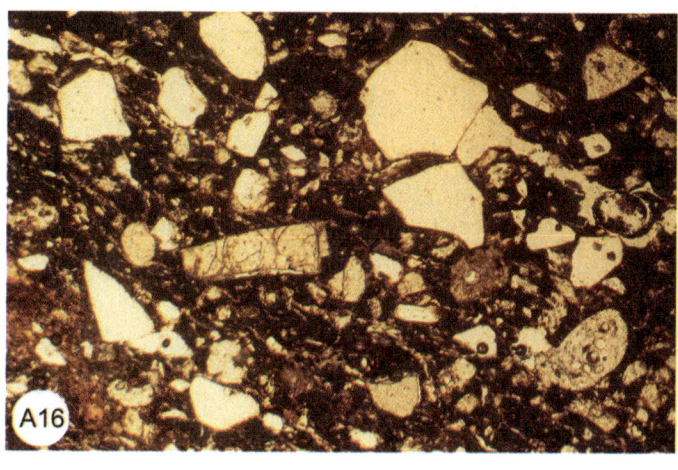

Figure A16. Andesitic arc temper class (quartzose variant), Sabanan Fiang site, Saipan Island, Mariana Islands: PL, FoV 1.0 × 1.5 mm. Temper grains are dominantly clear volcanic quartz mineral grains and internally more heterogeneous vitric volcanic lithic fragments composed of pale (nearly colorless) felsic volcanic glass (one glass particle near center displays internal microperlitic cracks).

Nonplacer Temper Aggregates (*continued*)

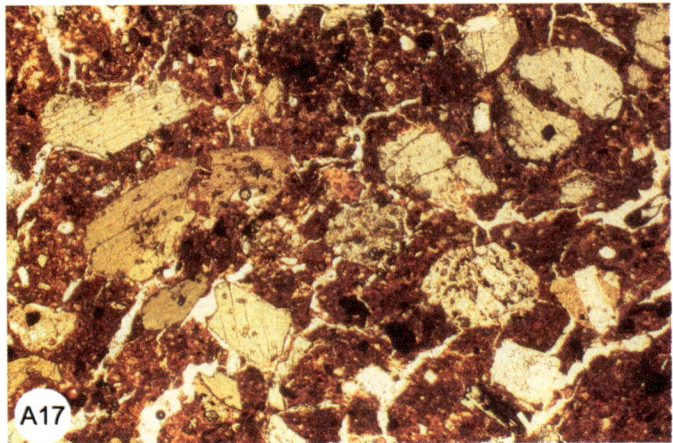

Figure A17. Postarc temper class, Natunuku site (Vatia temper), north coast of Viti Levu Island, Fiji: PL, FoV 1.0 × 1.5 mm. Temper grains include volcanic lithic fragments (polycrystalline microlitic), and clinopyroxene (pale green), hornblende (green-brown), and plagioclase (colorless) mineral grains.

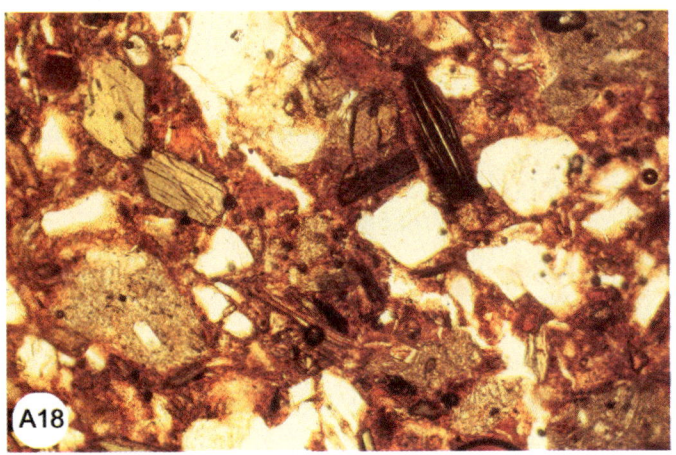

Figure A18. Andesitic arc temper class, exotic sherd from calcareous Trobriand Islands, Papua New Guinea (Solomon Sea): PL, FoV 0.4 × 0.6 mm. Note microlitic volcanic lithic fragment containing a prominent plagioclase microphenocryst on left below center, two green hornblende mineral grains in upper left, two dark biotite mica flakes (toward upper right from center), and multiple clear (colorless) plagioclase mineral grains.

Figure A19. Dissected orogen temper class (heterogeneous sand derived from diverse bedrock sources), Sigatoka dune site, south coast of Viti Levu Island, Fiji: PL, FoV 1.0 × 1.5 mm. Visible temper grains include pale (colorless) quartz and feldspar mineral grains, colored pyribole mineral grains (pale green clinopyroxene and darker green hornblende), opaque iron oxide grains, and sparse polycrystalline-polymineralic lithic fragments.

Placer Temper Aggregates

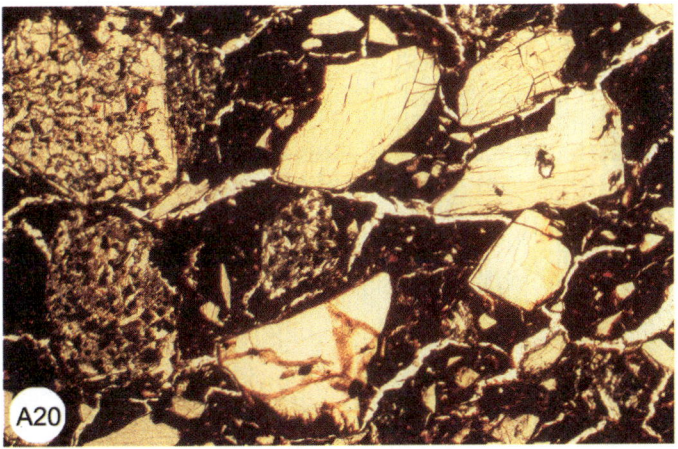

Figure A20. Oceanic basalt temper class, To'aga site, Ofu Island, American Samoa: PL, FoV 1.0 × 1.5 mm. Temper sand includes three clinopyroxene mineral grains (greenish and cleaved) in upper right, two olivine mineral grains (untinted except for partial reddish brown discoloration by growth of deuteric iddingsite) toward lower right, and lathwork volcanic lithic fragments (polycrystalline-polymineralic sand grains of heterogeneous intergranular to intersertal texture), two on left and a smaller one near center.

Figure A21. Andesitic arc temper class, Paubake-Buin site, Bougainville Island, Papua New Guinea (but part of Solomon island chain): PL, FoV 1.0 × 1.5 mm. Subequal proportions of pleochroic (pale to dark green) hornblende and colorless plagioclase mineral grains (subordinate clinopyroxene mineral grains of intermediate tone also visible).

Figure A22. Andesitic arc temper class, Manus Island sherd from M'Buke Island, Admiralty Group, Bismarck Archipelago, Papua New Guinea: PL, FoV 2 × 3 mm. Note abundant green-brown hornblende, common pale green clinopyroxene (see Fig. A24 for truer color), and sparse clear (colorless) plagioclase feldspar mineral grains.

Placer Temper Aggregates (*continued*)

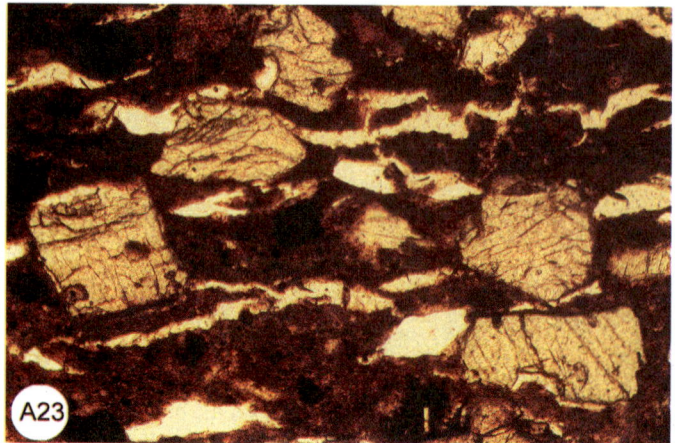

Figure A23. Andesitic arc temper class (placer sand), Nendö Island sherd from calcareous Ngangaua Island, Reef Islands of Santa Cruz Group (politically part of Solomon Islands, but geologically part of New Hebrides island arc of Vanuatu): PL, FoV 0.4 × 0.6 mm. Temper sand is dominantly pale green clinopyroxene mineral grains (see Fig. A24 for truer color).

Figure A24. Andesitic arc temper class (placer sand), Tongatapu Island, Kingdom of Tonga: PL, FoV 1.0 × 1.5 mm. Temper sand is dominantly pale green clinopyroxene mineral grains.

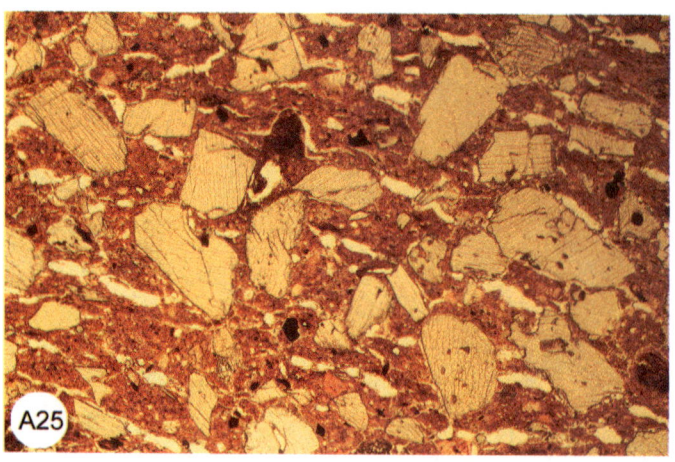

Figure A25. Postarc temper class (placer sand), Natunuku site (Tavua temper), north coast of Viti Levu Island, Fiji: PL, FoV 1.0 × 1.5 mm. Temper sand is dominantly pale green clinopyroxene mineral grains (see Fig. A24 for truer color).

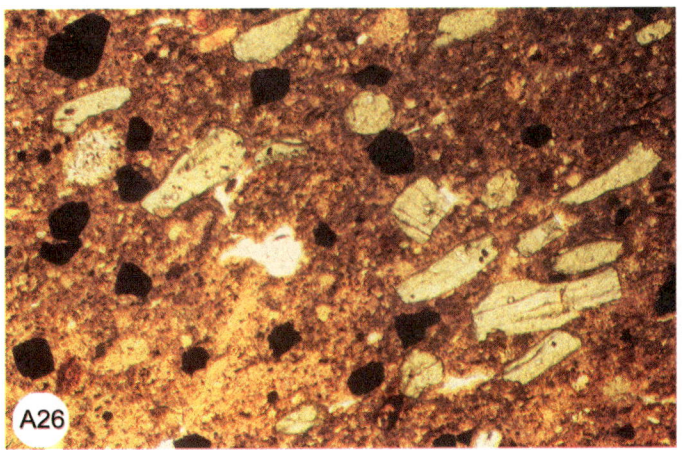

Figure A26. Postarc temper class (placer sand), Amfuli site, islet of Malendok (Tanga Islands), TLTF chain, Bismarck Archipelago, Papua New Guinea: PL, FoV 1.0 × 1.5 mm. Subequal proportions of transparent (green) clinopyroxene (aegirine-augite) mineral grains and opaque iron oxide grains.

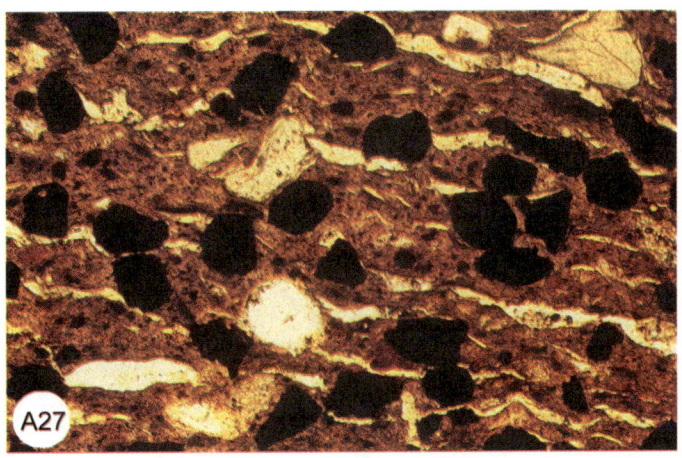

Figure A27. Andesitic arc temper class (placer sand), New Britain sherd from Apalo site, calcareous Arawe Islands off New Britain, Bismarck Archipelago, Papua New Guinea: PL, FoV 1.0 × 1.5 mm. Temper sand is dominantly opaque iron oxide grains.

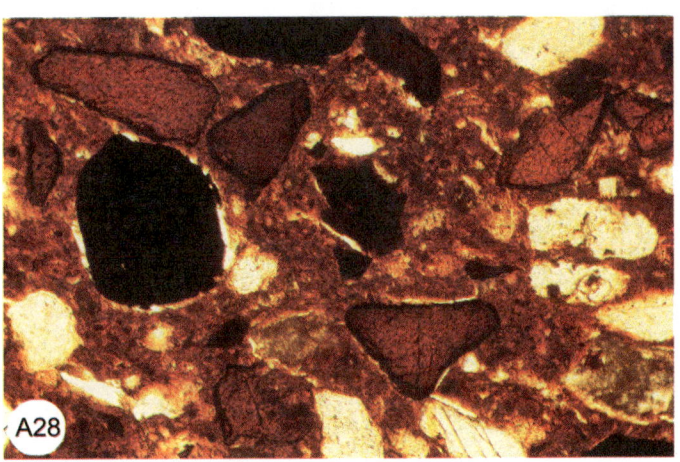

Figure A28. Tectonic highland temper class (ophiolitic variant), exotic sherd of New Caledonia origin from Avnitare site, Malo islet off Santo, Vanuatu: PL, FoV 0.4 × 0.6 mm. Visible temper grains include transparent plagioclase and clinopyroxene mineral grains, opaque (black) iron oxide grains, and translucent red chrome-spinel mineral grains (the latter derived from the peridotite massif of New Caledonia).

Anomalous Temper and Paste

Figure A29. Exotic quartz-calcite temper of unknown derivation in sherd from Roviana Lagoon, New Georgia Group, Solomon Islands: XN, FoV 1.0 × 1.5 mm. Internally clear quartz grains display white to yellow first-order interference tints, and internally murky calcareous grains of reef detritus display extreme birefringence (high-order interference tints).

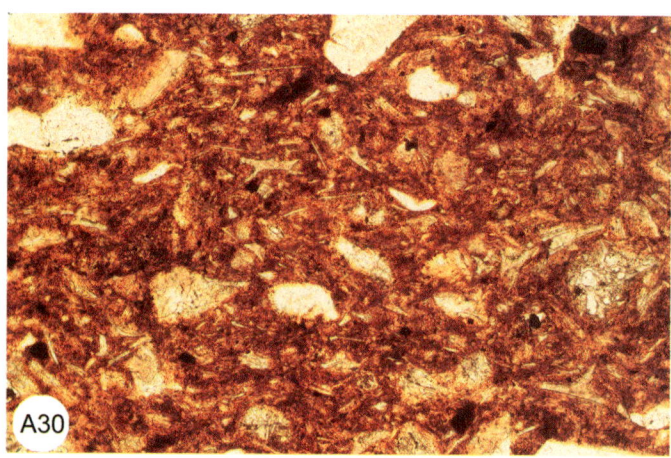

Figure A30. Shard-bearing clay paste in exotic Banda sherd from Wangil site, Wammar Island, Aru Islands (Arafura Sea): PL, FoV 0.4 × 0.6 mm. Colorless to tan glass shards (branching or arcuate in shape) of volcanic ash embedded in murky clay paste.

APPENDIX 3. INDEX TO ISLANDS, ISLAND GROUPS, AND ARCHAEOLOGICAL SITES

Typography: (1) islands, compact island clusters, and key peninsulas in normal typescript, major regional island groups in **boldface**, archaeological sites in *italics*; (2) text page numbers in normal typescript, figure numbers (not page numbers) in **boldface** (prefix **A** denotes figures in Appendix 2), text table numbers (not page numbers) in *italics (fn* refers to footnotes of selected tables); (3) letter codes used for individual islands and sites of the sherd catalogue (Table A1 of Appendix 1) given in parenthesis.

Index

A
Admiralty Group, 52, 75, 76, 104, 112, 113, 147, 148, **20, 31, 37, 38, A22**, *13A, 16A, 25CE*
Adwe (Adw), 55, 113, 139, 147
Aguijan (Agu), 41, 42, 107, 143, 147, **13B**, *25D*
Agrihan, 44, **13A**
Ahus islet, 113
Airai State (Air), 140, 147
Aitape coastline (Ait), 88, 139, **4**
Aiwa (Aiw), 71, 117, 142, 147, **27, 28**, *13D, 15, 25E*
Alamagan (Ala), 42, 107, 143, 147, **4, 13A**, *25B*
Alofi (Alo), 79, 80, 144, 147, **31, 38**, *16A*
Alu islet (Alu), 58, 144, 147, **21B, 22**, *25E*
Ambitle, see Anir
Ambon (Amb), 47, 50, 98, 110, 111, 143, 147, **4, 18, A10**, *25E*
Ambrym, 62, **23**
American Samoa, 2, 30, 32, 34, 36, 37, 38, 138, **8, 11, 12, A20**, *7, 8, 24A, 24fn*
Amfuli, **A26**
Anaio, 119
Anatom, 62, **23**
'Anatu (Ana), 145, 147, **29**
Angaur (Ang), 46, 47, 107, 140, 147, **16**, *12, 25D*
Anir (Ani), 76, 78, 79, 113, 139, 147, **2, 31, 37**, *16A, 24G, 25E*
Anuta (Anu), 79, 107, 146, 147, **4, 23**
'Aoa Valley (Aoa), 37, 144, 147, **11, 12**, *7*
Aoba, 62, **23**, *13fn*
Apalo (Apo), 55, 113, 139, 147, **A27**
Apansantate (Apa), 143, 147
Arawe Islands (Ara), 55, 56, 89, 112, 113, 139, 147, **2, 19, 20, 35, A27**, *13A, 18A, 24E, 25D*
Aru Islands, 2, 47, 96, 98, 109, 110, 139, **4, 6A, 17, 18, A30**, *25F, 26*
Asio, **19**
Asmatmos (Asm), 143, 147
'Ata (Ata), 73, 75, 145, 147, **8, 29, 30, 37**, *13D, 24B*
Atafu (Atu), 117, 145, 147, *25A*
Atchin, 66

Atiu, 107
Atuona Valley (Ato), 37, 38, 143, 147, *9*
Aunu'u (Aun), 36, 144, 147, **11**, *7*
Australia, 5, 7, 14, 88, 96, 98, 109, 111, 115, **1, 4, 7, 9**
Avnitare (Avn), 146, 147, **A15, A28**
Ay islet (Ay), 143, 147, *17*

B
Babase (Bas), 76, 78, 139, 147, **2, 31, 37**, *16A*
Babeldaob (Bab), 46, 47, 107, 140, 147, 148, **4, 16, 17, A2**, *12, 24D, 25D*
Balbalankin (Bal), 76, 113, 139, 147, **2**, *16A*
Baluan (Bal), 75, 139, 147, **19**, *16A, 25E*
Banda (Ban), 2, 14, 39, 47, 50, 103, 109, 110, 139, 143, 147, 148, **1, 4, 6A, 17, 18, A30**, *24D, 25F, 26*
Banda Naira islet (BN), 143, 147
Banks Islands (Bak), 63, 66, 107, 116, 146, 147, 148, **23, 24**, *13C, 13fn, 14, 24D, 25E*
Belau, see Palau
Bellona (Bel), 115, 144, 147, **4, 7, 22**, *25A*
Beqa (Beq), 80, 83, 85, 86, 93, 104, 116, 117, 141, 147, 148, **2, 26, 34, 36, 37, 38, 39**, *16C, 17, 18C, 24FG, 25DE*
Bismarck Archipelago, 2, 5, 7, 8, 10, 18, 39, 50, 56, 58, 63, 76, 88, 89, 101, 103, 104, 112, 113, 114, 115, 139, **1, 2, 3, 4, 5, 7, 9, 19, 20, 31, 35, 37, A22, A26, A27**, *13A, 16A, 18A, 24C-G, 25CE*
Boduna (Bod), 55, 56, 101, 113, 139, 147, **2, 19, 20, 37**, *13A, 24C, 25D*
Boeng islet (Boe), 139, 147
Bolipi (Bol), 140, 147
Bonin Islands, 14, 39, 40
Bougainville (Bou), 58, 60, 103, 114, 144, 147, **4, 19, 21A, 21B, 22, 37, A21**, *13B, 24FG, 25DE*
Bourewa (Bou), 93, 142, 147, **26**
Buin (Bui), 58, 60, 145, 148, **19, 21B, 22, A21**, *13B*
Buka (Buk), 58, 60, 104, 114, 139, 144, 147, 148, **2, 3, 4, 19, 21B**, *25AD*
Bulia (Bul), 70, 141, 147, **26, 27**, *13D*
Buru (Bur), 96, 97, 98, 143, 147, **4, 6A, 18**

C
Caroline Islands, 2, 10, 21, 30, 31, 107, 108, 140, 141, **1, 3, 5, 7, 10**, *25AD*
Casino (Cas), 143, 147
Chalan Piao (Cha), 143, 147
Chelab, 47, **17**
Chelbacheb (Che) Group, 140, 147, **16**
Choiseul (Cho), 91, 92, 104, 114, 144, 148, **19, 21A, 35, 37, 38, 39**, *18B, 19, 25E*
Chuuk, 30, 31, 147, **3, 4, 7, 10**, *A1*
Cikobia (Cik), 70, 85, 86, 116, 118, 141, 147, **2, 8, 26, 27, 32, 37, 38**, *13D, 16B, 17, 24D, 25D*
Cook Islands, 107, 119, 141, **1, 30, 40**, *13D, 25B*

D
Dauar (Dau), 143, 147, *25B, 27*
D'Entrecasteaux Islands, 2, 96, 98, 99, 104, 114, 141, 145, 148, **4, 7, 19, A14**, *25A*

Duff Islands, see Taumako
Duke of York Islands (DoY), 52, 56, 91, 112, 139, 147, **2, 19, 20, 35**, *13A, 18A, 25E*

E
Efate (Efa), 66, 67, 107, 115, 116. 146, 147, **2, 23, 25**, *13C, 13fn, 14, 24D, 25DE*
Eloaua (Elo), 113, 139, 147, **2, 19**
Emae islet (Ema), 67, 146, 147, **23**, *13C*
Emananus (Ema), 113, 139, 147
Emo (Emo), 146, 147
Epi, **23**
Eretoka, 67
Erromango (Err), 20, 62, 63, 67, 115, 146, 147, **2, 23, 25, 37**, *13C, 14, 24B, 25C*
Erueti, **2**
Etapakengaroasa (Eta), 139, 147
'Eua (Eua), 73, 75, 104, 119, 145, 147, **29, 30, 39**, *13D, 25E*

F
Fais (Fai), 47, 107, 108, 140, 147, **4, 6A, 40**, *11, 12, 21, 25AD*
Fakatafenga (Fak), 145, 147
Faleloa (Fal), 145, 147, **29**
Fana islet (Fan), 47, 107, 140, 147, **4**, *12, 25D*
Fauro, 58, **21B**
Fefan (Fef), 30, 31, 140, 147, **4**, *A1*
Feni Islands, 75, 76, 78, 79, 113, 147, **19, 31**, *16A, 24D, 24G, 25E*
Fergusson, 99, **19**
Feru (Fer), 115, 145, 147
Fiji (Fij; Lau indexed separately), 2, 5, 8, 20, 39, 67, 70, 75, 79, 80, 85, 86, 87, 88, 89, 92, 93, 95, 101, 103, 104, 107, 116, 117, 118, 119, 138, 141, 142, 145, 148, **1, 2, 3, 4, 5, 8, 9, 26, 27, 32, 34, 36, 37, 39, 40, A11, A12, A17, A19, A25**, *13D, 16B, 16C, 17, 18C, 24B-H, 25A-E, 28*
Fissoa (Fis), 139, 147, **19, 20**, *13A*
Foa (Foa), 145, 147, **29**
Fouha Bay (Fou), 143, 147
Futuna (Fut), 79, 144, 147, **2, 3, 4, 8, 31, 38**, *16A*

G
Gachpar (Gac), 141, 147
Gagil-Tomil, 96, 97, **15**
Garua (Gar), 55, 56, 101, 113, 139, 147, **2, 19, 20, 37**, *13A, 24C, 25D*
Giru Gaja (Gir), 143, 147, **26**
Goodenough (Goo), 98, 99, 141, 147, **19, A14**
Gorom (Gor), 97, 98, 109, 110, 143, 147, **4, 17, 18**, *25F, 26*
Graciosa Bay (Gra), 146, 147
Grand Terre, see New Caledonia
Guadalcanal, 92, **21A**
Guam (Gua), 40, 41, 42, 43, 44, 107, 143, 147, 148, **4, 6, 13, 14, 37, A8**, *10, 24B, 25DE*

H
Ha'ano (Hno), 145, 147, **29**
Ha'apai Group, 11, 73, 107, 119, 147, **2, 29, 30**, *13D*
Ha'atuatua, 107
Hahake, **29**

161

Hahula (Hah), 145, 147
Halmahera (Hal), 14, 39, 47, 110, 143, 147, **4**, **5**, **6**, **17**, **18**
Hane (Han), 37, 38, 143, 147, *9*
Haruku (Har), 47, 50, 111, 143, 147, **18**, *25E*
Hiva Oa (Hiv), 37, 143, 147, *9*
Holopeka (Hol), 145, 147, **29**
Ho'oumi (Hou), 37, 143, 147, *9*
Horne Islands, 2, 79, *16A*
Hunga (Hun), 145, 147, **29**
Huon Peninsula (Huo), 18, 19, 112, 139, 147, 148, **4**, **7**, **19**, **20**, *13A*, *24D*, *25D*
Hutumuri, **A10**

I

IBB operations (IBB), 143, 147
Ifo (Ifo), 67, 146, 147, *13C*, *14*
Île des Pins (IdP), 101, 115, 144, 147, **2**, **4**, **8**, *23*, *25D*
Indonesia, 5, 7, 14, 104, 109, **1**, **5**, **18**
Ipan Beach (Ipa), 143, 147, **A8**

J

Jane's Camp (Jan), 144, 147, **11**, **12**, *7*, *8*
Japan, 107, **5**
Jurlay (Jur), 139, 147

K

Kabara (Kab), 117, 142, 147, **28**, *16fn*, *25E*
Kadavu (Kad), 67, 70, 80, 104, 116, 117, 119, 141, 142, 147, 148, **8**, **26**, **27**, **37**, *13D*, *24H*, *25E*
Kahet (Kah), 143, 147
Kai Islands, **6A**
Kamgot (Kam), 76, 139, 147, **2**, *16A*
Kanacea, 117, 119, 148, **28**, *16fn*, *24B*, *25E*
Kapa (Kap), 145, 147, **2**, **29**
Kapingamarangi, 107, **7**
Kapume (Kap), 144, 147, *23*
Karkur (Kar), 139, 147
Karobo (Kao), 92, 95, 142, 147, **26**, **36**, **39**, *13fn*, *18C*, *25E*
Kayangel (Kay), 47, 107, 140, 147, **12**, *25D*
Kedekede hillfort, 87
Kieta (Kie), 58, 60, 144, 148, **19**, **21B**, **22**, *13B*
Kiki (Kik), 146, 147
Kiriwina islet (Kir), 145, 147
Kita Iwo Jima (KIJ), 40, 146, 147
Kolombangara (Kol), 61, 91, 92, 144, 147, **19**, **21C**, **35**, *18B*, *19*, *25E*
Koné, see *Lapita*
Koro, 80, **26**
Koror islet (Kor), 47, 140, 147, **16**, *12*
Kosrae (Kos), 31, 140, 147, **3**, **4**, **7**, **10**
Kotu (Kot), 145, 147, **29**
Kreslo (Kre), 89, 139, 147, **2**, **19**, **35**, **37**, **38**, *18A*
Kulu Bay (Kul), 85, 86, 104, 141, 147, 148, **2**, **34**, **37**, **38**, **39**, *16C*, *17*, *24F*
Kurin (Kur), 101, 142, 147, **2**
Kusaie, see Kosrae

L

Laguas (Lag), 42, 143, 147, **14**
Lakeba (Lak), 70, 71, 87, 93, 116, 117, 119, 142, 147, 148, **2**, **27**, **28**, **36**, *13D*, *13fn*, *16fn*, *18C*, *24D*, *25E*

Lamotrek (Lam), 107, 108, 140, 147, **4**, **7**, **10**, **40**, *11*, *21*, *25A*
Lapita (Lap), 99, 100, 144, 147, **2**, **A13**
Lasigi (Las), 52, 113, 139, 147, **19**, **20**, *13A*, *25C*
Late, 73, **29**
Lau (Lau), 2, 8, 20, 39, 70, 71, 75, 80, 87, 93, 103, 116, 117, 118, 119, 138, 141, 142, 145, 148, **5**, **8**, **9**, **27**, **28**, **34**, **37**, *13D*, *15*, *16C*, *24BD*, *25AE*
Laucala (Lac), 87, 141, 147, **26**, **34**, **37**, *16C*, *24B*
Lavaka (Lav), 146, 147
Lavongai, see New Hanover
Lease Islands (Lea), 47, 50, 104, 111, 143, 148, **6A**, **17**, **18**, **38**, **A10**, *25E*
Lelepa, 67
Lelu islet (Lel), 31, 140, 147
Lesu, see *Lossu*
Lewaki Levu (Lew), 141, 147
Lifuka (Lif), 107, 145, 147, **29**
Lihir Islands, 75, 76, **19**
Lomaiviti Group, 80, 86, 87, 93, 116, 117, 119, **26**, **34**, *16C*, *25E*
Lossu (Los), 52, 139, 147, **19**, **20**, **39**, *13A*
Lou (Lou), 75, 139, 147, 148, **19**, *16A*, *25CE*
Loyalty Islands, 20, 101, 115, 142, **2**, **4**, **8**, *23*, *25D*
Luangiua (Lua) islet, 144, 147

M

Maap, 96, **15**
Mabuiag (Mab), 88, 145, 147
Maer (Mer), 143, 147, *25B*, *27*
Maewo, 62, **23**
Mago (Mag), 87, 142, 147, **2**, **28**, **34**, **37**, *16C*, *24B*
Maka Bay (Mak), 144, 147
Makada, 91
Makira (Mak), 107, 115, 145, 148, **21A**, *25E*
Malae'imi Valley (Mae), 144, 147, **11**, *7*
Malai (Mai), 139, 147
Malaita, 19, 96, **21A**
Malakula (Mlk), 62, 63, 66, 103, 115, 116, 146, 147, 148, **23**, **25**, **37**, **A7**, *13C*, *13fn*, *14*, *24F*, *25DE*
Malau (Mal), 142, 147
Malekolon (Mao), 76, 113, **2**, *16A*
Malekula, see Malakula
Malendok (Mad) islet, 139, 147, **A26**
Malo (Mlo), 63, 66, 115, 116, 146, 147, **2**, **23**, **25**, **A15**, **A28**, *13C*, *14*, *25CD*
Malua Bay (Mau), 146, 147
Maluku (also see Molucca Islands), 96, 110, **38**, *24*
Mamalua (Mam), 144, 147
Mamanuca Islands, 67, **27**
Mangaasi (Mag), 146, 147
Mangai'a (Maa), 145, 147
Manono, **11**
Manu'a Group, 34, 36, 37, **11**
Manus (Man), 52, 75, 76, 112, 113, 139, 147, 148, **7**, **19**, **20**, **37**, **A22**, *13A*, *24F*, *25CE*
Maré (Mar), 101, 142, 147, **2**, *23*, *25D*
Mariana Islands, 7, 10, 11, 14, 20, 39, 40, 41, 42, 43, 44, 103, 104, 107, 143, **1**, **3**, **4**, **13**, **14**, **38**, **A8**, **A16**, *10*, *24*, *25BCE*
Marquesas Islands, 2, 30, 37, 38, 95, 107, 143, **1**, *9*, *24B*
Mask Cave (Mas), 88, 145, 147

Matanamuani (Mat), 86, 141, 147
Matuku (M/M; in Fiji), 86, 87, 117, 148, **26**, **37**, *16C*, *25E*
Matuku (Mat; in Tonga), 145, 147
Mauke (Mau), 119, 141, 147, **30**, *13D*, *25B*
M'Buke (MBu), 75, 113, 139, 147, **19**, **A22**, *16A*, *25E*
Meinakapa (Mei), 60, 144, 147, **19**, **21B**, **22**, *13B*
Melanesia, 5, 7, 10, 20, 75, 96, 112, 119, 144, **1**, **3**, **5**, **9**, *13fn*
Melanesian borderland, 2, 30, 75, 79, 92, 118, 138, **31**, *16A*, *24A*
Melekeok (Mek), 140, 147
Mele Plain (Mel), 67, 107, 146, 147
Micronesia, 5, 7, 39, 42, 96, 107, **1**, **3**
Moala (M/M), 86, 87, 117, 148, **26**, **37**, *16C*, *25E*
Mochong (Moc), 42, 104, 143, 147, **14**
Mogmog (Mog), 140, 147
Molucca Islands (also see Maluku), 14, 39, 47, 104, 109, 143, **6**, **17**, **18**, *A10*, *24D*, *25EF*
Mono islet (Mon), 145, 147, **21B**, **22**
Moturiki (Mot), 86, 87, 93, 116, 117, 141, 147, 148, **2**, **26**, **34**, **36**, **37**, *13fn*, *16C*, *16fn*, *18C*, *25DE*
Mouly (Mou), 142, 147, *23*
Mu'a (Mua), 145, 147, **29**
Mulifanua (Mul), 34, 118, 144, 147, **2**, **11**, **12**, **A4**, *7*, *8*
Murray Islands, 104, 111, 112, 143, **4**, **7**, *25B*, *27*
Mussau Group, 7, 52, 76, 89, 113, 147, **2**, **7**, **19**, **20**, **31**, **37**, *13A*, *13fn*, *16A*, *25E*
Muyua (Muy), 88, 91, 145, 147, **4**, **19**

N

Nabulei Lisa (NLi), 139, 147
Naigani (Nag), 86, 116, 117, 141, 147, **2**, **26**, **34**, **37**, *16C*, *25DE*
Naitabale (Nab), 86, 141, 147
Nananu-i-ra (Nan), 80, 141, 147, **26**, **32**, *16B*
Nanggu (Nan), 146, 147
Nan Madol (Nad), 31, 140, 147
Nanumanga (Nau), 117, 145, 147, **8**, **32**, *25A*, *28*
Nanumea (Nae), 117, 145, 147, **8**, **32**, *25A*, *28*
Nasilai (Nas), 92, 93, 95, 116, 142, 147, **26**, **36**, **39**, **A12**, *13fn*, *18C*, *25E*
Natunuku (Nat), 83, 85, 86, 93, 95, 116, 142, 147, **2**, **26**, **32**, **33**, **36**, **37**, **39**, **A17**, **A25**, *16B*, *18C*, *24B*, *24C*
Navaprah (Nvp), 146, 147
Navatu (Nav), 80, 83, 85, 86, 93, 95, 116, 117, 142, 145, 147, 148, **26**, **33**, **36**, **37**, **39**, **A11**, *16B*, *17*, *18C*, *28*, *24C*, *25A*
Navua Delta (Nvu), 70, 116, 142, 147, 148, **26**, **27**, **37**, *13D*, *13fn*, *24B*
Nayau (Nay), 70, 71, 117, 142, 147, **27**, **28**, *13D*, *15*, *25E*
Nendö (Nen), 62, 63, 67, 107, 115, 116, 119, 146, 147, 148, **2**, **4**, **23**, **24**, **30**, **37**, **A23**, *13C*, *14*, *24B*, *25D*, *29*
New Britain (NBr), 18, 20, 39, 52, 55, 56, 76, 88, 89, 91, 101, 103, 112, 113, 139, 147, 148, **2**, **4**, **7**, **19**, **20**, **35**, **37**, **38**, **A27**, *13A*, *18A*, *25C-E*
New Caledonia (NCa), 2, 5, 8, 20, 96, 99, 100, 101, 104, 115, 142, 144, 146, 148, **1**, **2**, **3**, **4**, **5**, **8**, **9**, **A13**, **A28**, *22*, *23*, *25CD*

Index

New Georgia Group (NeG), 58, 60, 61, 91, 92, 114, 115, 144, 147, 148, **2**, **4**, **19**, **21A**, **21C**, **22**, **37**, **A29**, *13B*, *19*, *24C*, *25AEF*
New Guinea, 2, 7, 8, 14, 18, 19, 20, 21, 39, 50, 52, 55, 88, 96, 98, 107, 109, 112, **1**, **4**, **5**, **6**, **9**, **18**, **19**, *25ABD*
New Hanover, 52, 76, 113, **7**, **19**, *13A*, *13fn*
New Hebrides, 8, 20, 61, 63, 66, 79, 103, 115, 116, 138, **8**, **9**, **23**, **24**, **25**, **A23**
New Ireland (NIr), 52, 75, 76, 91, 103, 104, 112, 113, 139, 147, 148, **4**, **7**, **19**, **20**, **37**, **39**, *13A*, *13fn*, *24F*, *25CD*
Ngambelipa (Ngm), 146, 147
Ngangaua (Ngn), 146, 147, **A23**
Ngaraard State (Ngr), 140, 147, **17**
Ngargasang, **A2**
Ngatpang State (Ngt), 140, 147
Ngchesar State (Ngc), 140, 147
Ngiwal State (Ngw), 140, 147
Ngulu (Ngu), 47, 107, 140, 147, **4**, **6A**, *11*, *12*, *21*, *25AD*
Nissan (Nis), 52, 91, 113, 114, 139, 147, **2**, **4**, **19**, **20**, **35**, **37**, *13A*, *18A*, *25D*
Niuatoputapu (Niu), 73, 119, 145, 147, **2**, **29**, **30**, **A9**, *13D*
Nomna Bay (Nom), 143, 147
Nomuka (Nom), 73, 119, 145, 147, **29**
Normanby, 99, **19**
Nuku (Nuu), 144, 147
Nuku Hiva (Nuk), 37, 95, 107, 143, 147, *9*
Nukuleka (Nuk), 119, 145, 147, **29**, **30**, *13D*, *25F*, *29*
Numbulevu (Num), 141, 147

O
Ofu (Ofu), 34, 36, 37, 144, 147, **11**, **12**, **37**, **A20**, *7*, *8*, *24A*
Ogasawara, see **Bonin Islands**
Olosega, 36, **11**
Oma (Oma), 143, 147
Ono (Ono), 70, 141, 147, **26**, **27**, *13D*
Ontong Java (OnJ), 107, 114, 144, 147, **7**, **19**, *25A*
Ordnance Annex (Ord), 143, 147
Ormi (Orm), 143, 147, *27*
Orrak islet (Orr), 47, 140, 147, *12*
Ouvea (Ouv), 101, 142, 147, **23**, *25D*
Ovalau (Ova), 86, 87, 117, 141, 148, **26**, **34**, **37**, *16C*, *24G*, *25D*

P
Pagan (Pag), 42, 44, 107, 143, 147, **4**, **13A**, **14**, *25B*
Pakea (Pak), 63, 66, 107, 146, 147, **23**, *13C*, *13fn*
Palau (Pal), 10, 14, 21, 39, 46, 47, 103, 107, 140, 147, 148, **3**, **4**, **6**, **16**, **17**, **37**, **40**, **A2**, *12*, *24*, *25AD*
Papakula (Pap), 139, 147
Papuan Peninsula (PP), 19, 20, 91, 98, 99, 104, 114, 145, 148, **7**, **9**, **19**, *25A*
Papua New Guinea (Pap), 18, 19, 88, 111, 138, 139, 147, **2**, **7**, **A14**, **A18**, **A21**, **A22**, **A26**, **A27**, *24fn*, *25A-E*
Paradise (Par), 144, 147, **11**, *7*
Patho, **2**
Paubake (Pau), 58, 60, 144, 147, **19**, **21B**, **22**, **A21**, *13B*
Pea (Pea), 119, 145, 147, **29**

Pelau islet (Pel), 144, 147
Peleliu (Pel), 46, 47, 107, 140, 147, **16**, *12*, *25D*
Pemrang (Pem), 141, 147
Pentecost, 62, **23**, *13fn*
Philippine Islands, 5, 7, 107, **1**, **5**
Pidia (Pid), 60, 144, 147, **19**, **21B**, **22**, *13B*
Pohnpei (Poh), 21, 31, 140, 147, **3**, **4**, **7**, **10**
Poi, 112
Polynesia, 5, 7, 8, 10, 38, 119, **1**, **3**
Ponamla (Pon), 67, 146, 147, *13C*, *14*
Ponape, see Pohnpei
Pukotala (Puk), 145, 147, **29**
Pulantat (Pla), 143, 147
Pulau Kumo, 47, 110, **17**
Pulemelei (Ple), 36, 144. 147, **11**, *7*

Q
Qaqaruku (Qaq), 142, 147, **26**, **32**, **37**, *16B*
Qaranicagi, 70
Qoqo, 93

R
Rakival (Rak), 139, 147
Rano, 66
Reber (Reb), 139, 147
Reef Islands (Ree), 62, 63, 116, 146, 147, **2**, **4**, **23**, **24**, **37**, **A23**, *13C*, *14*, *25D*
Regusa (Reg), 143, 147
Rendova, 60, **19**, **21C**
Rennell, 7
Rewa Delta (Rew), 93, 95, 107, 116, 141, 148, **26**, **A12**, *13fn*, *25E*
Rota (Roa), 40, 41, 42, 44, 104, 107, 143, 147, **4**, **6**, **13**, **14**, *25D*
Rotoapi (Rot), 146, 147
Rotuma (Rou), 30, 32, 37, 118, 144, 147, **8**, **32**, **37**, *16B*, *24A*, *25C*
Roviana Lagoon (Rov), 58, 60, 61, 92, 101, 115, 144, 147, **2**, **19**, **21C**, **22**, **35**, **37**, **38**, **A29**, *13B*, *18B*
Rumung, 97

S
Sabaig (Sab), 140, 147
Sabanan Fiang (Saf), 143, 147, **A16**
Saint Matthias Group, see Mussau Group
Saipan (Sai), 20, 40, 41, 42, 43, 44, 103, 104, 107, 143, 147, 148, **4**, **6**, **13**, **14**, **38**, **A16**, *10*, *24E*, *25BDE*
Samoa (also see **American Samoa**), 2, 8, 10, 30, 32, 34, 36, 37, 38, 75, 117, 118, 138, 144, **1**, **2**, **3**, **4**, **5**, **8**, **11**, **12**, **37**, **40**, **A3**, **A4**, **A5**, *7*, *8*, *24A*, *25C*
San Cristobal, see Makira
Sanmeima (San), 143, 147, **14**
Santa Ana (StA), 114, 145, 147, **4**, **21A**, *25E*
Santa Cruz Group, 8, 62, 63, 107, 116, 119, **1**, **2**, **3**, **8**, **30**, **A23**, *25D*
Santa Isabel, 19, 96, **21A**
Santo (San), 62, 63, 66, 103, 115, 116, 146, 147, 148, **23**, **25**, **37**, **A15**, **A28**, *13C*, *14*, *24F*, *25C-E*
Saparua, 47, 111, **18**
Sapota (Sap), 140, 147, **A1**
Satawal, 108

Sasoa'a (Sas), 34, 144, 147, **11**, **12**, **A3**, *7*, *8*
Saulevu (Sau), 86, 141, 147
Savai'i (Sav), 34, 36, 144, 147, **11**, *7*
Schouten Islands, 18, 112, **7**, **19**
Seputu (Sep), 141, 147
Seram (Ser), 47, 50, 96, 97, 98, 109, 110, 111, 143, 147, 148, **4**, **5**, **6A**, **17**, **18**, *25EF*, *26*
Shepherd Islands (She), 66, 67, 146, 147, 148, **23**, **25**, *13C*, *24D*
Shortland Islands (Sho), 58, 60, 103, 144, 147, **19**, **21B**, **22**, **37**, *13B*, *24F*, *25E*
Siassi Islands, 112, 139, 147, **19**, *25D*
Sigatoka (Sig), 92, 93, 95, 142, 147, **2**, **26**, **36**, **39**, **A19**, *18C*
Sikumango (Sik), 144, 147
Sinapupu (Sin), 146, 147
Solomon Islands, 2, 7, 8, 14, 18, 19, 39, 56, 58, 60, 61, 63, 88, 91, 92, 96, 101, 103, 104, 107, 112, 114, 115, 116, 138, 144, **1**, **2**, **4**, **5**, **7**, **8**, **9**, **21A**, **22**, **23**, **24**, **35**, **37**, **38**, **39**, **A21**, **A23**, **A29**, *13B*, *18B*, *24CF*, *25ADE*
Sonsorol Islands (Son), 47, 107, 140, 147, *12*, *25D*

T
Tabar Islands, 75, 76, **19**
Tafahi, 73, **29**
Taga (Tag), 143, 147
Talafofo (Taf), 143, 147
Talepakemalai (Tal), 139, 147
Talof, 114
Tanga Islands (Tan), 75, 76, 78, 79, 104, 113, 139, 147, **19**, **31**, **A26**, *16A*, *24D*, *25E*
Tanna, 62, **23**, *13fn*
Tarague (Tar), 143, 147
Tataga-matau (Tat), 144, 147, **11**, *7*
Ta'u, 36, **11**
Taumako (Tau), 62, 63, 107, 146, 147, **4**, **8**, **23**, **24**, **38**, *13C*, *14*
Tavai (Tai), 144, 147
Tavatava (Tav), 146, 147
Taveuni (Tav), 80, 87, 141, 147, **26**, **34**, *16C*
Temei (Tem), 117, 145, 147, *25A*
Tenmiel (Ten), 146, 147
Teop islet (Teo), 58, 60, 114, 144, 147, **19**, **21B**, *13B*, *25D*
Tikopia (Tik), 63, 66, 79, 116, 146, 147, **4**, **23**, **24**, *13C*, *13fn*, *24A*, *25E*
Tinakula, 63, **23**, **24**
Tinian (Tin), 41, 42, 107, 143, 147, **13B**, **14**, *25D*
TLTF chain, 75, 76, 103, 104, 113, 139, 147, 148, **19**, **31**, **37**, **A26**, *25E*
To'aga (Tog), 34, 36, 144, 147, **11**, **12**, **A20**, *7*, *8*
Tobi, 107, **4**
Tofua, 73, **29**
Tokelau, 117, 118, 145, **1**, **5**, **8**, **32**, **40**, *16B*, *25A*
Tonga (Ton), 5, 8, 10, 11, 20, 38, 39, 63, 70, 71, 73, 75, 79, 80, 95, 103, 104, 107, 117, 118, 119, 141, 148, **1**, **2**, **3**, **4**, **5**, **8**, **9**, **29**, **30**, **37**, **39**, **40**, **A6**, **A9**, **A24**, *13E*, *24B*, *24E*, *25BEF*
Tongatapu (Ttu), 73, 119, 145, 147, **2**, **29**, **30**, **A6**, **A24**, *13D*, *25EF*
Tongoa islet (Ton), 67, **23**, *13C*
Torres Islands (Tor), 66, 116, 146, 147, **23**, **25**, *13C*, *25D*

Torres Strait Islands, 2, 88, 111, 145, **4**, *7*
Totoya (Tot), 86, 87, 93, 116, 117, 141, 147, **2**, **26**, **34**, **36**, **37**, *16C*, *16fn*, *18C*, *24B*, *25E*
Treasury Group, 147
Trevanion islet (Tre), 146, 147
Trobriand Islands (Tro), 99, 104, 114, 145, 147, **4**, **7**, **19**, **20**, **A18**, *25A*
Truk, see Chuuk
Tuam (Tua), 139, 147
Tumon Bay (Tum), 143, 147
Tungua (Tun), 95, 107, 119, 145, 147, **29**
Tutuila (Tut), 30, 34, 36, 37, 144, 147, **11**, *7*
Tuvalu, 80, 117, 118, 145, **1**, **4**, **5**, **8**, **32**, **37**, **40**, *16B*, *17*, *25A*, *28*

U

Ua Huka (UaH), 37, 143, 147, *9*
Udu Peninsula (Udu), 67, 70, 116, 118, 141, 142, 144, 148, **26**, **27**, **38**, *13D*, *25BC*
Ugaga (Uga), 85, 93, 116, 117, 141, 147, **2**, **26**, **34**, **36**, *16C*, *17*, *18C*, *25BD*
'Uiha (Uih), 11, 145, 147, **29**
Ulithi (Uli), 107, 108, 140, 147, **4**, **6A**, *11*, *21*, *25D*
Ulong islet (Ulo), 47, 107, 140, 147, **16**, *12*, *25D*
Ulunikoro (Ulu), 142, 147
Umboi, 147
Unai Achugao (Una), 143, 147
Unai Bapot (Unb), 143, 147
Unai Chulu (Unc), 42, 143, 147, **14**
Upolu (Upo), 34, 36, 37, 118, 144, 147, **2**, **11**, **37**, **A3**, **A4**, **A5**, *7*, *8*, *24A*, *25C*
Uri, 66
Uripiv, 66, **A7**
Utuleve, 32

Uvea (Uve), 30, 32, 37, 144, 147, **2**, **3**, **4**, **8**

V

Vaga Bay (Vag), 85, 141, 148, **37**, *24G*
Vailele (Val), 144, 147, **11**, **12**, **A5**, *7*, *8*
Vaipuna (Vap), 145, 147, **29**
Vaitupu (Vai), 117, 145, 147, **8**, **32**, **37**, *17*, *25A*, *28*
Vanikolo (Vak), 63, 66, 107, 116, 146, 147, 148, **2**, **4**, **23**, **24**, *13C*, *13fn*, *14*, *24D*, *25E*
Vanua Balavu (VaB), 87, 117, 142, 147, **28**, **34**, *16C*, *16fn*, *25E*
Vanua Lava, 63, **23**
Vanua Levu (VaL), 67, 80, 85, 86, 116, 117, 118, 141, 142, 145, 147, 148, **26**, **32**, **37**, **38**, *13D*, *16B*, *16C*, *17*, *24E*, *25A-D*
Vanuatu (Van), 2, 8, 20, 32, 39, 61, 62, 63, 66, 67, 70, 79, 95, 103, 107, 115, 116, 118, 138, 145, 146, 148, **1**, **2**, **3**, **4**, **5**, **8**, **9**, **23**, **24**, **25**, **37**, **40**, **A7**, **A15**, **A23**, **A28**, *13C*, *14*. *24ABDF*, *25C-F*
Vao islet (Vao), 66, 146, 147, **2**
Va'ota (Vat), 34, 144, 147, **11**, **12**, *7*, *8*
Vatcha (Vac), 144, 147, **2**, *23*
Vatulele, 80
Vaturekuka (Vau), 142, 147
Vava'u Group, 73, 119, 147, **2**, **29**, **30**, *13D*
Veitongo (Vei), 145, 147
Vele (Vel), 144, 147
Vella Lavella, 61, **19**, **21C**, **37**, *24F*, *25E*
Vitiaz Strait, 21, 112, *25D*
Viti Levu (ViL), 2, 5, 20, 67, 70, 80, 83, 85, 86, 87, 88, 89, 92, 93, 95, 104, 107, 116, 117, 119, 141, 142, 145, 147, 148, **2**, **9**, **26**, **32**, **33**, **36**, **37**, **38**, **39**, **A11**, **A12**, **A17**, **A19**, **A25**, *13D*, *13fn*, *16B*, *17*, *18C*, *20*, *25ABE*, *28*
Volcano Islands, 39, 146, **1**
Votua (Vot), 87, 142, 147
Vuda (Vud), 83, 93, 95, 104, 116, 142, 147, **26**, **32**, **36**, **37**, **38**, **39**, *16B*, *17*, *18C*, *24F*

W

Wabao (Wab), 142, 147, *23*
Wala, 66
Wammar, **A30**
Wangil (Wan), 139, 147, **A30**, *26*
Waterfront Annex (Wat), 143, 147
Watom (Wat), 52, 55, 56, 91, 103, 112, 113, 139, 147, **2**, **19**, **20**, **35**, *13A*, *18A*, *24E*, *25C*
Waya, 70, 116, *25E*
West Irian, **6**, *7*
Willaumez Peninsula, 55, **19**
Witu Islands, 55, **19**
Woodlark, see Muyua
Wusi (Wus), 146, 147

Y

Yadua (Yad), 92, 95, 142, 147, **26**, **36**, **39**, *18C*
Yanuca (Yan), 92, 95, 116, 142, 147, **2**, **26**, **32**, **36**, **39**, *16B*, *18C*, *25E*
Yap (Yap), 2, 10, 14, 39, 44, 96, 97, 107, 108, 140, 147, 148, **3**, **4**, **6**, **15**, **40**, *11*, *12*, *21*, *25AD*
Yasawa Islands (Yas), 67, 70, 116, 142, 147, **26**, **27**, *13D*, *24D*, *25E*

Z

Zomar (Zom), 143, 147